Smart Innovation, Systems and Technologies

Volume 41

Series editors

Robert J. Howlett, KES International, Shoreham-by-Sea, UK
e-mail: rjhowlett@kesinternational.org

Lakhmi C. Jain, University of Canberra, Canberra, Australia, and
University of South Australia, Adelaide, Australia
e-mail: Lakhmi.jain@unisa.edu.au

About this Series

The Smart Innovation, Systems and Technologies book series encompasses the topics of knowledge, intelligence, innovation and sustainability. The aim of the series is to make available a platform for the publication of books on all aspects of single and multi-disciplinary research on these themes in order to make the latest results available in a readily-accessible form. Volumes on interdisciplinary research combining two or more of these areas is particularly sought.

The series covers systems and paradigms that employ knowledge and intelligence in a broad sense. Its scope is systems having embedded knowledge and intelligence, which may be applied to the solution of world problems in industry, the environment and the community. It also focusses on the knowledge transfer methodologies and innovation strategies employed to make this happen effectively. The combination of intelligent systems tools and a broad range of applications introduces a need for a synergy of disciplines from science, technology, business and the humanities. The series will include conference proceedings, edited collections, monographs, handbooks, reference books, and other relevant types of book in areas of science and technology where smart systems and technologies can offer innovative solutions.

High quality content is an essential feature for all book proposals accepted for the series. It is expected that editors of all accepted volumes will ensure that contributions are subjected to an appropriate level of reviewing process and adhere to KES quality principles.

More information about this series at http://www.springer.com/series/8767

Vladimir L. Uskov · Robert J. Howlett
Lakhmi C. Jain
Editors

Smart Education and Smart e-Learning

 Springer

Editors
Vladimir L. Uskov
Department of Computer Science
   and Information Systems
Bradley University
Peoria, IL
USA

Lakhmi C. Jain
Faculty of Education, Science,
   Technology and Mathematics
University of Canberra
Canberra
Australia

Robert J. Howlett
KES International
Shoreham-by-Sea
UK

ISSN 2190-3018   ISSN 2190-3026   (electronic)
Smart Innovation, Systems and Technologies
ISBN 978-3-319-36774-3   ISBN 978-3-319-19875-0   (eBook)
DOI 10.1007/978-3-319-19875-0

Springer Cham Heidelberg New York Dordrecht London

Printed on acid-free paper

Springer International Publishing AG Switzerland is part of Springer Science+Business Media
(www.springer.com)

Preface

Smart Education and Smart E-Learning (SEEL) is an emerging and rapidly growing area that represents an integration of (1) smart and intelligent systems, smart objects and smart environments, (2) smart technologies, various branches of computer science and computer engineering, (3) state-of-the-art smart educational software and/or hardware systems, agents, and tools, and (4) innovative pedagogy and advanced technology-based teaching strategies and learning methodologies. This is the main reason that in June 2013 a group of enthusiastic and visionary scholars from all over the world arrived with the idea to organize a new professional event that would provide an excellent opportunity for faculty, scholars, Ph.D. students, administrators, and practitioners to meet well-known experts and discuss innovative ideas, findings and outcomes of research projects, and the best practices in smart education and smart e-learning.

The main research, design, and development topics in SEEL area include but are not limited to (1) conceptual frameworks for Smart Education (SmE), (2) infrastructure, main characteristics, and features of smart universities and smart classrooms, (3) university-wide smart software, hardware, security, communication, collaboration, and management systems, (4) SmE analytics and economics, (5) main components and techniques of smart pedagogy, (6) Smart e-Learning (SeL) concepts, strategies, and approaches, (7) SeL environments, (8) smart e-learner modeling, (9) assessment and quality assurance in SeL, (10) social, cultural, and ethical dimensions and challenges of SmE and SeL, (11) applications of various innovative technologies—ambient intelligence, Internet-of-Things, smart agents, sensors, wireless sensor networks, context awareness technology, smart gamification, smart multimedia—and smart software/hardware systems in universities and classrooms, (12) mobility, security, access, and control issues in smart learning environments and numerous other topics. We hope that active and open discussion of those topics within SEEL research and academic communities will help us to (a) organize mutually beneficial partnerships, stimulate national and international research, design, and development projects in SEEL area, (b) propose innovative pedagogy, teaching, and learning strategies, standards, and policies in

SEEL, (c) identify tangible and intangible benefits, economic, technical, organizational, and financial feasibility and effectiveness of SEEL.

The inaugural international KES conference on Smart Technology-based Education and Training (STET) was held at Chania, Crete, Greece, during June 18–20, 2014. This book contains the contributions presented at the 2nd international KES conference on *Smart Education and Smart e-Learning*, which took place in Sorrento, Italy, during June 17–19, 2015. The book chapters, a total of 45 peer-reviewed chapters, are grouped into several parts, including: Part 1—Smart Education, Part 2—Smart Educational Technology, Part 3—Smart e-Learning, Part 4—Smart Professional Training and Teachers' Education, and Part 5—Smart Teaching- and Training-related Topics.

We would like to thank scholars who dedicated a lot of efforts and time to make SEEL international conference a great success, namely: Dr. Luis Anido (Spain), Dr. Elena Barbera (Spain), Dr. Claudio da Rocha Brito (Brazil), Dr. Dumitru Burdescu (Romania), Dr. Nunzio Casalino (Italy), Dr. Feng-Kuang Chiang (China), Prof. Melany Ciampi (Brazil), Prof. Adriana Burlea Schiopoiu (Romania), Dr. Michele Cole (USA), Dr. Pasquale Daponte (Italy), Dr. Natalia Dneprovskaya (Russia), Prof. Dimity Dornan (Australia), Mr. Marc Fleetham (UK), Dr. Mikhail Fominykh (Norway), Dr. Brian Garner (Australia), Dr. Jean-Pierre Gerval (France), Dr. Karsten Henke (Germany), Dr. Maung Htay (USA), Dr. Alexander Ivannikov (Russia), Prof. Vladislav Pirogov (Russia), Dr. Jung-Sing Jwo (Taiwan), Dr. Aleksandra Klasnja-Milicevic (Serbia), Dr. Marina Lapyonok (Russia), Dr. Greg Lee (Taiwan), Dr. Ezra Mugisa (Jamaica), Prof. Andrew Nafalski (Australia), Dr. Enn Õunapuu (Estonia), Dr. Elvira Popescu (Romania), Dr. Anitha S. Pillai (India), Mr. Valeri Pougatchev (Jamaica), Dr. Ekaterina Prasolova-Førland (Norway), Dr. Maria Riccio (Italy), Prof. Jerzy Rutkowski (Poland), Dr. Danguole Rutkauskiene (Lithuania), Dr. Demetrios Sampson (Greece), Prof. Richard Schumaker (USA), Prof. Boris Starichenko (Russia), Prof. Masanori Takagi (Japan), Dr. Wenhuar Tarng (Taiwan), Dr. Yoshimi Teshigawara (Japan), Dr. Vladimir P. Tikhomirov (Russia), Dr. Boban Vesin (Serbia), Prof. Natalia Gerova (Russia), Dr. Alan Weber (Qatar), Dr. Heinz-Dietrich Wuttke (Germany), Prof. Stelios Xinogalos (Greece), Prof. Chengjiu Yin (Japan), Prof. Shyan-Ming Yuan (Taiwan), and Dr. Larisa Zaiceva (Latvia).

We are indebted to many international collaborating organizations that made SEEL international conference possible, specifically: KES International (UK), InterLabs Research Institute, Bradley University (USA), Institut Superieur de l'Electronique et du Numerique ISEN-Brest (France), Silesian University of Technology (Poland), Institute for Design Problems in Microelectronics, The Russian Academy of Sciences (Russia), Science and Education Research Council COPEC (Brazil), Moscow State University of Economics, Statistics and Informatics (Russia), Tallinn University of Technology (Estonia), World Council on Systems Engineering and Information Technology—WCSEIT (Portugal), Multimedia Apps D&R Center, University of Craiova (Romania), Knowledge Networks PTY. Ltd. (Australia), Information Security Laboratory, Tokyo Denki University (Japan), and Technische Universitaet Ilmenau (Germany).

It is our sincere hope that this book will serve as a useful source of valuable data and information, and provide a baseline of further progress and inspiration for research projects and advanced developments in the SEEL area.

Peoria, USA	Vladimir L. Uskov
Shoreham-by-Sea, UK	Robert J. Howlett
Canberra, Australia	Lakhmi C. Jain
June 2015	

Contents

Part II   Smart Educational Technology

Part III   Smart e-Learning

Part I
Smart Education

The Ontology of Next Generation Smart Classrooms

Vladimir L. Uskov, Jeffrey P. Bakken and Akshay Pandey

Abstract Fast proliferation of various types of smart devices, smart systems, and smart technologies provides academic institutions, students and learners with enormous opportunities in terms of new approaches to learning technologies, education, learning processes and strategies, corporate training, user's personal productivity and efficiency, and faster and better quality of services provided. This paper presents the developed ontology of Smart Classroom systems - it helps to understand and analyze current smart classroom systems, and identify features, hardware, software, services, pedagogy, teaching and learning-related activities of the next generation Smart Classroom systems.

Keywords Smart classroom · Ontology · Hardware · Software · Technology · Pedagogy · Learning and teaching activities

1 Introduction

Modern sophisticated smart devices, smart systems, and smart technologies create unique and unprecedented opportunities for academic and training organizations in terms of new approaches to education, learning and teaching strategies, services to on-campus and remote/online students, set-ups of modern classrooms and labs. The performed research clearly shows that smart education market, in general, and

V.L. Uskov (✉) · A. Pandey
Department of Computer Science and Information Systems, Bradley University, Peoria, USA
e-mail: uskov@fsmail.bradley.edu

A. Pandey
e-mail: apandey@fsmail.bradley.edu

J.P. Bakken
The Graduate School, Bradley University, Peoria, USA
e-mail: jbakken@fsmail.bradley.edu

© Springer International Publishing Switzerland 2015
V.L. Uskov et al. (eds.), *Smart Education and Smart e-Learning*,
Smart Innovation, Systems and Technologies 41,
DOI 10.1007/978-3-319-19875-0_1

3

market of software and hardware for smart classrooms and smart universities, in particular, will exponentially grow in upcoming years.

Smart Education Market in 2013-2017. "The global smart education and learning market is expected to reach $220.0 billion by 2017 at a CAGR of 20.3 % between 2012 and 2017, including (a) services segment with projected $97.9 billion by 2017 with a CAGR of 26.6 %, (b) content segment - $72.9 billion in 2017, at a CAGR of 12.1 %, (c) software segment - $37.2 billion, and (d) hardware - $12.1 billion in 2017. Companies such as Ellucian, Inc. (U.S.), Smart Technologies (U.S.), Blackboard Inc. (U.S.), Kaplan Inc. (U.S.)., Promethean World Plc (United Kingdom), Pearson PLC (United Kingdom), and Informa Plc (Switzerland) are among key players on the smart education market" [1].

Smart Classrooms' Market in 2014-2018. "The global smart classroom market will grow at a CAGR of 31.25 % over the period 2013-2018. The two key factors contributing to this market growth are interactive display instruments and 3D education. Multiple global companies are among leaders in this area, including Apple, IBM, Microsoft, and SMART Technologies Inc." [2].

Therefore, it is necessary to perform research and get clear understanding of what specific technologies, software, hardware, services, learning-related activities and strategies will be required by next generation of smart classrooms in the near future.

2   Generations of Smart Classrooms: Literature Review

The concept of smart classroom was introduced several years ago; it is in permanent evolution and improvement since that time. "Smart Classrooms represents a focus on re-orienting our school structures and business processes around individual students and their learning needs. It is a transformative strategy to transition from traditional ways of working to a digital way of working that is meaningful, engaging and connected" [3].

We may identify several generations of implementations of smart classroom concept and corresponding software/hardware solutions in academic institutions.

The first (2001-2007) generation of smart classrooms. The early smart classroom implementations were primarily focused on synchronous delivery of learning content to local (i.e. students in actual physical classroom with face-to-face learning/teaching mode) and remote/online (i.e. students, in remote locations with online mode of learning/teaching) as well as synchronous teacher-students and local student-to-remote student communications. Shie, Xie, Xu, et al. in [4] showed that "…in the Smart Classroom, teachers can use multiple natural modalities while interacting with remote students to achieve same effect as a teacher in a classroom with local students. … In this type of tele-education, multimedia education systems let teachers and students in different locations participate in the class synchronously". Additionally, Xie, Shi, Xu et al. in [5] presented that "The Smart Classroom demonstrates an intelligent classroom for teachers involved in tele-education, in which teachers could have the same experiences as in a real classroom. … The magic of the Smart Classroom is the way teachers using the system - teachers are no longer tied up to the desktop computer, nor cumbersome keyboard and mouse".

V. Uskov and A. Uskov in [6] described a synchronous teaching of local students and remote/online students using the Internet-2 technology with 2-way full-scale synchronous high definition video, high quality audio, real-time 2-way discussions between local and remote students, active participation of remote students (who were as far as 2,400 miles away from actual physical classroom) in classroom activities, their questions and feedback.

The second (2008 - current) generation of smart classrooms. The second generation of smart classroom implementations is mainly based on active use of mobile technology, user/student/learner mobile devices and automatic communications between then and smart classroom environment.

Yau et al. [7] proposed Smart Classroom solution based on Reconfigurable Context-Sensitive Middleware (RCSM), connected situation aware PDA (i.e. with awareness about location, light, noise, and mobility) for each student, pervasive computing technology and collaborative learning.

O'Driscoll et al. in [8] described their version of Context Aware Smart Classroom (CASC). "It is a classroom that responds to lecturers and student groups based on preset policies and the lecture timetable. The pervasive nature of personal mobile devices permits the investigation of developing low-cost location and identification systems that support development of a smart classroom. The smart classroom CASC uses a central scheduling system to determine the teaching activity".

Huang et al. in [9] proposed "… a SMART model of smart classroom which characterized by showing, manageable, accessible, interactive and testing. … A smart classroom relates to the optimization of teaching content presentation, convenient access of learning resources, deeply interactivity of teaching and learning, contextual awareness and detection, classroom layout and management, etc.".

Pishva and Nishantha in [10] define a smart classroom as an intelligent classroom for teachers involved in distant education that enables teachers to use a real classroom type teaching approach to teach distant students. "Smart classrooms integrate voice-recognition, computer-vision, and other technologies, collectively referred to as intelligent agents, to provide a tele-education experience similar to a traditional classroom experience" [10].

Glogoric, Uzelac and Krco [11] addressed the potential of using Internet-of-Things (IoT) technology to build a smart classroom. "Combining the IoT technology with social and behavioral analysis, an ordinary classroom can be transformed into a smart classroom that actively listens and analyzes voices, conversations, movements, behavior, etc., in order to reach a conclusion about the lecturers' presentation and listeners' satisfaction" [11].

Slotta, Tissenbaum and Lui described an infrastructure for smart classrooms called the Scalable Architecture for Interactive Learning (SAIL) that "employs learning analytic techniques to allow students' physical interactions and spatial positioning within the room to play a strong role in scripting and orchestration" [12].

Koutraki, Efthymiou, and Grigoris developed a real-time, context-aware system, applied in a smart classroom domain, which aims to assist its users after recognizing any occurring activity. The developed system "...assists instructors and students in a smart classroom, in order to avoid spending time in such minor issues and stay focused on the teaching process" [13].

The Samsung Smart School solution has three core components: (1) the interactive management solution, (2) the learning management system, and (3) the student information system. Its multiple unique features and functions are targeted at smart school impact on education and benefits, including (1) increased interactivity, (2) personalized learning, (3) efficient classroom management, and (4) better student monitoring [14].

3 Research Project Goal and Objectives

The performed analysis of these and multiple additional publications and reports relevant to (1) smart classrooms, (2) smart technologies, (3) smart systems, (4) smart devices and meters, (5) smart universities, (6) smart environments, (7) smart cities, (8) ambient intelligence, (9) Internet-of-Things, and (10) smart educational systems undoubtedly shows that (a) smart classrooms, (b) smart labs, and (c) smart universities will be essential topics of multiple research, design and development projects in upcoming 5...10 years. It is expected that in near future smart classroom concept and hardware/software solutions will have a significant role and be actively deployed by leading academic intuitions – smart universities - in the world.

Based on our vision of smart classroom, smart university and up-to-date obtained research outcomes, we believe that the next generation of Smart Classroom systems should significantly emphasize not only software/hardware features but also "smart" features and functionality of smart systems (Table 1) [15, 16].

Table 1 Classification of levels of "smartness" of a smart system [15, 16]

Smartness levels (i.e. ability to ...)	Details
Adapt	Ability to modify physical or behavioral characteristics to fit the environment or better survive in it
Sense	Ability to identify, recognize, understand and/or become aware of phenomenon, event, object, impact, etc.
Infer	Ability to make logical conclusion(s) on the basis of raw data, processed information, observations, evidence, assumptions, rules, and logic reasoning
Learn	Ability to acquire new or modify existing knowledge, experience, behavior to improve performance, effectiveness, skills, etc.
Anticipate	Ability of thinking or reasoning to predict what is going to happen or what to do next
Self-organize	Ability of a system to change its internal structure (components), self-regenerate and self-sustain in purposeful (non-random) manner under appropriate conditions but without an external agent/entity

Therefore, next generation of smart classrooms should pay more attention to implementation of "smartness" maturity levels or "intelligence" levels, and abilities of various smart technologies.

The goals of performed research were to identify smart classroom's components, features, interfaces, inputs, outputs, limits/constraints, and develop ontology of Smart Classroom systems. The premise it that such taxonomy will enable us to identify and predict most effective hardware, software, services, pedagogy, teaching/learning activities for the next generation of Smart Classroom systems.

4 Research Outcomes: Smart Classrooms' Ontology

The proposed ontology of next generation Smart Classroom system is presented below in a tabular form. It is based on Systems Thinking approach that uses the following general descriptors: (1) goals, (2) components, (3) environment, (4) constraints/limits, (5) links (relations) between interrelated components, (6) interfaces between components and system-environment types of relations, (7) inputs, (8) outputs, and (9) boundary between Smart Classroom system and environment. Due to space limits of this paper, we present research outcomes for only the first four descriptors.

4.1 Goal and Objectives of the Next Generation Smart Classroom Systems

The main goal of next generation Smart Classroom systems is to demonstrate significant maturity at various "smartness" levels, including (1) adaptation, (2) sensing (awareness), (3) inferring (logical reasoning), (4) self-learning, (5) anticipation, and (6) self-organizations and restructuring (Table 1). The specific objectives of the next generation Smart Classroom systems are given in Table 2.

Table 2 Goals and objectives of the next generation Smart Classroom systems

Scope	Goals and objectives (main functions of the system)
	Goals
	• Goal 1: Next generation Smart Classroom systems must demonstrate significant maturity on various "smartness" levels, including (1) adaptation, (2) sensing (awareness), (3) inferring (logical reasoning), (4) self-learning, (5) anticipation, and (6) self-organizations and restructuring (Table 1)
	• Goal 2: Next generation Smart Classroom systems must have corresponding modern hardware (Table 3), software (Table 4), provide various services and activities to facilitate really smart teaching and learning strategies (Table 5), and deploy a great variety of learning strategies and pedagogies (Table 6)

(continued)

Table 2 (continued)

Scope	Goals and objectives (main functions of the system)
	Objectives
Common	• Facilitate learning through collaboration and interaction with students both remotely and locally using various types of interconnected wireless devices
	• Facilitate collaborative learning between local and remote students regardless students' native languages
	• Seamlessly connect several distant smart classrooms to share lectures and information via networking
	• Seamless connect various types of users' mobile smart devices and technical platforms; provide scalability and timely update of software systems and applications used by various users
	• Go well beyond the conventional static HTML-based presentation on the Web page or PPT slide show to multiple HD screens-based interactive learning experience (i.e. provide "smart learning cave" effect)
	• Automatically record all class activities and provide students with post-class review activities, for example, to review/learn content at student's own pace and comfort level
	• Accommodate, adapt and implement newest and emerging technologies and innovative trends, for example, computer vision, face recognition, speech recognition, noise cancellation, gesture recognition, etc.
	• Provide voice recognition, quality and fats automatic translation from English language to other languages, and visa versa
	• Provide ambient intelligence features, specifically, to adjust features like light intensity, temperature, humidity, safety, odor, etc. to suit and comfort local students in order to facilitate learning environment
Specific to instructor	• Allow instructor to communicate and express himself/herself naturally as in a traditional classroom without having significant experience in computer, information, and communication technologies and Web services
	• Allow teachers to host, join, form and evaluate group discussions on mobile devices and laptops for both local (in traditional classroom) and remote (online) students who present learning outcomes
	• Empower instructor with voice recognition, face recognition, gestures and smart pointing devices and boards to navigate, edit and display information on smart boards
	• Provide instructor with various analytical and recommender systems to maintain high quality and effective learning and teaching processes
	• Allow instructor to move freely and naturally without the need of a human cameraman to keep switching panoramic camera angles and views to display and present information to remote students (or, to use a smart cameraman component)
Specific to local student	• Provide students with quality automatic translation features (for example, from Spanish to English, or from English to Chinese)
Specific to remote student	• Provide a regular face-to-face learning like experience to online/remote students logging into a session in a smart classroom or smart lab
	• Provide students with quality automatic translation features (for example, from Spanish to English, or from English to Chinese)
	• Provide remote student with access to automatically recorded class activities for post-class review, for example, to review/learn content at student's own pace and comfort level

4.2 Components of Next Generation Smart Classroom Systems

Next generation Smart Classroom system's components include but are not limited to (a) hardware components, devices or equipment, (b) software systems, applications, and emerging technologies, (c) various activities related to learning and teaching, and (d) types of learning or pedagogy to used. Table 3 below contains detailed information about main hardware components, Table 4 – software components, Table 5 – learning activities related to learning and teaching, and Table 6 – types of learning and/or pedagogy to be used in next generation Smart Classroom systems.

Table 3 Proposed hardware components of next generation Smart Classroom systems

Scope	Hardware/equipment details
Common	• Array of video cameras installed to capture main classroom activities, movements, discussions, expressions, gestures, etc.
	• Ceiling-mounted projector(s) with 1 or 2 big size screen to display main activities in actual classroom; in some cases – 3D projectors
	• Student boards (big screen displays or TV) to display images of remote/online students from different locations
	• One or many (depending on class size, number of remote students, learning needs and workload) hidden computer systems to actually run the software and components of the Smart Classroom system
	• Bluetooth and Internet enabled devices like cell phones, smart phones, PDAs and laptops to facilitate communication and information/data/notes exchange
	• Network equipment (for example, Wi-Fi routers, zig bee transceivers, infrared, RFID readers and tags) to facilitate authorization and other forms of inter- device secure and reliable communication
	• Access to the Internet (mobile Web)
	• Wireless sensor network
	• Sensors (location detection, voice detection, motion sensors, thermal sensors, humidity, sensors for facial and voice recognition, etc.)
	• Robotic controllers and actuators to perform functions like intensity control, temperature control, movement, etc.
	• Devices: context aware devices, virtual mouse, biometric based login devices, automated zoom-in devices
	• Controlled and self-activated microphones(s) for instructor and students
	• Various type of speakers
	• Various types of lights
Specific to instructor	• Instructor's tablet PC (to write formulas, equations, run PPT presentations, video and audio clips, etc. in real time)
	• Big size smart board (to write formulas, equations, etc. in real time)
	• Document camera (connected to projector)
Specific to local student	• Array of mobile devices: smart interconnected mobile devices - smart phones, PDAs, laptops, smart headphones, etc.
	• (In some cases only): 3D goggles
Specific to remote student	• Desktop or tablet PC or laptop with connected or built-in microphone, speakers
	• Access to the broadband Internet

Bradley University contracted the Crestron company (http://www.crestron.com) to set-up multimedia top-quality Web-lecturing and capturing equipment for several classrooms including smart boards, HD video cameras, projectors, document camera, computer systems and software for instructor, microphones, speakers, etc. – this is the first step towards Smart Classroom establishment. A total cost of a full set-up (equipment + installation) of one classroom of this type is about $40,000 (as of May 2015).

Table 4 Proposed software components of next generation Smart Classroom systems

Scope	Software details
Common	• Agent-based systems to enable various types of communication between devices in the Smart Classroom system
	• Learning management system (LMS) or access to university wide LMS
	• Advanced software for rich multimedia streaming, control and processing
	• Software systems to address needs of special students, for example, visually impaired students (speech and gesture based writing/editing/navigation and accessibility tools to facilitate reading and understanding)
	• Smart cameraman software (for panoramic cameras)
	• Recognition software: face, voice, gesture
	• Motion or hand motion stabilizing software
	• Noise cancellation software
	• Security system for a secure log-in and log-out of registered student
	• Implementation of Internet-of-Things technology
	• Implementation of elements of various emerging technologies (for example, Smart Environments, Ambient Intelligence, Smart Agents)
Specific to instructor	• Smart drawing tools (for example, Laser2cursor) for drawing on smart boards, navigating and giving remote students floor to speak
	• Situation and/or context aware analytical system (that may generate hints and/or recommendations to instructor)
	• Analytical systems to analyze and rank class performance and outcomes
	• Systems to analyze presence, attendance, etc.
Specific to local student	• Smart notebook/laptop/tablet PC software
	• Main office software applications
	• Same view and smart view software
Specific to remote student	• Remote client programs to facilitate remote learning
	• Main office software applications
	• Same view and smart view software

Table 5 Types of activities related to learning/teaching to be actively used in the next generation of Smart Classroom systems

Scope	Activities (to …)
Common	• Communicate with other classmates as well as local and remote student project team members (a sub-group of students)
	• Share student team project documents on the Web or portal
	• Communicate with students in other smart classrooms at different locations
	• View learning content in a preferred language
	• Collect immediate feedback from students in terms of interest and likeability of an activity, session, or an overall subject and teacher
	• Recognize and classify the movement of any students in class
	• Automatically collect data from sensors and run analytics on students in terms of behavior, performance, interest, participation, etc.
	• Help special students, for example, visually impaired students (speech and gesture based writing/editing/navigation and accessibility tools to facilitate reading and understanding)
	• Adjust automatically classroom environment (lights, AC, temperature, humidity, etc.) or by voice commands
	• Agent-based systems to enable various types of communication between devices in the Smart Classroom system
Specific to instructor	• Give voice commands to the system to perform specific actions or to follow designated Web links
	• Initiate a classroom session with voice/facial/gesture commands
	• Give a floor to a remote student
	• Recognize each and every individual and his position/location in or outside physical classroom
	• Ability to suggest changes to the system
Specific to local student	• Learn and discuss presented learning content using reach multimedia and various communications tools and devices; participate in all class activities
	• Discuss presented learning content and assignments with remote students in real-time and using preferred language by each student
Specific to remote student	• Learn and discuss presented learning content synchronously with local students
	• Complete in-classroom assignments in real time and submit corresponding documents from remote locations synchronously with local students
	• Vote in a student team (with local and remote students) regarding an issue synchronously with local students
	• Ask teacher a question in real time during class session
	• Present in front of the local students from a remote area
	• Discuss and annotate any learning materials in real time

Table 6 Types of pedagogy to be used in the next generation of Smart Classroom systems

Scope	Types of pedagogy or learning
Common	• Smart classroom pedagogy (or, smart technology based teaching)
	• Learning-by-doing
	• Collaborative learning
	• Project-based learning
	• Advanced technology-based learning
	• e-Learning pedagogy
	• Games-based learning and pedagogy
	• Flipped classroom pedagogy

4.3 Environment and Constraints/Limits of the Next Generation Smart Classroom Systems

Various academic (schools, colleges, universities) and training (centers, businesses) organizations will primarily serve as the environment for the next generation Smart Classroom systems. Several examples of identified constraints for those organizations are presented in Table 7.

Table 7 Expected limits/constraints of the next generation Smart Classroom systems

Scope	Types of visible limits/constraints (several examples)
Common	• Limit on number of local and remote students in smart classroom
	• A combination of various sophisticated (cutting-edge) hardware and software systems and services may cause incompatibility problems
	• System can get "confused" by multiple voice-activated modules
	• Storage space (backdoor room) with servers, equipment, security, maintenance, etc. may be an issue both technologically and financially
Technology-related	• Technology constraints; potential problems with technology update, upgrade and maintenance at academic institutions
	• Too many interlinked devices, software systems, data exchange protocols used, etc.; as result, those devices and/or modules may lead to unexpected significant complexity and consequent failures of the entire system
	• Internet bandwidth constraints, especially for remote students in geographically isolated locations
	• Smart board technology is not yet as good and natural as unparalleled "white board-and-marker" or "chalk-and-board" methodology
Financial	• Financial constraints due to costs of high tech hardware, software, services, servers, maintenance, etc.
Student/learner-related	• Limit on number of hours a person can spend in front of computer screen or a mobile device
	• Absence of modern laptop, smart phone, software can be a constraint both financially and emotionally
	• Technological knowhow may be a constraint for older, not technology savvy students; a set of constraints will occur for disabled students
	• Non-social ("quite") student/learner issues

5 Conclusions. Future Steps

The performed research, identified evolution and development tendencies and obtained research findings and outcomes enabled us to make the following conclusions:

(1) Smart education market and market of software and hardware for smart classrooms will exponentially grow in upcoming years.
(2) Smart classrooms and smart universities will be essential topics of multiple research, design and development projects in upcoming 5...10 years.
(3) Leading academic intuitions in the will deploy smart classroom concept and hardware/software solutions in the near future.
(4) Next generation of Smart Classroom systems should pay more attention to implementation of "smartness" maturity levels and abilities of smart technologies.
(5) The proposed and developed ontology of Smart Classroom systems (components, functions, interfaces, inputs, outputs, limits/constraints, etc.) enabled us to identify and predict hardware, software, services, teaching/learning activities for the next generation Smart Classroom systems (Tables 2, 3, 4, 5, 6, 7).

Based on obtained research findings and outcomes, and developed ontology of the next generation Smart Classroom systems, the future steps in this research project are to (a) implement, test, validate, and analyze various identified learning strategies and pedagogies in smart classroom environment, (b) perform summative and formative evaluations of local and remote students and get sufficient data on quality of Smart Classroom main components - hardware, software, technologies, services, etc.), and (c) design and develop software systems for advanced Smart Classroom systems.

References

1. Smart Education and Learning Market. http://www.marketsandmarkets.com/Market-Reports/smart-digital-education-market-571.html (2013)
2. Global Smart Classroom Market 2014-2018 Report, http://www.researchandmarkets.com/research/2n7vrr/global_smart (2014)
3. e-Learning For Smart Classrooms, Smart Classroom Bytes journal, http://education.qld.gov.au/smartclassrooms/documents/strategy/pdf/scbyte-elearning.pdf (2008)
4. Shi, Y., Xie, W., Xu, G., Shi, R., Chen, E., Mao, Y., Liu, F. The smart classroom: merging technologies for seamless tele-education. In: Proceedings 4th International Conference on Multimodal Interfaces (ICMI 2002), pp. 429–434. IEEE CS Press (2002)
5. Xie, W., et al.: Smart Classroom - an intelligent environment for tele-education, In: Shum, H.-Y., Liao, M., Chang, S.-F. (eds.) PCM 2001, LNCS 2195, pp. 662–668 (2001)
6. Uskov, V.L., Uskov, A.V.: Streaming media-based education: outcomes and findings of a four-year research and teaching project. Int. J. Adv. Technol. Learn. 2(2), 45–57 (2005). ISSN:1710-2251

7. Yau, S., Gupta, S., Karim, F. et al.: Smart classroom: enhancing collaborative learning using pervasive computing technology. In: Proceedings of 2003 ASEE Conference. http://www. public.asu.edu/~bwang/publications/SmartClassroom-2003.pdf
8. O'Driscoll, C. et al.: Deploying a context aware smart classroom. In: Proceedings of the Arrow@DIR Conference, Dublin, Ireland. http://arrow.dit.ie/engschececon/53/ (2008)
9. Huang, R., Hub, Y., Yang, J., Xiao, G. The Functions of smart classroom in smart learning age. In: Proceedings 20th International Conference on Computers in Education ICCE. Nanyang Technological University, Singapore (2012)
10. Pishva, D., Nishantha, G.G.D.: Smart classrooms for distance education and their adoption to multiple classroom architecture. J. Netw. 3(5), 54–64 (2008)
11. Gligorić, N., Uzelac, A., Krco, S.: Smart classroom: real-time feedback on lecture quality. In: Proceedings 2012 IEEE International Conference on Pervasive Computing and Communications Workshops (PERCOM Workshops), Lugano, Switzerland, 19–23 March 2012, pp. 391–394 (2012). doi:10.1109/PerComW.2012.6197517
12. Slotta, J., Tissenbaum, M., Lui, M.: Orchestrating of complex inquiry: three roles for learning analytics in a smart classroom infrastructure. In: Proceedings of the Third International Conference on Learning Analytics and Knowledge LAK'13, pp, 270–274. New York, NY (2013). doi:10.1145/2460296.2460352
13. Koutraki, M., Maria, Efthymiou, V., Grigoris, A.: S-CRETA: smart classroom real-time assistance. In: Ambient Intelligence - Software and Applications, Advances in Intelligent and Soft Computing, vol. 153, pp 67–74. Springer, Berlin (2012)
14. The Next-Generation Classroom: Smart, Interactive and Connected Learning. http://www. samsung.com/es/business-images/resource/white-paper/2012/11/EBT15_1210_Samsung_ Smart_School_WP-0.pdf (2012)
15. Derzko, W.: Smart Technologies. http://archives.ocediscovery.com/discovery2007/ presentations/Session3WalterDrezkoFINAL.pdf (2007)
16. Uskov, A., Sekar, B.: Smart gamification and smart serious games. In: Sharma, D., Jain, L., Favorskaya, M., Howlett, R. (eds.) Fusion of Smart, Multimedia and Computer Gaming Technologies, Intelligent Systems Reference Library, vol. 84, pp. 7–36, Springer, Berlin (2015). doi 10.1007/978-3-319-14645-4_2, ISBN: 978-3-319-14644-7

Assessment in Smart Learning Environment – A Case Study Approach

Blanka Klimova

Abstract Current learning is becoming smarter thanks to the rapid development of mobile technologies. This, however, causes changes in traditional educational settings, such as learning takes place anywhere and at any time and in real-world context. The main emphasis is put on a learner itself and all is aimed at satisfying his learning needs in a new, smart learning environment. Therefore, the purpose of this article is to explain the current concept of smart learning environment, to explore its key criteria, and to show through a case study approach how assessment can be done in this smart learning environment. This is done at the example of an assignment carried out in the Course of Academic Writing taught at the Faculty of Informatics and Management in Hradec Kralove, Czech Republic. In conclusion the author of this article lists the skills which students need to succeed in their studies in this environment.

Keywords Smart learning environment · Assessment · Case study approach · Students' skills

1 Introduction

With the fast development of mobile technologies, current learning processes are becoming, on the one hand, more effective but, on the other hand, more complex because they require constant modifications and adjustments of teaching methods and materials so that learners' needs could be satisfied. One of the most popular buzz collocations nowadays in connection with learning is *smart learning environment*. There are ample definitions of smart learning environment (cf. [1–3], or [4]). The most recent has been provided by [5] who understands smart learning environment (SLE) as technology-supported learning environment that can make

B. Klimova (✉)
University of Hradec Kralove, Rokitanskeho 62, Hradec Kralove, Czech Republic
e-mail: blanka.klimova@uhk.cz

© Springer International Publishing Switzerland 2015   15
V.L. Uskov et al. (eds.), *Smart Education and Smart e-Learning*,
Smart Innovation, Systems and Technologies 41,
DOI 10.1007/978-3-319-19875-0_2

adaptations and provide appropriate support (e.g., guidance, feedback, hints or tools) in the right places and at the right time based on individual learners' needs, which might be determined via analyzing their learning behaviours, performance and the online and real-world contexts in which they are situated.

Furthermore, [5] also listed three essential criteria of SLE which are characterized as minimally *context-aware* (taking into account the learner's situation or the context of the real-world environment in which the learner is located); minimally *supportive* of the learner and his online and real-world contexts; minimally *adaptive* to the user interface (e.g., ways of presenting information), subject contexts to meet the personal factors (e.g., learning preferences), and learning status (e.g., learning performance). In addition, [2] provide more detailed descriptions of these three criteria:

- *Mobility*: The continuousness of computing while learners move from one position to another.
- *Location awareness*: The identification of learners' locations.
- *Interoperability*: The interoperable operation between different standards of learning resources, services, and platforms.
- *Seamlessness*: The provision of everlasting service sessions under any connection with any device.
- *Situation awareness*: The detection of learners' various situated scenarios, and the knowledge of what learners are doing with whom at what time and where.
- *Social awareness*: The awareness of learners' social relationship, including what do they know? What are they doing at a moment? What are their knowledge competence and social familiarity?
- *Adaptability*: The adjustability of learning materials and services depending on learners' accessibility, preferences, and need at a moment.
- *Pervasiveness*: The provision of intuitive and transparent way of accessing learning materials and services, predicting what learners need before their explicit expressions.

However, in such an environment learners must be self-directed, motivated to study and responsible for their own work [6]. Therefore, this article aims at demonstrating how these three criteria mentioned above can be reflected in assessing learners' skills of formal writing in the Course of Academic Writing which is taught at the Faculty of Informatics and Management (FIM) of the University of Hradec Kralove in the Czech Republic.

2 Background Information

The purpose of this article is to explore how students' assessment can be done in SLE. To attain this research objective, the author used a *case study approach* [7]; an empirical inquiry that investigates a contemporary phenomenon within its real-life context; when the boundaries between phenomenon and context are not clearly evident; and in which multiple sources of evidence are used. The author chose a representative assessment done in the blended Course of Academic Writing as the case study sample. Within this one-semester optional course for FIM students, which usually lasts 11 weeks, students learn how to write professionally. In addition, they focus on those features which are different in English and Czech, such as citations, compiling a bibliography or using appropriate English. As for the last aspect, there are independent sections on grammar structures in written English, lexical structures, and punctuation. The main topics are as follows: a summary of a lecture or a seminar; argumentative essay; professional essay 1 (including references and bibliography); professional essay 2 (including references and bibliography); writing an article for Wikipedia; and final consolidation and evaluation of the course by both the teacher and the students. There are usually five assignments/assessments during the course (see Fig. 1 below): a one–paragraph summary of a lecture/seminar; an argumentative essay without bibliographies and references; two essays including bibliographies and references and writing an entry for Wikipedia. In order to get a credit from this course, students have to do all assignments, including uploading the last article into the online version of Wikipedia.

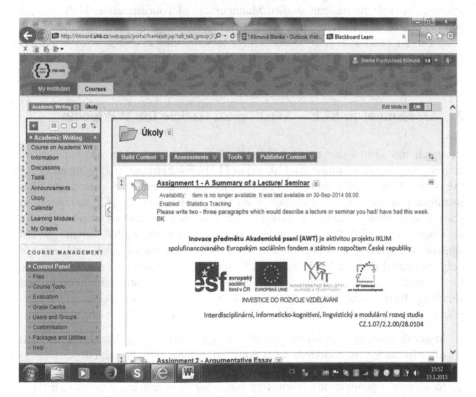

Fig. 1 An example of five continuous assessments in the course of academic writing [8]

All these assignments are formative assessments and thus they are carried out throughout the whole course every second week in order to help students to improve their writing skills. According to [9], formative assessment should be prevalent in writing courses for the following beneficial reasons:

- Formative assessment gives students a reason to read and understand the instructor's comments on their writing.
- Formative assessment aids students in applying the instructor's comments to the same or a very similar writing assignment, thus aiding them to become better writers.
- Formative assessment builds more time into the students' schedules for thinking and writing about assigned topics and results in better thinking and writing.
- Formative assessment helps students become better critics of their own writing, hence better revisers of their own writing.

For the purpose of this article the author chose just one assessment which corresponds to the SLE framework and that is the writing of an entry for Wikipedia and can be easily evaluated since it appears in the real-world settings in the end. Alltogether 14 students were working on this task. Out of 14 students only two were male students. And all the students studied Management of Tourism, either in the first or second year of study. Although the sample of respondents might seem small, for this course it is a maximum since students have to write long essays which the teacher must correct within a relatively short period to provide students with timely feedback.

3   Assessment in SLE – Writing an Article for Wikipedia

Research has shown that using Wikipedia for educational purposes, such as an assessment, is quite common (cf. [10, 11] or [12]). Furthermore, writing an entry for the online form of Wikipedia corresponds to the three essential criteria for SLE mentioned above. This assessment is context-aware since as [13] indicates, in producing a text for Wikipedia, students gain a real sense of audience and enjoy the satisfaction of seeing their work published on a high-traffic global website. When students are creating an authentic article for Wikipedia from scratch, they are not only motivated to write but begin to recognize the usefulness and necessity of the formal writing aspects of their course, e.g. the importance of attending to errors and checking facts when writing to be published. They are usually given a few guidelines as a minimum support by the course teacher [6]. These are as follows:

1. to get thoroughly acquainted with the website itself, e.g. to discover what kind of information is included, what kind of information is excluded, what is included in the footnotes . . . . . ;
2. to choose a genuine and interesting topic for their article, obviously, a topic which has not been covered in the wiki yet; if necessary to negotiate the topic with the person or institution that might be concerned;

3. to gather appropriate and relevant information on the topic and select only the most reliable and important facts;
4. to make an outline of the article;
5. to draft and revise the article a few times, preferably get someone to proofread it or to consult the facts in the article;
6. to format the source, make references and footnotes;
7. to submit the article and expect further revisions from the wiki reviewers.

Besides, the Wiki platform itself is SLE since it also provides instant support for the creation of a wiki article (for further information see [14]). And if students are still at a loss on how to edit their article, they ask their fellow students of computer science for help. In this was they also socially interact with other persons outside their group.

Furthermore, the course is run as a blended course. That means students meet a teacher once every two weeks to discuss and clarify the mistakes they made in their assignments, while at the same time, students are expected to undertake deep self-study of the materials that form their online e-Learning course. Thus, blended learning is perceived as an integration of face-to-face teaching and learning methods with online approaches [15]. This form of learning is quite popular among the students as several research studies conducted at FIM have already shown [16, 17] or [18]. Students have enough time for writing their assignment. Moreover, they can write it independently on time and location. That means it is adaptive to personal factors, such as students' learning styles and behaviour. Figure 2 below then demonstrates which days during the semester students accessed the online course and were the most active and how many times they accessed the course on the particular day. In addition, Fig. 3 is then more specific about the days of the week during which students were active at most. As Figs. 2 and 3 below illustrate, students were the most active on Mondays and Tuesdays. The reason is that Tuesday was the deadline of submitting their assignments. Figure 4 also indicates that students usually studied and wrote their assignments between 7p.m. and 10p.m. This is the time when students habitually finish their daily routines and can concentrate in the cosiness and peace of their home at more complex issues which require more time.

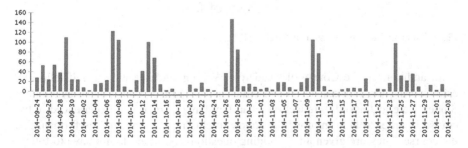

Fig. 2 Overall students' access in the course of the whole semester [19]

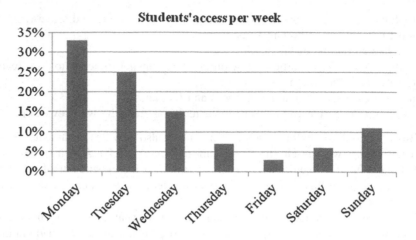

Fig. 3 Students' access per day of the week (author's own source)

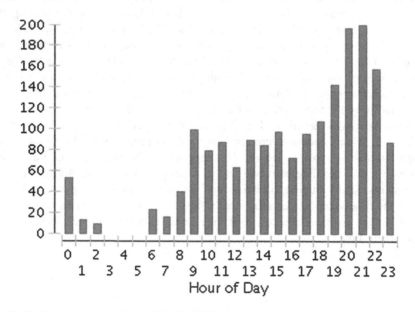

Fig. 4 Students' access per hour of the day [19]

In addition, in the Course of Academic Writing students write a self-reflective essay during the last contact lesson which is not part of their course evaluation. Students are just asked to reflect on what they have learned and experienced in the course in order to help improve the course, including the teacher's performance in the course. They are given a few guiding questions, but they do not have to follow them if they do not want to (see Appendix 1). They sometimes also mention the assignments. Thus, writing an entry for Wikipedia seems to be the biggest challenge and also the most difficult task for them. As one of the students put it:

> The greatest challenge for me was editing the Wikipedia article, it was quite difficult and I did not find much help on the Internet. However, in the end I was more proud of myself that I have completed this homework.

Nevertheless, no matter how difficult this task was, all students succeeded in completing and submitting it as a text into the online course as Fig. 5 partially shows below. After having their article checked and approved by their teacher, they had to upload into the online version of Wikipedia. See Fig. 6 for an example.

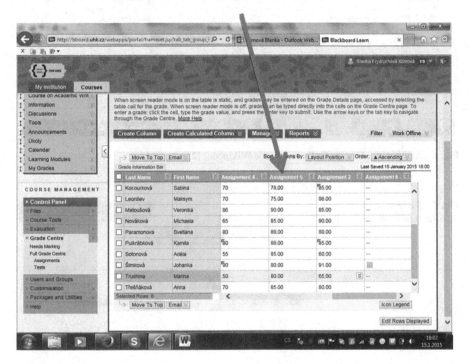

Fig. 5 Assignment being submitted into the online course to be checked by the course teacher [20]

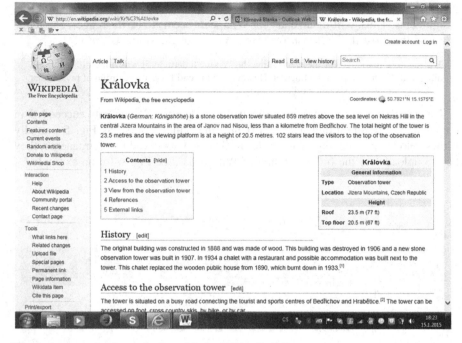

Fig. 6 An example of a student's article for Wikipedia [21]

Fig. 7 Students' skills in
SLE (author's own source]

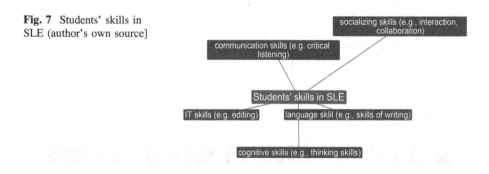

4   Conclusion

As this case study shows, technology-supported learning environment, appropriate support and adaptations can generate smart students who are able with a minimum support submit real-world assignments, such as an article for the online version of Wikipedia. Figure 7 below then summarizes relevant skills students need to succeed in solving SLE assignments with a minimum support from the teacher.

Appendix 1 – Guiding Questions for a Self-Reflective Essay

1. What did I actually achieve while attending this course? Which were the most difficult parts, and why were they difficult for me? Which were the most straightforward parts, and why did I find these easy?
2. How have I developed my knowledge and skills?
3. What were three greatest challenges in doing this course?
4. What was the most boring or tedious part of doing this course? What are your recommendations for a possible improvement?
5. In what way did this course help me in my future learning and career?
6. Did you find the online course useful, not useful?
7. What advice would I give a friend about to enrol in this course?
8. What have I learned about myself while doing this course? [22]

References

1. Winters, N., Walker, K., Rousos, D.: Facilitating learning in an intelligent environment. In: The IEEE International Workshop on Intelligent Environments, pp. 74–79. Institute of Electrcal Engineers, London (2005)
2. Yang, S.J.H., Okamoto, T., Tseng, S.S.: Context-aware and ubiquitous learning. Educ. Technol. Soc. **11**(2), 1–2 (2008)
3. Mikulecky, P.: Smart environments for smart learning. In: Proceedings of the 9th International Scientific Conference on Distance learning in Applied Informatics, pp. 213–222. Nitra, UKF (2012)
4. Hwang, G.J., Tsai, C.C., Yang, S.J.H.: Criteria, strategies and research issues of context-aware ubiquitous learning. Educ. Technol. Soc. **11**(2), 81–91 (2008)
5. Hwang, G.J.: Definition, framework and research issues of smart learning environments – a context-aware ubiquitous learning perspective. Smart Learn. Environ. **1**(4), 1–14 (2014)
6. Frydrychova Klimova, B.: Teaching Formal Written English, UHK, Gaudeamus (2012)
7. Yin, R.K.: Case study research: design and methods. Sage, Newbury Park (1984)
8. An example of five continuous assessments in the Course of Academic Writing. http://bboard.uhk.cz/webapps/portal/frameset.jsp?tab_tab_group_id=_2_1&url=%2Fwebapps%2Fblackboard%2Fexecute%2Flauncher%3Ftype%3DCourse%26id%3D_299_1%26url%3D (2014)
9. Trupe, A.L.: Formative assessment of student writing. http://www.bridgewater.edu/WritingCenter/Resources/sumform.htm (2001)
10. Cole, M.: Using wiki technology to support student engagement: lessons from the trenches. Comput. Educ. **52**(1), 141–146 (2009)
11. Deters, F., Cuthrell, K., Stapleton, J.: Why wikis? Student perceptions of using wikis in online coursework. MERLOT J. Online Learn. Teach. **6**(1), 122–133 (2010)
12. Kear, K., Donelan, H., Williams, J.: Using wikis for online group projects: student and tutor perspectives. Int. Rev. Res. Open Distance Learn. **15**(4), 70–90 (2014)
13. Tardy, M.: Writing for the world: Wikipedia as an introduction to academic writing. English Teach. Forum **48**(1), 12–19 (2010)
14. Wikipedia: about. http://en.wikipedia.org/wiki/Wikipedia:About (2015)
15. Littlejohn, A., Pegler, C.: Preparing for blended e-Learning. Routledge-Falmer, UK (2007)

16. Hubackova, S., Semradova, I.: Comparison of on-line teaching and face-to-face teaching. Procedia Soc. Behav. Sci. **89**, 445–449 (2013)
17. Frydrychova Klimova, B., Poulova, P.: Forms of instructions and students´ preferences – a comparative study, hybrid learning, theory and practice. In: Proceedings of the 7th International Conference (ICHL 2014), pp.220–231. Springer, Berlin (2014)
18. Frydrychova Klimova, B., Poulova, P.: ICT in the teaching of academic writing. Lect. Notes Manag. Sci. **11**, 33–38 (2013)
19. Evaluation reports. http://bboard.uhk.cz/webapps/portal/frameset.jsp?tab_tab_group_id=_2_ 1&url=%2Fwebapps%2Fblackboard%2Fexecute%2Flauncher%3Ftype%3DCourse%26id%3D_2 99_1%26url%3D (2014)
20. Assignments, Full grade centre. http://bboard.uhk.cz/webapps/portal/frameset.jsp?tab_tab_ group_id=_2_1&url=%2Fwebapps%2Fblackboard%2Fexecute%2Flauncher%3Ftype%3DCours e%26id%3D_299_1%26url%3D (2014)
21. An example of a student's article for Wikipedia. https://en.wikipedia.org/wiki/Kr%C3% A1lovka (2014)
22. Race, P.: Evidencing reflection: putting the 'w' into reflection. http://escalate.ac.uk/resources/ reflection/02.html (2006)

The Design Research of Future Informal Learning Space

Constructing the "Smart Space" of Beijing Normal University Library

Yun Zhang, Anan Liang, Huiping Sun, Lan Liu
and Fengkuang Chiang

Abstract With the development of modern technology, learning space has not only broken through the limits of traditional classrooms, but also extended to various places of human activities. Informal learning space has become an important part of learning space. Based on related literature, the analysis of the user's demands and design philosophy, this study reconstructs the spatial planning of the multimedia center of Beijing Normal University Library by means of graphic design and 3D modeling, evaluates the model, and gives some suggestions for the future design of informal learning space.

Keywords Informal learning space in future · Design · Library

Y. Zhang · A. Liang · H. Sun · F. Chiang (✉)
School of Educational Technology, Beijing Normal University, Beijing, China
e-mail: fkchiang@bnu.edu.cn

Y. Zhang
e-mail: Zyun@mail.bnu.edu.cn

A. Liang
e-mail: LiangAa@mail.bnu.edu.cn

H. Sun
e-mail: SunHp@mail.bnu.edu.cn

L. Liu
Library, Beijing Normal University, Beijing, China
e-mail: liul@lib.bnu.edu.cn

© Springer International Publishing Switzerland 2015   25
V.L. Uskov et al. (eds.), *Smart Education and Smart e-Learning*,
Smart Innovation, Systems and Technologies 41,
DOI 10.1007/978-3-319-19875-0_3

1 Introduction

1.1 Background

Generally speaking, "learning space" means all the places where learning activities occur; a complex of the physical environment between students and teachers, formal or informal, physical or virtual dimensions with resource services available. Nowadays, learners propose new requirements that learning space should be more flexible, efficient, intelligent and integrated. Concept of learning space in this paper is extended to the informal learning field. The model design should comprehensively take into consideration of the learners' perspective and learning activities from the following factors: the physical spaces of the integrated environment, hardware facilities, software resources and the study of space design.

As an important informal learning space, in recent years, some scholars have proposed the concept of a modern library. They consider that the library should be a place where people can communicate with each other, have academic discussion, and enjoy entertainment and other services. The building is considered as an integrated function space, including a lecture hall, multimedia presentation room, bookstore, music halls, readers' restaurant, underground parking, etc., embodying the comprehensiveness of a library [1]. In order to meet the new demands of learners, Beijing Normal University started the renovation project of Library Multimedia Learning Centre in 2014 which aims to build the future informal learning space to support the proactive exploration of learners and the information sharing of learning community.

1.2 Problems

The Multimedia Learning Centre in BNU (Beijing Normal University) Library built in 2010 is faced with several questions now, such as not satisfying the individual needs of the students. Therefore, our study is based on the renovation project of Multimedia Learning Centre in BNU, summarizing the basic concepts, functional characteristics and design principles of the future informal learning space in literatures. Incorporating both theoretical and practical, it also provides a viable solution by designing the 3D model of the future learning space. It focuses on designing a flexible, user-friendly, full technical support informal learning space to meet the demands of collaborative communication, self-exploration, and extra-curricular interest expansion to enhance the learning concepts and modes.

2 Research Status

- **Valuing the function and the design theory of learning space.** Considering the physical environment, Zehra Ahmed [2] pointed out that physical space is often defined by the lighting, furniture, and technology in your classroom. It can be somewhat flexible and faculty can arrange or control for certain aspects of it by maximizing its use to benefit student learning. The report of JISC (Joint Information Systems Committee) described the technological development of learning space and better cost-return usage of space. In addition, it also presented that the 21st century learning space design should follow the basic principles such as human -centeredness, personalization, comfort and flexibility, self-adaption, energy conservation and so on [3].

- **Concerning the future learning space construction of informal learning place like libraries.** As an important informal learning space, Walton G [4] indicated that "University libraries have massive experience in providing and developing different learning spaces and, at the same time, have the skills and expertise in their evaluation." Besides, Jamieson P [5] considered that "The library has been the spatial laboratory where many universities have explored the possibility of other formal and informal learning settings (e.g., IT training rooms, video conferencing centers). In addition, the library has been the place for initiating the use of new technologies, developing dedicated training centers for staff and students, and even incorporating cafés into formal working areas. Importantly, it is in the library that students often encounter the university's rich historical, cultural, and art collections that add an important dimension to the on-campus learning experience." Case B M C [6] pointed out that library should Have the following features: "Quiet zones for students who prefer quiet; social zones for students working collaboratively; a large collection of books at a wide range of reading levels, covering subjects, the students need or enjoy; computers with high speed Internet access and up-to-date hardware and software; well-qualified, helpful, and approachable librarians."
- **Design of informal learning space in the Library.** The library incorporates modern knowledge communicating, processing, spreading and accumulating into a dynamic academic center, and it gradually develops and transfers into the future learning space. Especially, the universities' libraries combine study and academic research together and their spatial layouts pay more attention to multi-information and multi-functional discussion area. Saltire Centre of Glasgow Caledonian University is a five-story facility incorporating the campus library, "Users can freely choose or move in the quiet area or active area based on their own will. In addition, the library provides a relatively close wall for persons and groups with inflatable and mobile equipment and the hall in the first floor can, according to the requirements of users, change its physical parameters, such as temperature [7]."

3 Design

3.1 Need Analysis

In order to understand the views and requirements of teachers and students about the multimedia learning center renovation project in BNU library, the research organized two special symposiums of the teachers and students for feeback by means of focus group interviews. The research chose 10 teachers from 9 departments and institutes of BNU as well as 6 undergraduates, 10 master students, and 5 Ph.D. candidates who attend the focus group interviews. These 31 interviewees often use the library's services (data from the users information of library database), and representatively reflect the library's users' needs. The integrated opinions of teachers and students' representatives are shown in Table 1, coded according to function.

Table 1 The integrated recommendations of teachers and students' representatives

Item	Description	Code
Function division	The entire learning space is divided into several different functional regions	A
Topic research zone	Semi-open, independent research unit for discussion and communication	A1
BNU gallery	Showing academic achievement, e-books, the history of BNU, to build a cultural atmosphere	A2
Creative experience zone	Support for exhibitions, lectures, salons and other diverse informal learning activities	A3
Personal learning zone	Independent personal self-study environment	A4
Leisure services zone	A comfortable sitting area with self-service	A5
Technical support	Advanced technical support, such as touch-screen presentations, holographic images, high-performance computers, and mobile applications	B
Resource sharing	Educational cloud platform to offer various, latest learning resources and tools	C
Flexible facilities	Facilities, such as tables, chairs, and sofas, can be remodeled to adapt to different forms of learning activities	D

3.2 Space Design

This study uses a series of software, such as 3Dmax2014, Sketchup2014, Photoshop CS6, to design the future informal learning space model.

Design Principle. Based on theoretical principles, and drawing on existing case studies of learning spaces, the design concepts of the "Smart Space" are put forwarded as follows.

Dynamic and Static Balance. By means of teachers and students' needs analysis, we know that the discussion areas (A1) and personal learning space (A4) in the library are greatly needed. The concepts of "Dynamic" and "Static" representing different learning activities characteristics should be integrated. "Dynamic" learning activities, such as discussions, debates, lectures, salons, games etc., emphasize socialization and generality, and require the interaction with outsiders and larger environment space; while "Static" learning activities, such as self-study, reading, exhibition, experiencing, etc., focus on the personalization and independence, require personal independent space and personalized preferences.

Virtual and Real Combination. According to distributed cognition theory, learning space design must be supported by a variety of modern products. Therefore, learning space should integrate the online virtual environment and offline physical space effectively and realize the seamless connection of online and offline learning. Online environment includes virtual platform, intelligent tutor, abundant digital resources and software application, etc., while offline space has the support of the physical environment, such as, communication space, self-service equipment, various collections of ancient books and plentiful face-to-face learning activities, such as salons and training.

Flexible Features. Future learning space needs to have extensibility and re-configurability, in order to adjust hardware facilities and update software resources to adapt to the needs of different learning activities. For example, using reconfigurable tables and chairs, make it possible to quickly change the spatial layout depending on the number of people and scope of activities; wireless devices can dynamically adapt to learning activities, promoting ubiquitous learning.

Design Elements and Functions. Based on the needs of A, A1, A2, A3, A4, A5 (Table 1), the space is divided into five functional areas shown in Fig. 1.

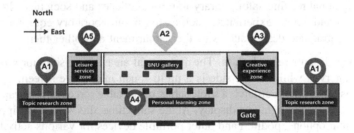

Fig. 1 The division of five functional areas

Topic discussion area: collaborative, building the learning community. Topic discussion area, located on both sides of the space, is the core zone. It includes 14 miniature seminar rooms accommodating 4–8 people and 2 large ones accommodating 10–20 people.

- Semi-open privacy design reflecting a humanistic perspective. The glass door, whose top is transparent and bottom is frosted glass decorated with ginkgo leaf, is convenient for staff to check the situation in the seminar room and create private spatial sense both at the same time.
- Screen wall. Seminar rooms are separated by the wall, which meanwhile is an interactive screen. The demonstration area could be adjusted.
- High-tech equipment. Seminar room is equipped with moveable white board, recording system and small platform and so on, to facilitate the discussion effectively.

- Triangular table. Tables in the area are light and flexible. The desktops are triangular, and they can be assembled into different shapes according to the seat arrangement and the number of students.

BNU gallery: interactive, reflecting characteristics of the library. It is an important area reflecting BNU's construction and development, the academic achievements of teachers, students and outstanding graduates.

- Corridor Learning. The region's body is a more-than-30 m central wall in the north, which transformed into a multi-display and multi-touch, interactive screen, displaying development history, legendary figures, achievements and so on in virtual and digital ways. The adaptive light skylight makes it convenient for people to learn interactively when walking along the gallery. Based on the perspective of the environmental psychology, "a wide aisle is helpful to promote people to communicate while walking, to slow down walking speed" [8]. Therefore, the corridor is spacious and comfortable, to effectively promote the informal learning.
- Entity books. The region combines virtual interactive digital resources and tangible resources, such as books, and pamphlets, which are placed in bookshelf and portable display board.

Creative experience zone: Flexible, supporting various learning activities. Creative experience zone is characterized by elasticity and plasticity. It is a platform for the occasional reading salon, library training, colleges and societies' show, new technologies and device experience, such as the brain laboratory equipment of the Department, graduate design shows of the Department and other theme activities.

- Arc-shaped double-sided glass. The functional areas using soundproofed glass as the interval, while on the side is a multi-touch interactive screen.
- Curtains screen. The curtain is a stretchable, projectable screen, which can be used to block out the sunlight and display. At the same time, this zone is equipped with audio, microphone, podium and other portable devices for various activities.
- Equipment display. Top of the wall is a dual display screen for showing digital resources, such as library lectures, campus events news, community videos and so on. Below that, there are a variety of advanced technology and equipment for students to experience and lend, such as 3D imaging holographic projection, IPad, SLR camera and electronic schoolbag etc.
- Chairs and tables supporting diverse activities. This zone features reconstruction and flexibility, which can meet the needs of different activities like exhibition, sharing, training, book salon. Chairs and tables are designed with stitching, multi-functional characteristics, following ergonomics comfort requirements, a need for a chair backrest, height and width consistent with human's using habits. Thus, the design of the bench is dual purpose: it can be used as a display platforms and it can connect with others, while the height can be adjusted for display.

Leisure service area: comfortable and intelligent perception service. Leisure service area faced the area in the direction of french windows, which is beneficial to relax oneself and alleviate eye fatigue.

- Semi-open division. To protect privacy and distinguish from the personal learning area and corridor learning, a semi-open clapboard and mobile bookshelf are used as intervals.
- Splicing-style sofa and self-adaption windows. It provides splicing-style sofas and coffee tables for relaxing. French window can intelligently sense light and adjust the degrees of sunlight reflection.
- Self-service equipment. It provides one-stop service and students can copy, print, inquire and borrow books by themselves. It also supplies a coin-operated coffee maker, updated newspaper and magazines, time-limited password storage and charging service, etc.

Personal Learning Area: personalized, meeting diversified self-study demands. Personal study area meets the need of self-study and reading, where tables are equipped with computers and laptops, supplying many kinds of extracurricular resources and software.

- Splicing-type personal desks. Around the pillars, splicing-type personal desks can be spliced into semi-circular or circular style flexibly and the seats can be increased or decreased according to the number of people.
- Wave-shape long desk. The long desk cross the whole middle space and the raised platforms are designed at the regular distance as seat intervals to create suitable sense of personal space. Besides, raised platforms can be placed with small pot plants to enhance greening effect and relax people's eyes and emotions.
- Diversified bookshelves against the wall. The south wall is uneven and has several narrow corners, so it is hard to be reconstructed. To make full use of space, the wall that is uneven can be equipped with book mobile shelves and exhibit shelves to provide more books for available reading.

Three-dimensional Modeling. According to the design of functional module division, 3D models are made in Google Sketch up, and shown in the Figs. 2 and 3.

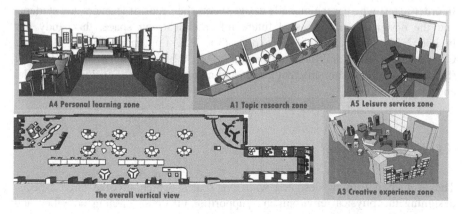

Fig. 2 Functional module division

3.3 Innovations

Diversity of regional function. Break through the limitations of single library function, the model can be dynamic transformed, such as the creative experience area can always adapt to different activities; one room is multi-purpose.

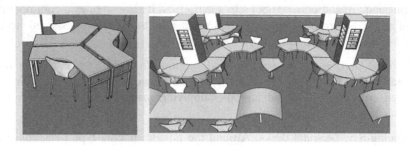

Fig. 3 Part of the splicing-type desk models

Fusion of "move" and "static" elements. Breaking the traditional library design which gives priority to static parts, the model integrates activities factors such as technology experience, corridor learning, to effectively promote the informal learning.

Feature of the informal learning. Set up the topic discussion area for students to custom topics. Students in different professions and grades can join the discussion according to their personal interests at any time.

4 Evaluation

In order to test whether the model's design meets the needs of teachers and students and the design specifications of future informal learning space, this study has invited 2 educational technology experts (T1, T2), 1 interior designer of Urban Construction Design and Research Institute (D1), 1 library renovation project staff (L1) and 6 students from different grades and majors (S1–S6) to evaluate the model and make comments.

4.1 Evaluation Tool

The evaluation form is designed with five scales, which involve three aspects: designing the physical environment, supporting informal learning activities and planning library renovation project. The items of designing the physical environment include seven secondary indicators: the overall design, functional layout, humanity, comfort and so on. The items of supporting informal learning activities include five secondary indicators: technical support, independent learning, collaboration and communication, etc. The items of planning library renovation project include two secondary indicators: the school's cultural atmosphere and site environmental features. In addition, two open topics for evaluators to give suggestions are provided.

4.2 Evaluation Result

After we collected the evaluation sheets and statistics of evaluation data, we found that each item got an average score above 4 (on a five-point scale), except for the 13 indicator (The model can show the characteristics and humanist approach of BNU) got 3.5 in average. Therefore, estimators are satisfied with the model on the whole. Especially, on the flexibility(Mean = 4.6, SD = 0.66), interactivity(Mean = 4.4, SD = 0.8), and function layout(Mean = 4.7, SD = 0.46).

The result indicated that students generally showed higher satisfaction on the model (Mean > 4.3). Students are the main user community in the library, so the model can meet the users' demand to a large extent. However, some students gave suggestions on the space utilization and management issues in open topics. S2 pointed out that: *"the seats are not enough, and it's a waste of space"*, while S3 was concerned about seat management: *"The mechanism of management and supervision should be concerned"*.

The interior designer agreed that the design model was in line with the basic principles of interior design, and the model had high availability, flexibility and intelligence. Meanwhile, D1 point out that the model should improve the guidance for users: *"A plan introduction should be set in the Entrance, at the same time, signs and instructions should be clearly set up in every regions"*, *"pay attention to the stream of people guidance and to set up the recommended routes design"*.

In educational technology experts' opinion, the current model is in line with the needs of the library. As for the whole, to improve the model's design, T1 suggested that creative elements or symbol marks should be integrated into the overall design of the space: *"Some philosophical slogans and classical Chinese ink paintings can be added on the metope; sofa should be set in the personal learning area to create reading space anytime and anywhere"*. Besides, he also concerned about whether the stream of people will make interference: *"People walking in and out between the BNU gallery and leisure service area may interfere with users who would like to study in quiet individual learning areas"*. Besides that, the library's staff paid more attention on the humanities environment design. L1 point out that:*" The construction of space should be based on Campus culture and on behalf of the school image; this principle should be strengthened in this study"*.

Based on above comments and feedbacks, the current model is considered to meet users' requirements and design specification with innovation ideas. However, it also has insufficiencies, such as a lack of route planning and humanistic atmosphere. Some modification and improvement should be carried on.

5 Summary

At present, informal learning space design and practice is inadequate. This research designed the informal learning space of BNU library by puttng related theory into practice, with needs analysis, and 3D modeling and evaluation, to form a set of specific space design patterns. This study can provide some reference as follows:

- Consider the physical environment features and the actual requirements of users together. This model is based on the data of local investigation at the scene and users' demand from the focus group interviews.
- Pay attention to the partitioning and transitions of different areas. According to the results of needs analysis, users are fond of space partitions, which can support various informal learning activities. In addition, the evaluation result showed that the model is expected to be improved on the route signage, space transition and area guidance. In related research, the transition between different areas is rarely discussed and designed, which is worthy of further research.
- Focus on creating a humanitic atmosphere and reflect the library culture. The evaluation result shows that the library characteristics and cultural atmosphere is insufficient in this model. Some creative elements and decorative symbols, such as quotes by famous people, lighting and greenery should be considered.

Besides, based on the literature review and evaluation result, furniture restructuring, technical application and theoretical support are important. Reconfigurable furniture and multi-purpose design is useful; new technology application are expected, but avoiding purely technical "stuffing"; synthesize cognitive tendencies, communication habits, environmental color principles, sensory laws, etc., so that the design conforms to informal learning characteristics and mechanisms.

Acknowledgments Our work is supported by the Fundamental Research Funds for the National Engineering Research Center of Educational Technology, and Faculty of Education in Beijing Normal University (CXTD201401).

References

1. Hao, L.: The study on the design of environment and art of Library Building (Doctoral dissertation, Xi'an University of Architecture and Technology) (2006)
2. Zehia, A., Nick, B., Amy, B., Petra, D., Sumiao, Li, Rich, S.: An online workshop, Center for teaching and learning, New York Institute Technology. J. Learn. Space (2014)

3. Jisc, J.: Designing spaces for effective learning: a guide to 21st century learning space design. JISC, Bristol (2006)
4. Walton, G., Matthews, G.: Evaluating university's informal learning spaces: role of the University Library. J. New Rev. Acad. Librariansh. **19**(1), 1–4 (2013)
5. Jamieson, P.: The serious matter of informal learning: from the development of learning spaces to a broader understanding of the entire campus as a learning space. J. Plann. High. Educ. 37(2), 18 (2009)
6. Case, B.M.C.: An open door: exploring how high school students with special needs view and use the high school library. Fielding Graduate University, Diss (2011)
7. Watson, L.: The Saltire Centre at Glasgow Caledonian University. SCONUL Focus, no. 37:. Retrieved 15 Oct. 2008, http://www.sconul.ac.uk/publications/newsletter/37/2.pdf, 4–11 (2006)
8. Pijir, P.M.: Investigation of the pedestrian movement in the streets in the downtown of the largest cities. Diss, Moscow (1971)

Moodle-Based Computer-Assisted Assessment in Flipped Classroom

Jerzy Rutkowski

Abstract Nowadays, Information and Communication Technologies (ICT) make great strides in computer assisted education. They enable development of new, Self-Directed Learning (SDL) driven, models of teaching and learning, and among them, Flip Teaching (FT) seems to be the most promising one. The flipped classroom is based on e-materials, then, the associated assessments have to be fully based on e-tests, FT-Computer-Assisted Assessment (CAA) program. Design of such program is the main motivation of the described research. Matching of this program to FT characteristics, taking into account both specific structure of the learning content and students behavior is the main objective. After a short introduction to FT, the e-test organization in the flipped classroom is briefly discussed. Then, Electric Circuit Analysis course case study is presented: organization of Moodle-quizzes based assessment program, taking into account common students behavior, their good and bad habits, is shown, conclusions are drawn and some guidelines are given.

Keywords Technology enhanced learning · Flip teaching · Computer assisted assessment

1 Introduction

In the era of dynamic development of Information & Communication Technologies (ICT), their use in education is ubiquitous. The ICT, properly used, may significantly contribute to the quality of education, enable development of new forms of Technology Enhanced Learning (TEL). Thanks to ICT development we are at the heart of a global learning revolution. The act of learning itself is no longer seen as simply a matter of information transfer, but rather as a process of dynamic

J. Rutkowski (✉)
Silesian University of Technology, Akademicka 16, 44-100 Gliwice, Poland
e-mail: jrutkowski@polsl.pl

© Springer International Publishing Switzerland 2015   37
V.L. Uskov et al. (eds.), *Smart Education and Smart e-Learning*,
Smart Innovation, Systems and Technologies 41,
DOI 10.1007/978-3-319-19875-0_4

participation in which students cultivate new ways of thinking and doing through active discovery and discussion [1]. They migrate toward Self-Directed Learning (SDL) experiences on computer and Internet. Then, an immediate attention to the conduct of teaching has to be given, such that both students' and teachers' expectations are met, worlds of "digital natives" (students) and "digital immigrants" (educators) merge. From different forms of TEL (MOOC, CDIO, PBL, SDL), SDL, Flip Teaching (FT) in particular, seems to be the most promising one [2–5] - it is predicted that by year 2020, all learning will be based on principles of SDL. In the FT, the student first studies the topic by himself, using e-materials created by the teacher. In the classroom, the student then tries to apply the knowledge by solving problems, together with the teacher and peers. That way, the traditional classroom evolves into discussion forum direction. In general, a course content may be divided into two components:

- learning content (knowledge delivery);
- assessment program (knowledge assessment).

In both traditional classroom and flipped classroom these two components have to be strongly correlated. In FT, the learning content is based on e-materials. Then, FT-Computer-Assisted (Aided) Assessment program should be fully based on e-materials as well and take into account the FT characteristics, the modular structure of a course in particular. The student is measured, both during the course (formative assessment) and the final exam (summative assessment). In FT-CAA program, the assessments are completed by the student at a computer, at home (formative assessment) or in a computer lab (exam), without the intervention of an academic. The following problem has been formulated: "How to assess large groups with minimum amount of resources but preserving quality?" [7]. Some extensive research has already been done in this field and effectiveness of CAA systems have been reported [8–10]. However, further extensive studies are necessary to make the assessment system fully computer-automated and reliable, such that traditional assessments can be replaced by e-assessments. In fact, traditional assessments have to be practically eliminated from the flipped classroom. Following Author's experience, it has to be clearly stated, that it is impossible to "assess large groups with minimum amount of resources but preserving quality"! To preserve high quality of learning in the flipped classroom, comprehensive bank of questions has to be created and this is a very time consuming job! Modularity of the learning content is essential for the flipped classroom development. Then, modularity of such bank of questions, of the FT-CAA program in general, is essential as well. In Sect. 2, different approaches to e-test organization are briefly discussed. In Sects. 3 and 4, Electric Circuit Analysis course case study is presented: organization of FT-CAA program, taking into account common students behavior, their good and bad habits, is shown, conclusions are drawn and some guidelines are given.

2   Organization of e-Tests

There are many types of tests that can be used at both stages of a course, formative and summative stage. In general, they can be divided into two categories:

- open-ended tests;
- close-ended tests.

Open-ended and close-ended tests differ in many characteristics, especially in regards to the role of a student when solving the test tasks [6]. Close-ended test limits the student to the numerical answer or the set of possible alternatives offered. Open-ended test allows the student to express the way of reasoning and own presentation of the obtained solution. Advantages of the open-ended tests include the possibility of discovering responses that individuals give spontaneously, and thus avoiding the bias that may result from suggesting responses [6]. Unfortunately, the traditional open-ended test can't be used in fully automated CAA, as the open-ended test marking has to be done manually. The manual marking of open-ended test has also other drawbacks, such as:

- lower reliability, as number of the tasks/questions is limited by marking time effort practically to a few, maximum ten tasks, and consequently, they can't cover all the course topics;
- subjectivity of marking.

Completion of the comprehensive Question Bank is the introductory step leading to creation of the assessment program, three attributes that describe each question character have to be assigned [11]:

1. Descriptor: LLx.y, where LL describes Part of a course, of x is lecture number within LL, y is question number within LLx;
2. Easiness: easy or moderate or difficult;
3. Reasoning: forward or reverse.

After completing such Question Bank, each question has to be transferred into one of Moodle formats, Moodle allows the following formats of computer-marked questions [12]:

1. Multiple Choice Question (MCQ): Allows the selection of a single or multiple responses from a pre-defined list. Each answer can have its own mark (positive or negative);
2. True/False: A simple form of MCQ;
3. Short Answer: Allows a response of one or a few words that is graded by comparing against various model answers, which may contain wildcards;
4. Numerical: Allows a numerical response, possibly with units, that is graded by comparing against various model answers, possibly with tolerances;
5. Calculated: Calculated questions are like numerical questions but with the numbers used selected randomly from a set when the quiz is taken;

6. Matching: The answer to each of a number of sub-question must be selected from a list of possibilities;
7. Random Short Answer Matching: Like a Matching question, but created randomly from the short answer questions in a particular category;
8. Calculated Multichoice: Like MCQs which choice elements can include formula results from numeric values that are selected randomly from a set when the quiz is taken;
9. Calculated Simple: A simpler version of calculated questions which are like numerical questions but with the numbers used selected randomly from a set when the quiz is taken;
10. Embedded Answers: Questions of this type are very flexible, but can only be created by entering text containing special codes that create embedded multiple-choice, short answers and numerical questions.

In Moodle, creating a new quiz is a two-step process. In the first step, the teacher creates the quiz activity and sets its options which specify the rules for interacting with the quiz. In the second step, the teacher adds questions to the quiz from the Question Bank. The teacher can:

- allow students between one and an unlimited number of attempts;
- restrict access to specific time window(s);
- restrict access by a variety of criteria including user id, previous performance and IP address;
- protect the assessment by a password, for example a password only given out at one time in an exam room;
- determine the amount of feedback, review options.

When planning formative assessments, the teacher has to select a strategy, decide whether quizzes are obligatory for students or not, whether some bonus for diligent students can be granted. Generally, "Carrot and Stick" approach seems to be the best one.

3 Case Study: Electric Circuit Analysis Course

The Electric Circuit Theory (ECT) course has been fully redeveloped into the flipped classroom model in the academic year 2012/2013 [5]. The assessment program supporting the flipped teaching has been introduced in the academic year 2013/2014, after some ten years of preparations. The Question Bank of around 500 questions has been created and it is revealed to students, filed on the course website. These questions are used:

- during lectures and classroom exercises,
- in pre-lab quizzes,
- in formative tests,
- in final examination.

Modularity of the learning content is essential for the flipped classroom development. The Electric Circuit Analysis course consists of three Parts:

- DC Analysis;
- Transient Analysis;
- AC Analysis.

The whole content, all three Parts are divided into lectures, basic units with fixed content. Each lecture is divided into modules supported by the video-podcasts. The block diagram for the first four lectures of DC Analysis, with hyperlinks to video-podcasts, is presented in Fig. 1. Such structure of the learning content requires the same structure of formative assessments and they have been organized in the following manner:

- six quizzes cover DC Analysis, with 20 questions per quiz,
- ten quizzes cover Transient & AC Analysis (4 Transient, 5 AC, 1 miscellaneous), with 15 questions per quiz.

The first quiz covers all modules of the 1st lecture (Intro) and the 2nd lecture (DC1), the second quiz covers one module of the 3rd lecture (DC2.1 – single module lecture) and the 1st module of the 4th lecture (DC3.1), the third quiz covers next two modules of the 4th lecture (DC3.2, DC3.3), and so on. There are two summative assessments:

- the 1st covers DC Analysis, with 6 questions, each selected from each formative quiz;
- the 2nd covers Transient & AC Analysis, with 10 questions (4 Transient, 5 AC, 1 miscellaneous), each selected from each formative quiz.

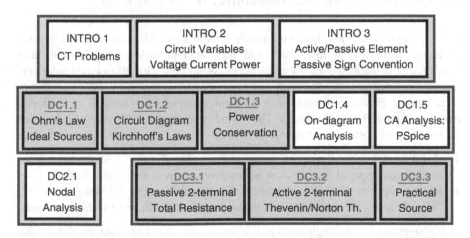

Fig. 1 Modular structure of ECT lecture, first lectures of DC analysis

3.1 Distribution of Question Attributes in Quiz (Formative and Summative)

1. Descriptor: LL = DC or TRansient or AC (see above);
2. Easiness: Gauss distribution is applied;
3. Reasoning: some 80 % of questions are forward reasoning.

3.2 Distribution of Question Formats in Quiz (Formative and Summative)

Moodle quizzes have been first used in the academic year 2005/2006 [11]. After seven year experience with quizzes, built of MC, Numerical and Matching Questions, it has been decided to redevelop most of the questions into Calculated format. Students must learn that in engineering calculations the accuracy of the final result is as important as the correctness of the process they follow to obtain it [13] and today all **formative quizzes** are built only of Calculated Questions. The **summative quiz** contains only 1(2) non-Calculated Question(s). In Calculated Questions, students are often making "silly" mistakes in calculations, mainly due to carelessness. Such a mistake results in "0" mark. This mark can be manually upgraded by the teacher from "0" to "0.5", after post-exam discussion with the student, based on scratchpad with calculations left before leaving the exam-room (computer lab).

3.3 Quiz (Formative and Summative) Activity Settings

For formative tests, multiple (unlimited) attempts are allowed and this setting excludes questions other than Calculated. Each attempt builds on the last, i.e. each new attempt contains the results of the previous attempt. This allows the student on the new attempt to concentrate on just those questions that were answered incorrectly on the previous attempt, uses the same questions in the same order, independent of randomization settings. Time window is set (see "Carrot & Stick") and no special password is required. Other settings:

- Question behavior: "Adaptive mode (no penalties)", allows students to have multiple attempts at the question before moving on to the next question;
- Review option: "Immediately after the attempt", then, "Whether correct", "Mark" (apply for the first two minutes after 'Submit all and finish' is clicked).

For summative test, single attempt is allowed, time window is set on 90 min (the test duration) and quiz is password protected, access restricted by student id and IP (lab) address. Other settings:

- Question behavior: "Deferred Feedback", student must enter an answer to each question and then submit the entire quiz, before anything is graded;
- Review options: same as for formative tests.

3.4 "Carrot & Stick"

The Author's first experience with formative e-tests included the application of LMS for non-obligatory quizzes [14–16], as part of the blended classroom. Unfortunately, there was very little feedback from the students – not more than 15 % tried to solve these quizzes. This is unfortunately consistent with the Author's experience with the traditional classroom – if an activity is non-obligatory, such as attending lectures, consultations and homework, the student interest is almost none, up till the final examination time. Similar conclusions were presented by Rodanski [13]. The students were surveyed and most of them appreciated the idea of having opportunity to learn using the stored repository, but somehow they happened not to take it.

Then, since the academic year 2008/2009, the formative quizzes are obligatory. If the student didn't pass the quizzes before the deadline, didn't solve 1/3 of all questions, then he/she is not admitted to the summative assessment and this is "**Stick**". As the result of this change, students have been forced to more systematic learning, evident improvement of pass/fail ratio has been observed. Today, practically 95 % of students pass quizzes, only some 10 % of them fail the exam – some 20 % failed before.

There is also "**Carrot**" for the top students, students that solve more than 95 % of all questions – the final mark of these students is upgraded by "0.5" and number of this students oscillates around 20 %. The fact that the summative assessment questions are drawn from the formative assessment quizzes is an additional motivation to solve these quizzes.

3.5 Exemplary Questions

Three exemplary questions are presented in Figs. 2, 3 and 4. The question of Fig. 2 is the Calculated one and has the following attributes:

- Descriptor: Part LL = DC, lecture number x = 1, question number y = 9;
- Easiness: moderate;
- Reasoning: reverse.

The question of Fig. 3 is the Calculated one and has the following attributes:

- Descriptor: Part LL = TRansient, lecture number x = 1, question number y = 1d;
- Easiness: easy;
- Reasoning: forward.

The question of Fig. 4 is the MCQ one and has the following attributes:

- Descriptor: Part LL = AC, lecture number x = 4, question number y = 14a;
- Easiness: moderate;
- Reasoning: forward.

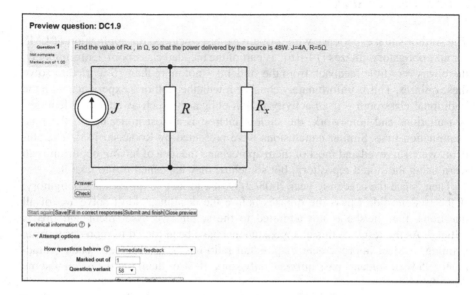

Fig. 2 Exemplary question: DC1.9

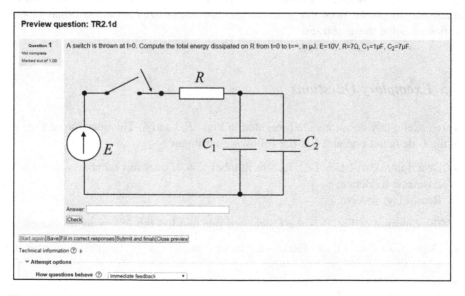

Fig. 3 Exemplary question: TR2.1d

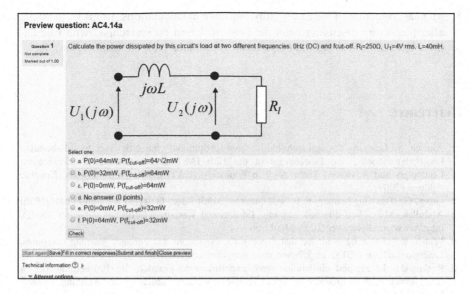

Fig. 4 Exemplary question: AC4.14a

4 Final Conclusions and Guidelines

The complete flipped classroom has to supported by both e-materials supporting knowledge delivery and e-materials supporting knowledge verification by formative and summative assessments. After some ten year experience with Moodle-quiz based e-assessments, one year experience with e-assessments in the flipped classroom, the following guidelines, recommendations, for an engineering course, can be given:

1. Quizzes, both formative and summative, should match modular structure of a course, questions of summative test should be drawn from formative tests, one question from each formative test;
2. Moodle quizzes, both formative and summative, should be built mostly of Calculated Questions, 80 % for exams, 100 % for formative tests, with the correct answer marked "1", incorrect answer marked "0" or "−0.5" for MCQ, Numerical and Matching;
3. Formative tests have to be obligatory and "Carrot & Stick" approach is recommended, e.g. upgraded exam mark for the top students (Carrot) and lack of exam admission (Stick);
4. Quiz, both formative and summative, easiness distribution should be the Gauss one;
5. Ratio of reverse-to-forward reasoning questions around 20 % per quiz is recommended, for both formative and summative tests;
6. In formative quizzes, setting of the pass/fail ratio on 50 % level is recommended;
7. In summative quiz, setting of the pass/fail ratio at 35 % to 40 % level is recommended,

8. Manual correction of the exam "silly" mistake in calculations, from "0" to "0.5", after post-exam discussion with the student, based on scratchpad with calculations left before leaving the exam-room (computer lab), should be possible.

References

1. Aldrich, S.: Learning through technology: from institutional course delivery to collaborative knowledge creation. In: Proceedings of the 12th IASTED International Conference on Computers and Advanced Technology in Education (CATE), US Virgin Islands, Keynote Address (2009)
2. Grow, G.: Teaching learners to be self-directed. Adult Educ. Q. **41**(3), 125–149 (1991/1996)
3. Abdullah, M.H.: Self-directed learning. Educational Resource Information Center (2002). http://www.ericdigests.org/2002–3/self.htm
4. Walsh, K.: Gathering evidence that flipping the classroom can enhance learning outcomes. EmergingEdTech (2013). http://www.emergingedtech.com/2013/03
5. Rutkowski, J.: Flipped classroom—from experiment to practice. In: Proceedings of 1st International KES Conference on Smart Technology Based Education and Training, Chania, Greece (2014)
6. Reja, U., Lozar, K., Hlebec, V., Vehovar, V.: Open-ended vs. close-ended questions in web questionnaires. Developments in Applied Statistics. Metodoloski zvezki, 19, Ljubljana: FDV, pp. 159–177 (2003). http://mrvar.fdv.uni-lj.si/pub/mz/mz19/reja.pdf
7. Starkings, S.: How to asses large groups with the minimal amount of resources but preserving quality. In: Proceedings International Statistical Institute—ISI 52nd session, Helsinki, Finland (1999). http://www.stat.fi/isi99/proceedings.html
8. Fielding, A., Bingham, E.: Tools for Computer-Aided Assessment, Learning and Teaching in Action, vol. 2, Issue 1. Learning and Teaching Unit of Manchester Metropolitan University (2003). http://www.celt.mmu.ac.uk/ltia/issue4/fieldingbingham.shtml
9. Chalmers, D., McAusland, D.M.: Computer-assisted assessment. In Houston, J., Whigham, D. (eds.) Handbook for Economics Lecturers. Glasgow Caledonian University, Glasgow (2002)
10. Sim, G., Holifield, P., Brown, M.: Implementation of computer assisted assessment: lessons from the literature. ALT-J, Res. Learn. Technol. **12**(3), 215–229 (2004)
11. Rutkowski, J.: Development and evaluation of computer assisted exam—Circuit Theory example. In: Proceedings of 9th IASTED International Conference on Computers and Advanced Technology in Education (CATE), Lima, Peru, pp. 333–338 (2006)
12. Moodle 2.7 Documentation: https://docs.moodle.org/27/en/Main_page
13. Rodanski, B.S.: Web-based tutorials and assessment in technical subjects. In: Proceedings of the International Conference on Signals and Electronic Systems (ICSES), Lodz, pp. 141–144 (2006)
14. Rutkowski J., Moscinska, K., Jantos, P.: Web-based assessment and examination system—from experiment to practice. In: Proceedings of the 10th IASTED International Conference on Computers and Advanced Technology in Education (CATE), Beijing, China, pp. 206–211 (2007)
15. Moscinska, K., Rutkowski, J.: Introduction of summative and formative web-based assessment to the Circuit Theory course. In: Proceedings of the International Conference Web-Based Education (WBE) 2007, Chamonix, pp. 395–400
16. Rutkowski, J., Moscinska, K., Grzechca, D.: Students' attitude to formative web-based assessment. In: Proceedings of the IASTED International Conference on Web Based Education (WBE), Innsbruck, pp. 139–144 (2008)

Three Dimensions of Smart Education

Vladimir Tikhomirov, Natalia Dneprovskaya
and Ekaterina Yankovskaya

Abstract Many principles of smart education are not explained now because of ambiguity of this concept. The system of key features of smart education is proposed in this article. It is one of possible ways to create the well-rounded paradigm of smart education. It means that it is necessary to take into account different aspects of this phenomenon, not only ICT components. Educational projects, which represent the vast majority of key components of smart education, can be regarded as the parts of smart educational trend. Three of the most important components – main dimensions of smart education - are identified and analyzed in this paper: educational outcomes, ICT, and organizational dimensions.

Keywords Smart education · Dimensions of smart education · Educational outcomes · E-learning · Instructional design · 21th century skills

1 Introduction

Nowadays progress of e-learning, new pedagogical techniques and approaches contribute to the formation of a new educational trend, which calls "smart education". This trend provides outstanding opportunities to acquire different professional skills, competences and knowledge through active use of various information and communication technologies (ICT). Therefore, contemporary ICTs have the enormous influence in the area of education. For example, big data analytics and

V. Tikhomirov (✉) · N. Dneprovskaya · E. Yankovskaya
Moscow State University of Economics, Statistics and Informatics (MESI),
Moscow, Russia
e-mail: VTikhomirov@mesi.ru

N. Dneprovskaya
e-mail: NDneprovskaya@mesi.ru

E. Yankovskaya
e-mail: EAJankovskaja@mesi.ru

© Springer International Publishing Switzerland 2015   47
V.L. Uskov et al. (eds.), *Smart Education and Smart e-Learning*,
Smart Innovation, Systems and Technologies 41,
DOI 10.1007/978-3-319-19875-0_5

technologies could be applied for collecting multiform information concerning individual learning experience of many students. These data help to create exiting personalized courses full of actual knowledge in the context of smart education. Thus, smart education can make different forms of e-learning and blended learning as qualitative as face-to-face learning but more quick, flexible and available.

Despite the importance of smart education, many principles of this new trend are not clarified now because of ambiguity of this concept. It has many different interpretations and the most of them stress the technological aspects of this significant trend and do not take into consideration other sides of this phenomenon. Such situation causes a conceptual uncertainty so it is difficult to understand special peculiarities of smart education. But it is necessary to formulate what the smart education is because sometimes we face with the different educational and learning projects, which are called "smart", but do not represent anything innovative that we really can regard as smart educational project or approach.

The goal of this paper is to offer a systematic approach to key features of smart education. It is the one of possible ways to create the well-rounded paradigm of smart education. It means that we try to take into account different aspects of this phenomenon, not only ICT components. We contend that only such educational projects, which represent the most of key components of smart education, can be regarded as the parts of smart educational trend.

2 Previous Research

The first publications on smart education was published just a few years ago. The key trends in development of education and futurological forecasts of further changes of the educational system were presented in these papers. However, these publications presented just descriptions and examples of system solutions and technologies in the field of education, but did not formulate yet a paradigm of smart education. The exception is publication [1], where many key features of smart education are discussed, for example, smart competences.

Some researchers are focused on perspectives of contemporary higher education and tendencies that correspond to the concept of smart education. For instance, in [2, 3] different authors discuss the problems of contemporary universities and their future perspectives in the context of applications of smart ICT in education. A comprehensive study of the role of universities and MOOC's in the context of smart education is presented in [4].

The other group of researches is focused on the problem of educational outcomes in the contemporary educational systems. The concept of outcomes-based education is proposed and this approach can be regarded as a part of the smart education paradigm. The main part of such outcomes includes the studies of cognitive abilities, needs, skills and their training through e-learning. Authors propose the concept of 21th century skills in [5]. There are also resources available on instructional design and cognitive science, which considers the problem of learning, structuring material, communication, and forming cognitive competence in this group, for example, the fundamental publication [6].

One more group of researches is devoted to the problem of smart learning and smart e-learning. There is attempt to analyze different possible ways of using this concept in [7]. The several other publications in this area (especially [8]) explore the problem of implementation of principles of smart e-learning in different educational services. Some authors propose possible models and schemes of smart educational systems.

The other resources include different attempts to make conceptual model of ICT infrastructure of smart educational and e-learning systems; for example, the questions concerning databases, standards (like SCORM), gadgets and learning equipment are discussed [9, 10].

There are some research projects (for example, [11]) that analyzed organizational aspects of the smart education such as educational trajectories, learning strategies, etc. These researchers usually emphasize the fact that many aspects of contemporary education need new flexible organizational structures, which we can be called "smart".

3 Methods and Theoretical Framework

The main method of our research is the conceptual analysis of the area of smart education. We summarized different aspects of this field and proposed to denote three main dimensions of the smart education: (1) educational outcomes, (2) ICT and (3) organizational aspects. We analyzed specific features of each dimension.

The theoretical framework of our research is based on approach of outcomes-based education, some aspects of instructional design, knowledge management, socio-cognitive paradigm, e-learning theories. We used several aspects of these theoretical concepts.

The "smart education" as a phenomenon of the modern e-learning should be split off on three dimensions: the educational outcomes, the ICT and the organizational dimension (Fig. 1).

The smart education should be considered from different points of view. Every dimension is obligatory for education development, and final goal of in in educational outcomes, which can be gained by society, company or person.

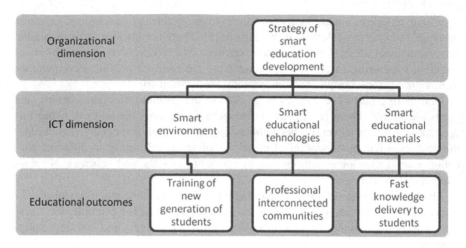

Fig. 1 Three dimensions of smart education

4 Results

4.1 Smart Education: The Educational Outcomes Dimension

The first and the most important dimension of smart education is dimension of educational outcomes. This dimension is so important because there are many new and complex requirements to each well-educated person in contemporary social and information context. Therefore, there are several important results which necessary to achieve during education process. Because of contemporary context, these results could be called "smart" if they help to get adaptation advantages.

Education outcomes are the purposes of students upon which educational programs are build. They can be presented as the sets of different skills, competences, knowledge contents, cultural background, values that students should get to be successful in different kinds of their lifelong activities. Outcomes-based approach corresponds to the contemporary requirements to each educated person. It allows to know, understand and be able to do according to demands of the complex, diverse, and globally interdependent world of the 21st century.

The central part of educational outcomes is the set of content, knowledge and cognitive skills or competences. It is obvious that nowadays the requirements of cognitive skills have changed because of social and technological changes. There is concept of 21th century skills that denotes several key sets of learning and innovation skills of 21th century. We will pay our attention to the several cognitive skills that are the most important for learners.

The first important cognitive skill is the cognitive self-organization, which means possibility to set cognitive goals, change cognitive strategies and tools if it is necessary. Using various types of reasoning as appropriate to the situation is the effective reasoning. System thinking as another cognitive skill is a possibility to analyze how parts of a whole interact with each other. This skill helps to integrate different and sometimes contradictory information and knowledge into whole system. The skill of critical thinking allows to reflect critically on learning experiences and processes. This skill is especially important in the changeable social and cognitive environment.

So the most important cognitive skills are educational outcomes are cognitive self-organization, system thinking, logical and analytical thinking and critical thinking. We can see that all of these skills enhance the learners' independence and their ability to adopt to the changeable environments. It is critically important because of variable changes of social structure, professional needs and other contexts.

Hence there are some approaches in instructional design, which suggest different instructional tools for enhancing these skills. These approaches should take into account difference between learners, for example, their cognitive styles, individual learning paces, motivation etc. Besides it is necessary to take into account the role of communication in learning and such phenomena as distributed cognition and knowledge. There are some instructional design requirements that are used in smart learning as a part of smart educational approach; for example, these are active learning, collaborative learning, person-centered learning, etc.

Besides several important cognitive skills there are many another educational outcomes, which smart education should provide. For instance, such fundamental skills as computer literacy and possibility to work with different kinds of information.

Also the one of the most important part of the education outcomes is the knowledge, which students acquire. Contemporary students need not only fundamental, but also actual knowledge.

Smart life skills are the sets of abilities, which help to live and achieve success in the contemporary dynamical and complex worlds. The examples of such skills are social skills, creativity, flexibility, leadership and so on.

Therefore, we can see that different outcomes of education need to be enhanced by appropriate structure of learning. It means that they should be formed by using appropriate instructional design and learning systems. Hence, the educational outcomes dimension of the smart education is the complex of necessary outcomes, which need appropriate ways of achieving them by structures, forms and methods of learning. We contend that this is the most important dimension of the smart education and other dimensions are just tools realizing its demands.

4.2 Smart Education: The ICT-Dimension

The ICT dimension is a set of ICT tools being used for providing different aspects of smart education. There are several groups of these tools. The first group includes the tools organizing and managing educational process, for example Learning Management Systems. Such systems corresponding to the principals of the smart education should provide an intellectual big data processing concerning educational process and automatize management of educational process. Actually, such systems should be virtual learning environments integrating all tools necessary for the e-learning and located in the "cloud". These systems should support specialized standards such as Tin Can API, which is replacing SCORM. Tin Can API allows to collect data about the wide range of learning experiences a person has.

Another group of tools consists of specialized software designed for the development of educational content. This group includes, for example, such software as Articulate Storyline or Course Lab. This group may include any tools to work with text and multimedia content being used for course development. Tools for development smart learning content should provide high interactivity and create an effect of virtual presence or telepresence. Software used for the development of educational content should make it more efficient transmission than the traditional transfer of teaching material. For example, interactive courses should provide possibility to form a hypothesis and then testing it out with special instruments such as virtual labs.

The third group of tools based on social interaction tools particularly social networking and software for webinars. This ICT tools allow to involve students into communication with teachers and with each other. Communication is necessary aspect of learning and it helps to improve learning. It is the obvious part of smart approach to learning and education. It is expected that social interaction tools are already actively used in the learning and will be integrated into the educational online services.

Finally, all educational services built on the principles of smart education should be implemented in the form of mobile software. Currently there are several main mobile platform such as Android, iOS, Windows Phone/Windows8. Using the mobile application will allow to teach and to learn regardless to the spatial localization of the desktop computer.

Therefore, ICT component of smart education should have the following key features: integrity, interactivity, social interaction tools, mobility. Integrity means that different ICT educational tools are integrated with each other by using common standards or by the other way, ideally this approach allows to make integrated learning environment. Interactivity means that student actively participate in learning because of special ICT tools such as virtual labs or simulators. Social interaction tools should be integrated into educational service. Mobility means that smart educational services should be presented as mobile application too.

4.3 Smart Education: The Organizational Dimension

The organizational dimension of smart education is as important as ICT component. We will consider some aspects of organizational component below. These aspects are educational programs, forms of learning, and principles of teaching.

Smart educational programs should be formed to be able to create a flexible educational trajectory for each student. This means customization of educational programs. Each student should be able to select specific courses to satisfy his/her professional and other needs. The smart education concept correlates with the principle of lifelong learning.

The smart educational systems should combine various forms of learning, for example, formal and informal education. Because of such combination, students can remain a kind of educational freedom. They can get cognitive skills and knowledge through informal communication without strict regulation. Besides combination of different learning forms allows using such learning approaches as project approach. This approach combines fundamental and applied knowledge. From the point of view of a smart approach, implementation and combination of different forms of training should be maintained using appropriate ICT infrastructure, innovative educational technologies and instructional design.

From the point of view of the smart approach, one of the key features of organizational component is openness of education. Currently there are many discussions concerning the term "open education". Necessary to mention that "open education" doesn't mean "free education". It means that enrollment is easier than in traditional education. Openness is ambivalent because it can prevent to control the quality of education, but at the same time, it provides great educational opportunities. From another point of view "openness" means that all learning content can be freely distributed for re-use on the basis of open licensing.

The one of the most important feature of smart education is using knowledge management in the educational systems. This aspect is very significant because of difficulties with using complex and changeable information and knowledge during education.

The last but not least is such feature as individualization and customization. This feature can have the form of the learner profile. The learner profile is the key element in the smart educational process. This profile allows taking into account different learners needs and creating flexible educational trajectory according to them. Also it is possible to combine different learning modules according to the individual educational needs.

We have denoted several important features of organizational component of the smart education: flexibility, combination of different forms of education, "openness", using knowledge management, individualization and customization. Flexibility means that smart education allows to create adaptive and flexible educational trajectory. Combination means that different forms of learning are combined in one program. Openness means that educational programs can be free and/or they have no admissions requirements. Using knowledge management helps to rule dynamic and changeable information and knowledge during learning. Individualization and customization help to create special profile to each learner and take into account his or her special needs, goals and achievements (Table 1).

Table 1 Key features of smart education and their implementations

Dimension	Key features	Implementations
Educational outcomes	21 century cognitive skills, special knowledge, literacy, smart life skills	Cognitive self-organization, system thinking, logical and analytical thinking and critical thinking, computer literacy, actual knowledge, social skills, creativity, flexibility, leadership
ICT	Integrity, interactivity, social interaction tools, mobility	Special standards and specifications (like Tin Can API), entire and smart educational environment with set of necessary instruments, LMS based on intellectual data analyze, virtual labs, special content development systems and authoring tools, social media, mobile applications and other
Organizational dimension	Flexibility, combination of different forms of education and learning, "openness", using knowledge management, individualization and customization	Flexible combinations of learning modules, sets of electives, special learning approaches (such as project learning, informal learning and so on), open enrollment, using academic knowledge management systems, individual learning profile, lifelong learning, individual recommendations system and other

5 Obtained Results

The new educational paradigm called "Smart Education" can be regard as the concept, which fixes the current tendencies in the field of education, and as a project of education of the near future. From the point of view of some approaches, this paradigm with its key concepts reflects the transformation from the fourth to the fifth generation of e-learning. So it is possible to denote the main aspects (dimensions) of the smart education and its key features, but rather difficult to give brief and clear definition of this phenomenon. So we have tried to point out the features of smart education, which can be used to regard different educational projects, services, systems, solutions as "smart" or "not smart'. But such instrumental approach to the phenomenon of the smart education can be argued.

According to the paradigm of smart education, the educational process and learning should be adapted to technical, cognitive, psychological, educational, professional and other needs of students, as well as to the changing socio-economic environment (Information Society and the economy based on knowledge). This adaptation is possible because of the introduction of a wide range of specialized ICT in the education and learning. Using ICT in the educational process initially involved in the paradigm of smart education because creating of flexible and well-adapted educational trajectories is possible only by applying automated intelligent big data processing and internet technologies of distance education.

We contend that the main dimension of smart education is not the ICT component. This component is just tool helping to achieve specific educational outcomes. But without this tool it is impossible for example to make the appropriate individualization and customization of educational process. So all dimensions of smart education are impossible without each other, but there is another approach regarding the ICT dimension as the most important component of smart education.

6 Conclusions

We denoted three main dimensions of smart education and explicated their special features and the main contents. These features indicate that a particular educational project applies the principles of smart education. These are educational outcomes, ICT and organizational dimensions.

Educational outcomes dimension includes different kinds of cognitive skills such as critical thinking and the different aspects of learning, which help to enhance these skills. ICT features are integrity, interactivity, social interaction tools, mobility. Organizational features are flexibility, combination of different forms, "openness", using knowledge management, individualization and customization. We can see that similar features have different implementations at the different dimensions of the smart education.

Probably we will face with different problems when we will try to implement all principles of smart education in the real educational projects. May it is impossible to make "perfect" smart education system. But the concept of smart education allows to catch the most significant trends in contemporary education.

References

1. Coccoli, M., Guercio, A., Maresca, P., Stanganelli, L.: Smarter universities: A vision for the fast changing digital era. J. Vis. Lang. Comput. **6**, 1003–1011 (2014)
2. Barnett, R.: The Future University. Routlege, New York (2012)
3. Temple, P. (ed.): Universities in the Knowledge Economy: Higher Education Organisation and Global Change. Routledge, New York (2011)
4. Tikhomirov, V.: The World on a Way to Smart Education: New Opportunities. Open Educ. **3**, 22–28 (2011)
5. Hilton, M.: Exploring the Intersection of Science Education and 21st Century Skills: A Workshop Summary. National Research Council (2010)
6. Richey, R.C., Klein, J.D., Tracey, M.W.: The Instructional Design Knowledge Base: Theory, Research, and Practice. Routledge, New York (2010)
7. Spector, J.: Conceptualizing the emerging field of smart learning environments. Smart Learn. Environ. **1** (2014)
8. Gamalel-Din, S.A.: Smart e-Learning: A greater perspective: from the fourth to the fifth generation e-learning. Egypt. Inf. J. **11**, 39–48 (2010)

9. Ke, C.-K., Liu, K.-P., Chen, W.-C.: Building a Smart E-Portfolio Platform for Optimal E-Learning Objects Acquisition. Mathematical Problems in Engineering
10. Learning media in the age of e-Learning 2.0 (2013)
11. Komleva, N.K.: Educational environment of the University. Econ. Taxes Right **11**, 209–212 (2011)

To Flip or Not to Flip: A Critical Interpretive Synthesis of Flipped Teaching

Virginia N. L. Franqueira and Peter Tunnicliffe

Abstract It became almost fashionable to refer to the term "flipped" in higher education. Expressions like flipped learning and flipped classroom are often used interchangeably as an indication of innovation, flexibility, creativity and pedagogical evolution. We performed an exploratory study on this topic following the Critical Interpretive Synthesis methodology for analysis of the literature. Our findings indicated that the term "Flipped Learning" is misleading and that, in fact, the synthetic concept behind it is "Flipped Teaching". We derived a synthesising argument, in the format of two synthesis models, of the potential benefits promoted by flipped teaching and the potential issues which affect its success in practice. Those models allow STEM course tutors not only to make informed decisions about whether to flip teaching or not, but also to better prepare for flipping.

Keywords Flipped learning · Teaching · Pedagogy · Higher education

1 Introduction

"Flipped" has become the pedagogical buzzword of the day: flipped learning, flipped classroom, flipped course, and even flipped university currently seem to be the preferred terminology to demonstrate that educators are up-to-date with latest developments, have a vision for the future and embrace changes. It is gaining incredible visibility and being subject of numerous debates not only within academia but also in the media. However, lots of questions remain unanswered about its novelty and effectiveness in practice. This paper reports on an exploratory study of flipped learning in higher education using Critical Interpretive Synthesis (CIS) [14]. This methodology, popular in the domains of Social Science and Health research, aims

V.N.L. Franqueira (✉)
Department of Computing and Mathematics, University of Derby, Derby, UK
e-mail: v.franqueira@derby.ac.uk

P. Tunnicliffe
Department of Education, University of Derby, Derby, UK
e-mail: p.tunnicliffe@derby.ac.uk

© Springer International Publishing Switzerland 2015
V.L. Uskov et al. (eds.), *Smart Education and Smart e-Learning*,
Smart Innovation, Systems and Technologies 41,
DOI 10.1007/978-3-319-19875-0_6

at analysing evidence collected from the literature reviewed to develop concepts and theories using induction and critical thinking [14]. Interpretive approaches, such as CIS, contrast with integrative approaches aimed at aggregation and summarization of evidence [13]. The former is interrogative and iterative in nature, thus, suitable for exploratory research starting from a broad topic without a specific research question [27] – these were reasons why CIS was selected as the methodological backbone for this study about flipped learning.

1.1 Methodology

CIS [14] is a methodology which uses literature as primary source of evidence of different kinds; e.g., qualitative and qualitative evidence collected from multidisciplinary or multi-method sources [3]. It incorporates some elements of Meta-ethnography (i.e., Lines-Of-Arguments as analysis strategy) and Grounded Theory (i.e., inductive approach for emergent theory generation). As such, CIS aims at the development of *synthetic constructs* which derive from new interpretations of existing concepts and constructs directly collected from the literature, and of a *synthesising argument* which relates existing and emerging concepts/constructs [14]. Table 1 summarises the key characteristics of CIS.

Table 1 Key characteristics of Critical Interpretive Synthesis (CIS) [3, 14, 15, 27]

Purpose	CIS is a process of review; it aims to explore a topic and develop a synthesising argument which critically integrates the literature reviewed
Process	CIS follows a cyclic approach where iteration, reflexivity and refinement coexist. Searching, sampling and analysis happen in parallel within iterations
Procedure	There are no pre-defined procedures and CIS recognises the "authorial voice" for the development of a synthesising argument grounded in evidence collected from sources critically analysed – reproducibility is not a requirement
Search	Search of literature is flexible and draws from both keyword search in databases, and researchers' awareness of relevant material. Exhaustive searches are outside the scope of CIS
Sampling	CIS uses a purpose-based sampling where sources are chosen according to the emergent theoretical framework allowing the selection criteria to evolve. Sampling saturation establishes coverage
Analysis	Interrogation rather than aggregation drives the analysis of sources; what is included in the review derives from a critical approach to the material selected
Results	Analysis allow the development of a synthesising argument which connects existing constructs &concepts to new ones derived from synthesis – synthetic constructs

We followed several iterations starting from the broad intent of understanding flipped learning and its foundations up to the more specific intent of analysing experiences with flipped learning compared to the traditional approach. Along the process, we restricted our review to empirical studies evaluating the implementation of a same STEM (Science, Technology, Engineering, and Mathematics) course in both modalities – flipped and traditional.

2 Concept & Foundation of Flipped Learning

The idea of class flipping is attributed to Baker [2] and Lage et al. [24]. The former focused on Web-based tools as an essential enabler of flipped classrooms, and the latter focused on inverted classrooms as a promoter of *inclusion* [9] better accommodating different learners' styles and abilities. Lage et al. [24, Page 32] describes the idea as: "Inverting the classroom means that events that have traditionally taken place inside the classroom now take place outside the classroom and vice versa". Such early definitions were too fuzzy, and allowed false interpretations [7]. For example, distance learning, empowered by the advent of MOOCs (Massive Open Online Courses), could be considered as a flipped approach but is not since "classroom" is completely redundant in this case.

The foundation of flipped learning is *Active Learning* [18, 32]. Active learning builds over constructivism – a student-centred approach which emphasises "learning by doing" [20]. It is anchored on the principles of (i) intentional learner, i.e., students are actively responsible for and owners of their learning, (ii) reciprocal teaching, i.e., learning is a collaborative process where students benefit from social interactions with peers and tutor, and (iii) anchored instruction, i.e., learning requires the application of knowledge to complex, contextualised, and real problems, case-studies or scenarios [10].

Although flipped learning draws from active learning practices, it goes further and completely moves passive and individual activities, such as assimilation of content and concepts, to outside contact time. In fact, flipped learning shifts learning activities which fall on lower levels of the Bloom Taxonomy [8] (e.g., knowledge and comprehension) to outside the classroom and focuses on higher levels activities (e.g., application, analysis, evaluation and synthesis) inside the classroom [21].

Although there is no consensus on a definition of Flipped Learning, the following recent definition seems to capture its essence [38, Page 5]:

Flipped Learning is a "pedagogical approach in which direct instruction moves from the group learning space to the individual learning space, and the resulting group space is transformed into a dynamic, interactive learning environment where the educator guides students as they apply concepts and engage creatively in the subject matter."

This definition exposes an intrinsic issue with recent definitions of flipped learning – they do not refer to *learning* but rather to *teaching*. Teaching facilitates student learning [6]; learning being the delta between what a student knows/understands prior and after a teaching intervention. Therefore, the synthetic construct which emerges from this study is **Flipped Teaching** defined as above.

2.1 Implementation of Flipped Teaching

Advances in technology and the open source movement have largely enabled the flip
in teaching from a traditional approach, based on live lectures, to an online approach,
based on video-lectures and Web assessments [7].

The Flipped Learning Network [16] has published what they consider as the Four
Pillars (F-L-I-P) of the flipped approach [21]. Chen et al. [12] criticise this F-L-I-P
model in terms of its comprehensiveness for application to higher education. They
based their critique in three aspects. First, F-L-I-P focuses more on content planning
than on delivery, providing poor insights on types of activities and how they should
be conveyed. Second, it focuses on the educators' perspective leaving the students'
perspective unaccounted for. Finally, it lacks guidance for the individual learning
space. To address those gaps, they proposed three additional pillars (P-E-D). All
these seven pillars are reviewed next.

1. Flexible environments. Flipped teaching requires flexible environments to meet
   students' needs of studying content anywhere, anytime, and their expectations of
   flexibility in relation to assessment and to learning curve.
2. Learning Culture. Flipped teaching requires a shift from a instructor-centred
   approach, where students are passive, to a student-centred approach, where students are active and owners of their learning. This aims to promote deep learning [26], and cooperative learning targeted at the *Zone of Proximal Development* [37]; this means that tutors should assist and challenge students up to the
   limit of their capacity, but not beyond since it would demotivate them [21].
3. Intentional Content. Flipped teaching requires that tutors evaluate (and prepare)
   content and activities appropriate for the individual learning space, and for the
   group learning space. They can draw from constructivist techniques such as
   problem-based learning [4] and peer-based learning [36].
4. Professional Educators. Flipped teaching demands more from tutors than traditional teaching. They have to be constantly reflecting on how to maximise contact
   time and how to assess students understanding of content absorbed on their own.
5. Progressive Networking Learning Activities. This feature emphasises the social
   ingredient of active learning delivery, i.e., the need for "learning by networking", achieved by activities centred in collaboration and teamwork, complement-
   ing "learning by doing". It also suggests the adoption of a progressive strategy of
   low-to-high-risk activities to gradually allow students to adapt. Low-risk activities tend to have short duration, be considerably planned, structured, not controversial, and familiar to students and tutors [12].
6. Engaging and Effective Learning Experience. This feature expands the role of
   "professional educators" and proposes the monitoring of *transactional distance*
   to improve learning. Transaction distance is the psychological or communication distance – disconnected from physical distance – between students and
   tutors [30]. It fluctuates in a flipped setting, therefore, should be managed by
   tutors with the purpose of decreasing the distance. Chen et al. [12] propose
   two ways to achieve that: increment dialogue and reduce preset structure. For

example, learners' autonomous activities (like watching video-lectures) increase the distance and should be balanced with activities which enhance student-tutor communication and allow tutors to monitor learning (like quizzes or personalised formative feedback via email or learning platform).
7. Diversified and Seamless Learning Platforms. This feature extends "flexible environments" and regards the need of digital platforms to fulfil requirements of individualization, differentiation, personalization, reliability and consistence [12].

3 Empirical Evaluations of Flipped Teaching

Empirical evidence were collected from the literature comparing STEM courses delivered via both traditional and flipped teaching. Traditional teaching is an approach where tutors present new content (concepts, facts, theories) in class while students take notes; practical sessions exercise the content to some extent, students consolidate their knowledge through homework, and address challenges via assessments. This section discusses findings.

Content coverage. Experience with a same course delivered in traditional and in flipped mode (e.g., [29, 39, 41]) indicates that the latter tends to run in a faster pace compared to the former. Flipped teaching allows more topics to be covered by students on their own, and more individual feedback and guidance to be provided in class.

An interesting aspect of flipped teaching is the possibility to cover and assess more learning outcomes [39]. This is achievable in part because it "increases [opportunity] levels of problem solving structure and practice" [1, Page 229]. It is also partially achievable because in-class group activities make it very convenient to assess learning outcomes related to teamwork, communication, and students' ability not only to solve but also to identify and formulate problems [7]. These are all soft skills regarded as valuable in industry.

Perception from tutors. Benefits of self-paced, asynchronous learning outside the classroom is a strong point often recognised by students who experienced flipped teaching [25, 29]. They can follow their own schedule to cover content at home or on the move, and can watch video-lecture passages repeatedly.

Increased motivation and student engagement were also observed when flipped teaching was compared to traditional teaching [39]. For example, students perceived in-class active learning as fun, and easier to remain focused [25]. Many students also perceived that intensive hands-on activities they were experiencing extensively in every class would allow them to acquire practical skills sought by the industry [28]. Students felt more at easy to ask questions and participate in the less structured and more cooperative flipped environment [11, 24, 29].

On the other hand, tutors reported that flipped teaching translated into an increased interest on the course subject area because students engaged into a variety of realistic problems and case studies [25, 39]. Kim et al. [23] observed significant better rates of students retention in their flipped version of an engineering course. They believe this is the result of the cooperative learning component of flipped teaching, and better performance decreasing the number of students who abandoned the course due to the lack of hope in a "pass" grade.

Academic performance. Empirical evidence from flipped teaching indicates an improved performance of students in assessments, compared to traditional teaching. Yarbro et al. [38] refer to examples where it either incurred in a significant or a marginal improvement in students' performance, but never the opposite. In addition, experience suggests that flipped teaching enables the acquisition of a richer skills set, such as higher-order thinking, innovation in problem-solving, cooperation, independence, collaboration and creativity [33, 35, 39, 41]. Cooperative learning and active learning, both embedded into flipped teaching, enable deeper understanding of concepts [39], therefore, promoting deep learning [6]. Love et al. [25, Page 322] provide insights in this respect: "over 70 % [of students] agreed that explaining a problem or idea to their partner helped them to develop a deeper understanding of it".

Time & effort for preparation. Often authors who engaged in flipped teaching mention a substantial effort for the transition from traditional mode. In particular, they refer to preparation of video-lectures. For example, 100 hours to generate and edit 45 video-lectures of 5–15 min [29]; 35 hours of recording to produce 48 video-lectures of 30–60 min [41]. The bright side of this substantial preparation effort is the possibility of reuse, and the shorter preparation required before each flipped class [11, 29]. The recommendation of short video-lectures (30 min or less) is echoed by many [7, 28, 40] – in this case, students would watch one or more videos per week. However, this strategy creates challenges. From the perspective of tutors, it creates the challenge of selecting and organising material in really small chunks [29]. From the perspective of students, it creates challenges regarding a lack of a clear module structure.

An alternative to avoid the dangers of poor quality teacher-created videos [22], and cut down on preparation investment, is to adopt off-the-shelf video-lectures [41]. A variety of educational videos is available (e.g., Khan Academy, MIT OpenCourseware, TED Talks, and YouTube), although quality may remain an issue.

Perception from students. Although the majority of students tend to be positive about flipped teaching, there seems to be always a consistent minority (15–25 %) who remain negative about it. For example, Butt [11] reported 25 % of students not seeing value in flipped teaching; Bates and Galloway's findings [5] showed that 8 % of students slightly/strongly preferred traditional teaching while 10 % were neutral; Kim et al. [23] obtained similar results: 15 % of students disliked or declared to be neutral.

Paradigm shift. STEM students throughout their academic life have been mainly exposed to traditional teaching. When the approach is turned up-side-down by flipped teaching, some students find it hard to adapt because they are required to leave their comfort zone to become active learners [34]. While some students succeed to adapt after a short transition period, some do not [12]. This also depends on students' readiness to self-directed learning [19]. Moreover, Gajewski and Jaczewski [17] suggest that attitude towards flipped teaching may be affected by cultural differences since some cultures may be more open to changes and innovations than others.

Pre-class preparation. Preparation for class assumes a more vital importance in flipped teaching; unprepared students reportedly feel strongly behind [12]. Therefore, preparation represents a point for adjustment required from students. Mason et al. [29, Page 434] observed that in the first weeks of their flipped course, students reported to be "frustrated" in class. However, "by the fourth week, students seemed to have realized that they would learn more during class time if they came prepared". This perception is corroborated by other authors (e.g., [39]). The lack of pre-class preparation, if substantial, may cause a burden for those who prepared and may feel demotivated [11]. It also causes a burden for tutors in terms of extra assistance required by those unprepared and the need of keeping the whole class busy. Despite the stereotype that learning in flipped mode is more demanding in terms of study time, preliminary results indicate that this is not the case [29].

Learning management. One crucial aspect of asynchronous content delivered outside class, as in flipped teaching, is learning management. Tutors need to specifically check students understanding of the material *before each class*. This allows formative feedback to be provided to all or to individuals in the next class or remotely before it. Quizzes seem to be the most commonly used method of input for that monitoring – the majority of authors report the use of pre-class quizzes [7, 23, 40], while others prescribe pre-class *and* post-class quizzes [39]. An alternative approach for learning management is Just-In-Time-Teaching (JITT) [31]. In JITT, class activities and scope are adapted depending on results from assessment of students understanding [25, 40]. One way to implement that is to pose questions at the beginning of each class and collect answers via clickers. Depending on the results, the tutor uses micro-lectures and adjusts peer activities to solve misconceptions and gaps. Therefore, learning management in flipped teaching runs the risk of becoming a 24/7 task [12].

4 Reflection and Synthesising Argument

Flipped teaching is not really a new pedagogy. In fact, it leverages from practices anchored on existing, well known, educational models such as Active Learning, Problem-based Learning and Peer-assisted Learning. The element of novelty which can be attributed to Flipped Teaching is the *flip* of content delivery away from the classroom (replacing face-to-face lectures by video-lectures) and 100 % use of classroom time for "learning by doing" and "learning by networking" activities. However, this delivery shift seems confusing even for authors reporting flipped experiences. For example, Marwedel and Engel [28, Section III] argue that flipped teaching is a novelty for Engineering but not for Social Sciences. They regard seminar-style lectures, popular in the latter sciences, as flipped teaching – maybe because it is a student-centred approach. Our study indicates, however, that this is a misconception since the delivery of content prepared by students as homework and its delivery as seminar in class, although promoting independent learning, remains a traditional

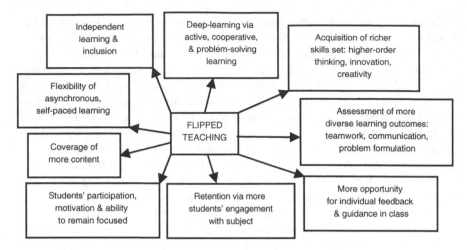

Fig. 1 Synthesis of potential benefits promoted by flipped vs. traditional teaching

setting. In flipped teaching, lectures content and delivery remain as a responsibility of the tutor. Figure 1 synthesises potential benefits promoted by flipped teaching when compared with traditional teaching.

The implementation of flipped teaching involves investment and raises challenges. Figure 2 synthesises factors that affect the success of flipped teaching in practice. One such factor is the effort-consuming preparation of high quality video-lectures to convey pre-class content. An alternative is to use off-the-shelf videos, but poll results among teachers [22, Page 63] suggest that finding those videos is rather a difficult task. Herreid and Schiller [22] propose standardisation and sharing of videos and case studies on a centralised repository such as the National Center for Case Study Teaching in Sciences[1] – a form of collaborative teaching – as a way forward.

According to Chen et al. [12], it typically takes 3 years to fine-tune a flipped course and achieve its maximum benefits. Questions remain unanswered regarding the return on investment of flipped teaching in a diversity of courses and cohorts. For example, there are STEM courses with stable content (e.g., "Foundations of Computer Sciences") while there are others with rather dynamic content (e.g., "Advances in Digital Forensics Research"). There has been no opportunity for longitudinal and realistic studies to answer such questions yet.

There is a paradox in flipped teaching. Some authors (such as [24]) claim that it accommodates well different learning styles, promoting inclusion through a variety of teaching methods which can be used. Others (such as [40]), however, raise the question about how different learning styles, different cultures [17] and different levels of students' readiness for self-study [19] adapt differently to it. The former's perception [40] is that active learners may fit best with flipped teaching although,

---

[1]http://sciencecases.lib.buffalo.edu.

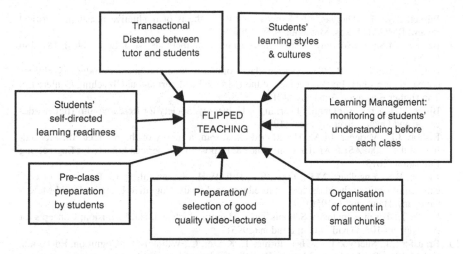

Fig. 2 Synthesis of potential issues that affect the success of flipped teaching

in the end, students with other learning styles may also benefit by developing skills via active-engaging tasks. Deeper and broader studies on factors affecting flipped teaching remains a gap which calls for empirical research.

5 Conclusion

This paper used Critical Interpretive Synthesis to review terminology, foundation, implementation and empirical evaluations of flipped teaching in STEM courses. "Flipped Learning", a widespread terminology, is misleading since it refers to a teaching strategy rather than to a learning strategy per se. Findings from this study allowed us to build a synthesising argument in the format of two models showing what flipped teaching promotes (i.e., potential benefits) and what affects its success (i.e. potential issues). Tutors considering to adopt flipped teaching should invest to minimise the latter to maximise the former. The paper also pointed to research directions needed to further improve our understanding of flipped teaching in practice.

References

1. Baepler, P., Walker, J., Driessen, M.: It's not about seat time: Blending, flipping, and efficiency in active learning classrooms. Comput. Educ. **78**, 227–236 (2014)
2. Baker, J.W.: The 'Classroom Flip': using web course management tools to become the guide by the side. In: Proceedings of the 11th International Conference on College Teaching and Learning. pp. 9–17. ERIC, Institute of Education Sciences (2000)

3. Barnett-Page, E., Thomas, J.: Methods for the synthesis of qualitative research: a critical review. BMC Med. Res. Methodol. **9**(59) (2009)
4. Barrows, H.S.: A taxonomy of problem-based learning methods. Med. Educ. **20**(6), 481–486 (1986)
5. Bates, S., Galloway, R.: The inverted classroom in a large enrolment introductory physics course: a case study. In: Proceedings of the HEA STEM Learning and Teaching Conference. The Higher Education Academy (2012)
6. Biggs, J.: Teaching for Quality Learning at University. Society for Research into Higher Education and Open University Press (1999)
7. Bishop, J.L., Verleger, M.A.: The flipped classroom: a survey of the research. In: Proceedings of the 120th ASEE Annual Conference & Exposition. American Society for Engineering Education (2013)
8. Bloom, B.S., Engelhart, M.D., Furst, E.J., Hill, W.H., Krathwohl, D.R.: Taxonomy of educational objectives: the classification of educational goals. Cognitive Domain. David McKay Company, Handbook I (1956)
9. Booth, T.: Towards Inclusive Schools, chap. Mapping Inclusion and Exclusion: Concepts for All?, pp. 96–108. David Fulton, London (1995)
10. Bransford, J., Sherwood, R., Hasselbring, T., Kinzer, C., Williams, S.: Cognition, Education, and Multimedia: Exploring Ideas in High Technology, chap. Anchored instruction: why we need it and how technology can help, pp. 115–141. Lawrence Erlbaum (1990)
11. Butt, A.: Students views on the use of a flipped classroom approach: evidence from australia. Bus. Educ. Accred. **6**(1), 33–43 (2014)
12. Chen, Y., Wang, Y., Chen, N.S.: Is FLIP enough? Or should we use the FLIPPED model instead? Comput. Educ. **79**, 16–27 (2014)
13. Dixon-Woods, M., Agarwal, S., Jones, D., Young, B., Sutton, A.: Synthesising qualitative and quantitative evidence: a review of possible methods. J. Health Serv. Res. Policy **10**(1), 45–53 (2005)
14. Dixon-Woods, M., Cavers, D., Agarwala, S., Annandale, E., Arthur, A., Harvey, J., Hsu, R., Katbamna, S., Olsen, R., Smith, L., Riley, R., Sutton, A.J.: Conducting a critical interpretive synthesis of the literature on access to healthcare by vulnerable groups. BMC Med. Res. Methodol. **6**(35) (2006)
15. Entwistle, V., Firnigl, D., Ryan, M., Francis, J., Kinghorn, P.: Which experiences of health care delivery matter to service users and why? A critical interpretive synthesis and conceptual map. J. Health Serv. Res. Policy **17**(2), 70–78 (2012)
16. Flipped Learning Network: www.flippedlearning.org
17. Gajewski, R.R., Jaczewski, M.: Flipped computer science classes. In: Proceedings of the 2014 Federated Conference on Computer Science and Information Systems. pp. 795–802. IEEE (2014)
18. Grabinger, R.S., Dunlap, J.C.: Rich environments for active learning: a definition. J. Assoc. Learn. Technol. **3**(2), 5–34 (1995)
19. Guglielmino, L.M.: Development of the self-directed learning readiness scale. Ph.D. thesis, University of Georgia, U.S.A. (1977)
20. Guney, A., Al, S.: Effective learning environments in relation to different learning theories. Procedia Social Behav. Sci. **46**, 2334–2338 (2012)
21. Hamdan, N., McKnight, P., McKnight, K., Arsfstrom, K.M.: A review of flipped learning. Flipped Learn. Netw. http://www.flippedlearning.org/research (2013)
22. Herreid, C.F., Schiller, N.A.: Case studies and the flipped classroom. J. Coll. Sci. Teach. **42**(5) (2013)
23. Kim, G., Patrick, E., Srivastava, R., Law, M.: Perspective on flipping circuits I. IEEE Trans. Educ. **57**(3), 188–192 (2014)
24. Lage, M.J., Platt, G.J., Treglia, M.: Inverting the classroom: a gateway to creating an inclusive learning environment. J. Econ. Educ. **31**(1), 30–43 (2000)
25. Love, B., Hodge, A., Grandgenett, N., Swift, A.W.: Student learning and perceptions in a flipped linear algebra course. Int. J. Math. Educ. Sci. Technol. **45**(3), 317–324 (2014)

26. Lublin, J.: Deep, surface and strategic approaches to learning. Centre for Teaching and Learning. UCD Dublin, nd (2003)
27. Mackenzie, M., Conway, E., Hastings, A., Munro, M., O'Donnell, C.: Is candidacy a useful concept for understanding journeys through public services? A critical interpretive literature synthesis. Soc. Policy Adm. **47**(7), 806–825 (2013)
28. Marwedel, P., Engel, M.: Flipped classroom teaching for a cyber-physical system course - an adequate presence-based learning approach in the internet age. In: Proceedings of the 10th European Workshop on Microelectronics Education. pp. 11–15. IEEE (2014)
29. Mason, G., Shuman, T., Cook, K.: Comparing the effectiveness of an inverted classroom to a traditional classroom in an upper-division engineering course. IEEE Trans. Educ. **56**(4), 430–435 (2013)
30. Moore, M.G.: Theorectical Principles of Distance Education, chap. Theory of Transactional Distance, pp. 22–38. Routledge, New York (1993)
31. Novak, G.M., Patterson, E.T., Gavrin, A.D., Christian, W., Forinash, K.: Just in time teaching. Am. J. Phys. **67**(10) (1999)
32. Prince, M.: Does active learning work? A review of the research. J. Eng. Educ. **93**(3), 223–231 (2004)
33. Roehl, A., Reddy, S.L., Shannon, G.J.: The flipped classroom: an opportunity to engage millennial students through active learning strategies. J. Family Consum. Sci. **105**(2), 44–49 (2013)
34. Strayer, J.F.: The effects of the classroom flip on the learning environment: a comparison of learning activity in a traditional classroom and a flip classroom that used an intelligent tutoring system. Ph.D. Thesis, Ohio State University, Ohio, USA (2007)
35. Strayer, J.F.: How learning in an inverted classroom influences cooperation, innovation and task orientation. Learn. Environ. Res. **15**(2), 171–193 (2012)
36. Topping, K., Ehly, S. (eds.): Peer-assisted Learning. Routledge (1998)
37. Vygotsky, L.S.: Mind and Society: The Development of Higher Mental Processes. Harvard University Press, Cambridge (1978)
38. Yarbro, J., Arfstrom, K.M., McKnight, K.: Extension of a review of flipped learning. Flipped Learn. Netw. http://www.flippedlearning.org/research (2014)
39. Yelamarthi, K., Drake, E.: A flipped first-year digital circuits course for engineering and technology students. IEEE Trans. Educ. PP(99) (2014)
40. Zappe, S., Leicht, R., Messner, J., Litzinger, T., Lee, H.W.: "Flipping" the classroom to explore active learning in a large undergraduate course. In: Proceedings of the 2009 ASEE Conference. American Society for Engineering Education (2009)
41. Zarestky, J., Bangerth, W.: Teaching high performance computing: lessons from a flipped classroom, project-based course on finite element methods. In: Proceedings of the Workshop on Education for High-Performance Computing. pp. 34–41. IEEE (2014)

Smart Edutainment as a Way of Enhancing Student's Motivation (on the Example of Board Games)

Vera Novikova and Ludmila Beskrovnaya

Abstract The paper examines specific features of smart edutainment on the examples of board games, particularly features that operate board games in teaching a foreign language, identifies the criteria and the requirements for the usage of Russian and English variants of board games in English lessons. The paper focuses on smart edutainment as a way of effective teaching method. The purpose of this study is to examine the effect of smart edutainment methods with young learners in an ESL class and to have a closer look at different games and activities to support teaching and learning process.

Keywords Edutainment · Smart learning · Board games

1 Introduction

We are living in the 21st century, in diverse, complex society. We have changed student's teaching methods because our major objective is to help all students and prepare them for work and life. ESL students are adults who are able to communicate, act and create changes. The teacher's first task is to teach creatively using some innovative techniques. This is no small challenge, and it is our responsibility to prepare our young people for the unique demands of a 21st century world.

National research Tomsk polytechnic university offers the opportunity to university students to undertake undergraduate studies in universities of The Czech Republic, Germany etc. for one semester. Increasing student mobility is a strategic priority for the university. English as a Second Language (ESL) programs are emerging as an important pathway for students who want to study abroad.

V. Novikova (✉) · L. Beskrovnaya
National Research Tomsk Polytechnic University, Tomsk, Russia
e-mail: veranovikova@sibmail.com

L. Beskrovnaya
e-mail: Blv82@yandex.ru

© Springer International Publishing Switzerland 2015
V.L. Uskov et al. (eds.), *Smart Education and Smart e-Learning*,
Smart Innovation, Systems and Technologies 41,
DOI 10.1007/978-3-319-19875-0_7

Developing the course for ESL students, we have taken into account the latest trends in teaching English as a foreign language. The purpose of this study is to examine the effect of smart edutainment methods with young learners in an ESL class and to have a closer look at different games and activities to support teaching and learning process.

2 Game-Based Learning

Edutainment was chosen as an effective, creative and innovative teaching method. English term "edutainment" is a combo word that combines two regular words into one term that really isn't a word. "Edutainment" is a blend of education and entertainment. This term is used to describe various forms of entertainment that also educate. Games can work well requiring minimal preparation and minimal technological resources or equipment.

The scope of application of this term is blurred and to describe it we split it into two subscopes to apply it differently to absorb some language skills. The first method focuses on the way of information transmission and skills' development helping learners who are not well motivated. At this point entertainment is considered as a sugarcoat. The second method focuses on entertainment-education process when ESL skills are acquired, developed and improved; the information is taken from multiple sources including those that are not regarded as educational materials. The first method implies passive perception of information, the second one – active search. Computer games, entertainment and educational programs, games as a part of academic process are related to the first category (passive perception); video podcast that teaches or discusses one particular aspect of life and society, television show, creating a blog are related to the second one (active search) [1].

Games have become a major component of edutainment practice. Edutainment games from board games to computer ones have become a phenomenon. They are an integral part of academic process and we can not imagine an educational process without games. It is difficult to get the fundamental knowledge or employability skills but using edutainment students can become more well-rounded, intelligent and well-read, they can develop and improve their skills and abilities in different spheres, particularly, in the sphere of learning foreign languages. Nowadays this technology is highly demanded in education. D. Perushev, who is one of the founders of edutainment establishment in Russia, claims that edutainment is the technology of knowledge transmission, a fantastic opportunity to learn something new from reliable sources, to find new information. It is not an alternative to academic education. It works among people of different ages and changes due to fashion trends. Entertainment or educational component of edutainment dominates taking into consideration a particular event. At the same time he emphasizes the mixture of components [2].

If we focus on educational component of the games, they may enhance students' skills or help to practice what they have learnt. If we focus on entertainment component, they provide enjoyment, emotions, adrenaline and sometimes students can turn their brain off.

Different games and methods are aimed at various aspects of teaching ESL. They are designed to help practicing important grammatical structures, train fluency or accuracy, improving communicative skills and they offer teachers enjoyable ways of letting their students practice what they have learnt. They are useful supplements to a broad program that is alive with authentic, interesting, and stimulating learning opportunities.

It is important to identify the following features of edutainment methods:

• availability of interactive teaching methods;
• availability of the process of two-way related activities within an educational process (subject-subject relations);
• setting goals in the organization of the learning process;
• comfortable learning environment;
• complex application of didactic and technological learning tools [3].

3 Smart Edutainment

Smart edutainment integrates edutainment with smart learning. Modern teachers are increasingly using a huge amount of resources: electronic, online and paper (book) in teaching a foreign language. Modern devices, smart boards, high-speed Internet are evolving rapidly. The objective of teaching is to increase motivation through engagement process, so to play with students, where learning potential raises sharply is our priority. In our research we take into consideration the attraction of modern devices and Internet in EFL (English as a foreign language) lessons. Many well-known teachers draw attention to the effectiveness of the use of board games in the learning process. "It is more than a game. It is an intuition" Thomas Hughes (1822–1896).

Our practical part of the research is focused on the implementation of smart technologies (mobile telephones and Internet) in the lesson. To carry out it we use different board games at the first stage of the learning process for knowledge and introduction of the game. Accordingly, the group of learners is involved into educational entertainment. Secondly, we engage students into the process of using smart devices (mobile telephones, Internet etc.). With the help of smart edutainment we provide learners with building and activating background knowledge, deep comprehension, goal achievement. In addition, the learners study to process information, monitor and transform learning into demonstration and understanding of personal skills.

Board game is very organized activity, requiring emotional and mental strength. The game itself involves making decisions that sharpens mental activity of students.

Everyone has equal rights. Moreover, the student of elementary language level may become the first in the game, resourcefulness and ingenuity here are sometimes more important than the knowledge of the English language. The atmosphere of enthusiasm gives an opportunity to overcome shyness; the linguistic material is mastered, there is a sense of satisfaction. The use of a board game on English lessons requires considerable preparation by the teacher and a guide of the games.

As a result teacher should consider the following factors when using board games:

- language proficiency;
- the studied material;
- the specific objectives and conditions of lesson;
- time management;
- consideration of certain psychological atmosphere;
- Wi-Fi, access to the Internet (computer class), smart devices.

Now we consider the most appropriate with smart learning board games during EFL lessons.

3.1 The Board Game "Dixit"

The board game "Dixit" was designed by Jean-Louis Roubira in 2008 as a party game for 4–6 players for 30 min spending time. There are expansions of "Dixit" most preferably we practice "Dixit Odyssey" (2011), because number of players in this version – 7–12 persons and the game itself becomes very useful for many students within the framework of learning English language (Fig. 1).

Fig. 1 The board game "Dixit"

How to play "Dixit Odyssey". The version includes: a folding game board, 12 rabbits scoring tokens, 84 picture cards (deck), 12 voter cards, 24 voter markers. One player in this game becomes the storyteller for the turn and looks at 6 picture cards with different images in his (her) hand. From one of these cards, he or she makes up association with the image on the card - a sentence or phrase, some words (for instance from poem, song, film etc.) and says it in a loud voice, not showing the card to the other players. You should not make up something very obvious. If the game is played from 7 to 12 players, each takes two voter markers. Players select the card in their hands which best matches the sentence (phrase, word) and gives this chosen card to the storyteller. Then, the storyteller mixes all the received cards (images) and lays out the cards on a folding game board according to numbers from 1 till 12. All cards with images are shown face up and every player has to bet upon which image was the storyteller's one. In "Dixit Odyssey" players can vote for the second card trying to improve the chances of success. Two voter markers are used accordingly.

Scoring.

- If all the players guessed the card of storyteller, or no one had guessed, the storyteller gets no points, and all the other players get two points.
- In any other cases, the storyteller, as well as all players guessed his (her) card receive three points.
- Each player, except the storyteller, gets one extra point for each vote for their image card.
- Players who guessed the card and voted for one card only get one extra point.

Thus, each player gets the cards from the deck again so that in their hands always six. If the deck is not enough cards, discarded cards are shuffled and they form new deck. The game ends when one player gets 30 points. The winner is the player with the highest number of points at the end of the last round.

Benefits of playing the game. Accordingly, this board game helps students to embody all their creative ideas, to use language material on the various topics more actively. It should be emphasized that the board game affects motivation in learning a foreign language, as well as increases the cognitive interest of the students. It is undoubtedly important for the creation and self-development of the modern graduate. The task of the teacher as a moderator - to acquaint the student with the game, introducing the basic rules of the game, creating a favorable atmosphere in a class. Thus, attracting students to the learning activities, subsequently the role of the teacher is referred to the introduction online game Dixit. Further, in the computer class with direct access to the Internet students play with the app on their individual.

3.2 The Board Game "Mister X"

Among the other board games that we put into practice is "Mister X" that was designed by Gabriele Mari in 2009. The thematic board game is played among 2–6 players and lasts approximately 60 min. "Mister X" is a modern version of classic "Scotland Yard" which shows the action story of Mister X who escapes the Scotland Yard detectives in London. It is the adventure chase on the real London streets (Fig. 2).

Fig. 2 The board game "Mister X"

How to play "Mister X". The version includes: game board, ticket board, ticket bag, 6 pieces, 6 ticket racks, 120 tickets, 15 start cards, 5 color markers, 10 position rings, 22 investigation chips, 1 checklist, 1 visor for Mr. X, 1 tracker for Mr. X. One of the players takes on the role of Mr. X. He moves from one point to another point around the map of London borrowing taxis, buses or subways that determines main kinds of transport. The rest of the players (detectives) act in concert – they move around similarly in an effort to move into the same space as Mr. X. But while the criminal's mode of transportation is nearly always known, his exact location is only known intermittently throughout the game.

Benefits of playing the game. Particularly this board game helps to develop mainly active independent thinking of students, teaches them to focus on practice and seek an approach to solving the problem. "Mister X" is a game for the development of critical thinking and the development of cognitive skills of the students, their ability to develop their knowledge independently, to summarize information.

3.3 The Board Game "Pictureka"

The board game "Pictureka" was designed by Eugene and Louise and Arne Lauwers in 2006 as a party or family game for 2–7 players for 30 min spending time. There are versions of "Pictureka", most preferably we use "Pictureka. Card game" (2008), because the game includes expansions: 8-away, Alphabetti, Matchureka and Cow's creative combo. It should be noted that the board game published by Hasbro is used as online game too and most preferably is used in English learning activities (Fig. 3).

Fig. 3 The board game "Pictureka"

How to play "Matchureka". The version consists of 78 picture cards, 32 mission cards (double cards), three special cards with penguins (penguin is a joker in a deck). "Matchureka" is an exciting and colorful board game for students and it tests their alertness and responsiveness. They should collect the most mission cards to win. Players take turns selecting a mission card and trying to find objects fast and first. Cards with pictures are placed face-down on the table in a 4 × 5 grid. The deck of mission cards is placed face down too. Students take mission cards in turns (for example, it might be a card to find an instrument and things that can fly) and try to find objects choosing cards with pictures that match the description. A student takes two picture cards and turn them face up. If the picture cards match the description of the mission card, student can collect the mission card. If one picture does not match the description of the mission card, the student place two picture cards face down in a grid. And watch out – special mission cards instruct players to turn over picture cards.

Scoring. The player who collects enough cards becomes the Matchureka champion. Students can play the game in different ways.

Discard 8 cards! (8-away). The objective of the game is to get rid of the cards as quickly as possible. Each player takes 8 cards. Cards with pictures are placed face-up on the table in a 4 × 2 grid. A mission card is placed face-up and players start looking for a card to match the description of the mission card. The first player to find the card cries "Pictureka!" and explains why the picture card matches the mission card. If other players agree, he/she gets rid of the card. The next mission card is placed face-up. The game continues until one of the players gets rid of 8 cards.

Alphabet (Alphabetti). This game is for elementary level students. The objective of the game is to find objects that start with a specified letter of the alphabet. For instance, students can select letters from A to K. Cards with pictures are only used. The first letter is A. Cards are placed face-up in a line. As soon as the player finds a card with an object that starts with the letter A he cries "Pictureka!" and identifies this object (for instance, aquarium). If other players agree, he or she takes the card. The game ends when the object starting with the last letter is found (for example, the letter K). The player who collects enough cards becomes the winner.

In our research we pay the greatest value to this game, because it is one of the favorites. This online game "Pictureka" is used as an individual element of smart edutainment in our teaching process. Throughout the study, we created a social group on the Internet, which aims to coordinate the overall work, to create a variety of topics and discussions. Discussions are encouraged by teachers. You can visit the website to play. http://www.hasbro.com/mylittlepony/en_US/play/details.cfm?R= 0B92B938-F9A3-4CE2-B9E1-AAE3D9073A0A:en_US.

Benefits of playing the game. This board game promotes competition and enables the tutor to give students a break from the stress of ESL learning.

3.4 The Board Game "Erudite"

The board game "Scrabble" was invented by Alfred Mosher Butts, an unemployed architect from Poughkeepsie, New York. Analyzing games, he found they fell into three categories: number games, such as dice and bingo; move games, such as chess and checkers and word games, such as anagrams. Attempting to create a game that would use both chance and skill, Butts combined features of anagrams and the crossword puzzle. First called LEXIKO, the game was later called CRISS CROSS WORDS [4].

The version of Russian Erudite game was inspired by popular board game Scrabble. The board game "Erudite" shares the same field and overall idea with Scrabble, but there are differences in game rules, number of tiles.

How to play "Erudite". The chips are put down side on a table and are mixed. Each player takes 7 chips. The first made up word is placed horizontally or vertically so that one of its letters is placed in the center of the board (white place).The player calculates the points received for making up a word (or words) and then takes the missing number of chips to 7. Another player forms a new word by adding one or more letters to the word or letter placed on the board. The universal chip can replace any letter of the alphabet. If this chip is placed on the board, it can be replaced with a valid chip that has the alphabetic symbol by the next player and it must be used by the player when he takes a turn. A player who could not (or did not want to make up words for tactical reasons) make a single word, has the right to substitute any number of chips (1–7) taking new chips and skipping his or her turn. A player can also find a word in the dictionary "Erudite" which helps to make up a new word. The game is finished when the first person receives a designated number of points (Fig. 4).

Fig. 4 The board game "Erudite"

Scoring. The score received for a turn is based upon the point values that appear on each chip used to make a word, and is modified based upon any bonus squares that are covered by the letters that are newly-placed during that turn.

Bonus squares:

Green square - this means that a letter placed on this square receives double the number of points that are shown on the letter.

Bonus squares for words:

Yellow square - This means that a letter placed on this square receives three times the number of points that are shown on the letter.

Blue square - This means that a letter placed on this square double the number of points of the word.

Red square - This means that a letter placed on this square triple the number of points of the word.

If the word covers several bonus squares points are summed up: firstly the number of points of the letter is counted, then, the number of points of the word is counted. If a player places 7 chips on the board per one turn he scores 15 points. A player earns a bonus if he is the first to put his chips on bonus squares.

Most significantly "Erudite" focuses on using mobile devices for finding right, new words, for checking the meanings of the words in the online dictionaries. During the game students are allowed to use electronic handheld devices, electronic dictionaries on their iPads.

Benefits of playing the game. Years of practice have shown that students who play "Erudite" become more sociable and socialized. This is a powerful teaching method that focuses on critical thinking, logic and analytical skills which are so necessary for future engineers, managers, philosophers. According to the results of the survey, conducted by ESL teachers of Tomsk Polytechnic University, 85 % of students noted that their English pronunciation has improved, 90 % of respondents

claimed that the game motivated them to enrich their vocabulary and 94 % of
students proved that they try to remember words to use them in other games and
situations. The table below shows how to use the aforementioned board games
during ESL classes (Table 1).

Table 1 Board games in the classroom

Name of the game	Number of players	Skills category	Key vocabulary	Grammar/spelling	Use of electronic devices
"Dixit"	7–12	Association Logic Communication	Various phrases (nouns, adjectives, verbs), sentences related to any topic (people, education etc.)	Word formation, superlatives (the most ..., the least etc.); -ing and the Infinitive: verb + -ing (enjoy doing); phrasal verbs	Computer class (online Dixit), mobile telephones (electronic dictionaries)
"Mister X"	2–6	Bluffing Deduction Cooperation	Words related to the topic: In the town streets. Kinds of transport (underground, bus, taxi). Directions, routes, ways	Questions with Yes/No; Who-; Past Simple, Future tenses. Prepositions of place; articles of landmarks, and sights	Mobile telephones (electronic dictionaries)
"Pictureka"	2–7	Abstract strategy Memory	Various nouns, verbs and adjectives etc. (hot, scary, kitchen, music, ocean animals, tooth, leg)	Relative pronouns who, which, that: Find someone who... Find something that... Find an object which...	Computer class (online Pictureka)
"Erudite"	2 and more	Memory Vocabulary Spelling Time management Logic Analytical skills	Various nouns, verbs, adjectives etc. The game can be focused on words related to a particular topic (travelling, food, entertainment etc.).	Accuracy of spelling Word formation	Mobile telephones (electronic dictionaries)

4 Conclusion

Smart edutainment integrated in academic process makes learning an exciting and fun activity. It makes it possible to mix games and learning process for an effective and beneficial experience and for students the use of new technologies is engaging and they easily use them.

Moreover, using board games in English lessons teacher should take into consideration the next things: expansion of language material, selection of professionally-oriented language, provision of practical language study through communicative games and various games that stimulate student's interest and their creativity, improve vocabulary, English language skills, computer-based skills.

Smart edutainment allows students to:

- become more sociable and socialized;
- to enrich their vocabulary;
- to have a break from the stress;
- to develop cognitive skills and independent thinking;

Games for learning English contribute to the enrichment of students' vocabulary, they help to master quickly the rules of spelling, grammar and pronunciation of English words, learn to work with the dictionary. Through the use of the games, learning process becomes an interesting and effective way of creative thinking, it develops logical abilities, stimulates teamwork and enhances personal skills. It also learns to choose the right tactics and make decisions.

References

1. Sakoyan, A.: Science is fun. http://polit.ru/article//2011/07/29/edutainment (in Russian)
2. Perushev, D.: Word formation lessons// ConNews.ru. http://finditnow.osa.pl/atp/? (in Russian)
3. Kobzeva, N.: To the question of entertaining foreign language teaching technology// Uchenye zapiski Tavricheskogo Natsionalnogo Universiteta im. V.I. Vernadskogo. Series«Filology. Social communications». 25(64), 280–283 (2012) (in Russian, No 1. Part 2)
4. National scrabble association. http://www.scrabble-assoc.com/info/history.html

Study Materials in Smart Learning Environment – A Comparative Study

Blanka Klimova and Ivana Simonova

Abstract At present there is a visible paradigm shift from the traditional learning environment to the technology-enhanced learning environment, which is understood as smart learning environment. This new smart learning environment is distinguished by context-awareness, mobility, seamlessness or adaptability; the factors which, among others, have an enormous impact on the development and use of learning materials. Thus, the purpose of this article is to discuss what study materials students prefer in this smart learning environment. This is done by comparison of two surveys performed at the Faculty of Informatics and Management in Hradec Kralove, Czech Republic, in the course of two periods in order to examine whether there has been any radical shift in students' preferences for the use of a particular form of the learning materials or not.

Keywords Study materials · Smart learning environment · Students' preferences · Multimedia elements · Surveys

1 Introduction

In current smart learning environment (SLE), which is perceived as technology-supported learning environment that can make adaptations and provide appropriate support (e.g., guidance, feedback, hints or tools) in the right places and at the right time based on individual learners' needs, which might be determined via analyzing their learning behaviours, performance and the online and real-world contexts in

B. Klimova (✉) · I. Simonova
University of Hradec Kralove, Rokitanskeho 62, Hradec Kralove, Czech Republic
e-mail: blanka.klimova@uhk.cz

I. Simonova
e-mail: ivana.simonova@uhk.cz

© Springer International Publishing Switzerland 2015   81
V.L. Uskov et al. (eds.), *Smart Education and Smart e-Learning*,
Smart Innovation, Systems and Technologies 41,
DOI 10.1007/978-3-319-19875-0_8

which they are situated (cf. [1]), close attention is paid to learning materials in order to make students' learning more independently effective and efficient in a sense of their context-awareness, seamlessness, adaptability and pervasiveness (cf. [2]). The issue of study materials is of high importance nowadays because of the increasing use of mobile devices such as notebooks, tablets or smart phones for learning purposes (cf. [3]). However, their displays, particularly of smart phones and tablets, are much smaller so that the structure and the content of the study materials must adjust to this fact. Furthermore, the learning materials should be interactive to draw learner's attention and influence more of his senses. [4] list the key criteria for the successful use of the study materials in SLE as follows:

• Study materials should have a clear, concise, logical and simple structure (information in bullets is preferred).
• They should be well-balanced (i.e. there should be an adequate amount of relevant teaching matter including learning objectives and exercises/assign-ments/self-tests).
• The materials should be comprehensible and up-to-date.
• They should be easily navigated.
• They should be interactive with appropriate multimedia components.
• They should be linked to other suitable materials and relevant websites.

In addition, [5] provides a simple framework for the development of any technology-enhanced learning materials which would suit mobile devices in current SLE. They should consist of no more than two pages and their structure should be as follows (see Fig. 1 below):

• topic (a concise sentence or a phrase of the lesson content);
• learning goal (a short statement motivating the participants to study the particular lesson); prerequisites (previous knowledge required to master the lesson);
• skills (a description of the knowledge/skills to be gained in the particular lesson);
• explanation of the basic concept and ideas of the teaching matter discussed in the lesson (in the form of text and questions);
• conclusion with self-tests, tasks, quizzes (with keys), or an assignment; and
• bibliographical sources and or links to them.

Moreover, short sentences, involving 20 words at maximum should be formed [7] or [8]. For a further and a more detailed description of the production of any study materials see the evaluation checklists of [9] or [10].

The purpose of this article is to explore what study materials students prefer in SLE. This is done by comparison of two surveys conducted in 2010/11 and 2013/14 in order to examine whether there has been any radical shift in students' preferences for the use of a particular form of the study materials or not.

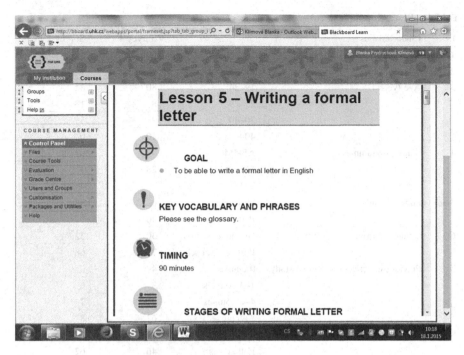

Fig. 1 An example of a smart learning material layout [6]

2 Materials and Methods

Data for this study were collected within two surveys which were held in 2011/12 (Survey 1 [11]) and 2013/14 academic years (Survey 2 [12]) at the Faculty of Informatics and Management (FIM), University of Hradec Kralove (UHK), Czech Republic. The data were collected by the method of questionnaire.

Two research samples are described by eight criteria. Some characteristics are rather close, other differ substantially. The data are displayed in Table 1 below.

The research samples have similar gender and age structures (items 2 and 3), they are close in the number of respondents who graduated from secondary grammar or professional school (item 6) and are rather equal in respondents' opinions on future use of online courses for education (item 8). However, they mainly differ in the amount of respondents (item 1), structure of full-time and part-time students (item 5) and previous experience in online study (item 7) – see Fig. 2 below.

In both surveys the questionnaires were available online to respondents within the LMS WebCT/Blackboard, which has been used to support both full-time and part-time forms of study at the institution since 2001. Totally, approximately 250 online courses have been already implemented in study programmes at FIM. This means students are familiar with working with questionnaires. Therefore limitations

Table 1 Research sample: a comparison (authors' own source)

Item	Characteristics		Survey 1 (%)	Survey 2 (%)
1	Respondents (n)		397	2,440
2	Gender	Male/Female	63/37	76/24
3	Age (years)	19–24	80	84
		25–29	8	13
		30–40	2	3
		40+	10	0
4	Study programmes	AI+IM	44	93
		FM	28	3
		TM	27	4
5	Form of study	Full-time	60	88
		Part-time	40	12
6	Graduated from	Grammar school	30	32
		Professional school	70	68
7	Previous experience in online study	0 course	4	0
		1–3 courses	3	6
		4–6 courses	31	7
		7+ courses	62	87
8	Future online study	Yes	32	23
		Rather yes	46	62
		Rather no	20	14
		No	2	1

Explanation: AI-Applied Informatics; IM-Information Management; FM-Financial Management; TM-Management of Tourism

were neither expected, nor detected within the process of the data collection. Both questionnaires contained multiple-choice questions with one or more choices and open-answer questions for providing additional information. In addition, the following research questions were defined:

1. In the period of SLE, what materials do students use more frequently – paper-printed or electronic?
2. What types of electronic materials do students prefer?

Furthermore, the authors focus on the following types of study materials:

- printed materials – in case of *yes* answer, students were asked whether they buy the study materials, or borrow them from libraries;
- electronic materials – in case of *yes* answer, the types of materials used were detected and students were asked whether they buy them, or use only those which can be downloaded from the Internet for free.

Questionnaire 1 (collecting data from Survey 1) consisted of 12 items; questionnaire 2 (Survey 2) included eight items. The data were processed by the method of frequency analysis by NCSS2007 statistic software.

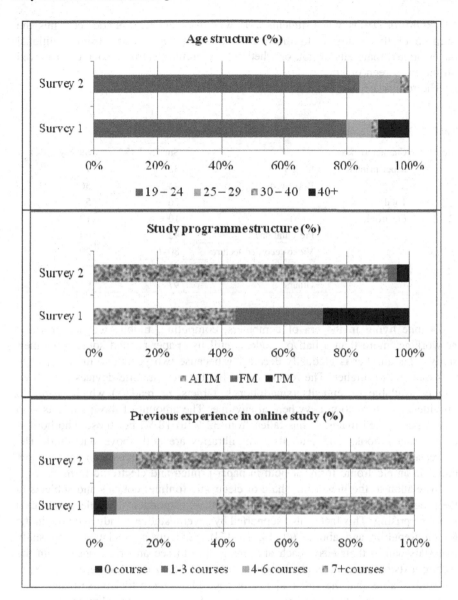

Fig. 2 An illustration of the research sample (authors' own source)

3 Results of the Survey and Their Discussion

Books and journals published in the paper form and materials printed from electronic sources were included under paper-printed materials whereas electronic materials were understood as those downloaded from the Internet or available on

CD-ROMs or DVDs which students work with from an electronic device. Thus, the purpose of this study is to discover whether students study from traditional paper-printed materials or not, or whether they prefer electronic sources displayed on the PC screen or not.

The results are displayed in Table 2 below.

Table 2 Research results: a comparison (authors' own source)

Item	Characteristics		Survey 1 (%)	Survey 2 (%)
1	Paper-printed	Bought	32	18
		Borrowed	44	36
	Total		76	54
2	Electronic	e-texts	100	100
		Presentations	92	69
		Video-recorded lectures	80	29
		Animations	57	26
		Others	97	97

Despite living in the era of computers, computing, mobile technologies and networking, more than a half of students still use paper-printed sources for their study. This number is gradually decreasing because two years ago the percentage was one quarter higher. The reason might be that new mobile devices have been widely available to students (smartphones, tablets, or readers) which enable to provide any information anywhere at anytime. The amount of those students who buy paper-printed materials has fallen from 32 % to 18 %, but those who borrow paper-printed books and journals from libraries are still above one third. The decrease could be caused by the fact that more and more sources are now available either in an electronic form, or both in paper-printed and electronic forms.

In relation to the increasing share of electronic (online) courses and subjects in the process of instruction, the fact that all (100 %) respondents use electronic texts is not surprising. This fact is also supported by several studies conducted both in the Czech Republic and abroad (cf. [4, 14–17]). Students appreciate having study materials within their easy reach and being able to see and read once again the lecture texts or other materials from their face-to-face classes. Furthermore, [18] in his study claims that the vast majority of page views were for course materials, a supplement for face-to-face contact on-campus. In addition, several research studies prove that eLearning materials have positive effect on student learning [19] or [20].

Besides the multiple choice of replies, the respondents were also able to express their own opinions in open-answer questions. The most frequent reasons for their satisfaction with the electronic study materials were as follows:

- access to study materials independently on time and location;
- a possibility of copying, downloading and printing;
- an easy access to study materials;

- low/no costs of learning materials;
- everything that is needed for study in one place;
- simplicity;
- clarity;
- relative conciseness;
- well-written and structured;
- comprehensiveness of materials and links to other suitable materials;
- graphics of VLA and its multimedia components;
- a sufficient amount of information;
- current information;
- a chance of self-study;
- individual pace of study;
- self-tests, a chance to check one's knowledge;
- a chance to return to study materials and complete one's learning
- interaction;
- ecological approach;
- assurance of correct learning materials;
- no need to take notes during the face-to-face lectures;
- no need to carry textbooks to classes or borrow them in the library;
- a chance to complete notes from the face-to-face classes; and
- a chance of searching.

Whereas in 2011/12 rather high values were detected in all observed criteria, two years later slightly decreased in presentations (from 92 to 69 %), but the decrease was sharp with video-recorded lectures (from 80 to 29 %) and with the use of animations (from 57 to 26 %). It is quite striking that students showed low preferences for the multimedia elements such video sequences and animation in Survey 2. There might be several reasons for this fact.

Firstly, the creation of these materials is quite demanding and therefore they are not so much common in the electronic materials. Consequently, students are not used to exploiting them.

Secondly, the creation of such materials is time-consuming. In comparison with the text or PowerPoint presentation which can be understood quickly, the video recordings or animation cannot since students would miss the whole point.

Thirdly, mobile technologies, as it has been already mentioned in the Introduction, possess different technical parameters. They often have a small display and a short battery life. Moreover, students predominantly use web pages whose access requires a relatively good Internet connection that is not always available everywhere.

Finally, students still like to highlight the key points they want to remember with different coloured marker pens, particularly for revising the material before sitting for an exam. These are just a few observations from the authors' point of view and they would definitely further investigations. But in spite of these pitfalls, multimedia elements should be incorporated into the development of smart learning materials since they represent a great potential for learning (cf. [21, 22], or [23]). In addition, [24] assigns the following attributes to the multimedia:

- they are modern/fashionable;
- they affect more senses;
- they are up-to-date, they can be usually easily modified;
- they are user-friendly;
- they are relatively inexpensive;
- they are eye-catching/appealing to students;
- they are stimulating;
- simply, they are natural means of student's everyday use.

According to [25], multimedia instructional materials have been recognized for enabling the understanding of complex engineering and IT decision-making situations. They have been also identified as an important tool for managers and students in their efforts to connect and apply classroom theory-based learning with the analysis of real-world problems. In addition, [26 and 27] claim that multimedia learning materials promote deeper learning and intercultural competences [23].

Although the same results were detected under the criterion *others*, this criterion contained some new items as respondents mentioned in the open answers. Whereas links to additional sources of information, applications such as dictionaries, glossaries, or wikipedia were monitored in Survey 1; networking, sharing sources, experience and opinions were mentioned in Survey 2. The latter issues in fact confirm the paradigm shift of traditional learning towards smart learning since learning in SLE is perceived as a social process that happens at a time and place of the learner's choosing, continuing throughout one's life. It is collaborative, evolving and informed by a process of self-paced development (cf. [28]). The learning collaborators together with learning contents and learning services are the three main learning resources of SLE (cf. [29 and 30]).

4 Conclusion

In conclusion, the comparison of the data collected in both surveys showed that a rather large amount of respondents welcomed having their study materials in electronic form (both online and offline). The most frequents reasons for their satisfaction with the electronic study materials were as follows:

- to have an anything/anytime/anywhere access;
- to re-check the information already mentioned in face-to-face classes;
- not to spend additional time in libraries and shops if other (electronic) sources are available;
- to become more confident with the new learning content as electronic sources can help them acquire new knowledge more easily and correctly.

Although more than half of students (54 %) still use in their studies traditional paper-printed materials, many more prefer various electronic sources with the multimedia components if available.

Finally, there is a visible paradigm shift from the traditional learning processes to those in smart learning environment which are distinguished by mobility, location awareness, interoperability, seamlessness, situation awareness, social awareness, adaptability, and pervasiveness (cf. [31]).

References

1. Hwang, G.J.: Definition, framework and research issues of smart learning environments – a context-aware ubiquitous learning perspective. Smart Learn. Environ. 1(4), 1–14 (2014)
2. Yang, S.J.H., Okamoto, T., Tseng, S.S.: Context-aware and ubiquitous learning. Educ. Technol. Soc. 11(2), 1–2 (2008)
3. Boranek, R.: Mobile devices soon overcome the number of personal computers (Mobilni zarizeni v poctu brzy prekonaji osobni pocitace). http://www.root.cz/zpravicky/mobilni-zarizeni-v-poctu-brzy-prekonaji-osobni-pocitace/ (2014)
4. Frydrychova Klimova, B., Poulova, P.: Analysis of online materials and their impact on learning. In: Proceedings of the Third International Conference on Digital Information Processing (ICDIPC 2013), The Society of Digital Information and Wireless Communications (SDIWC), USA, pp. 564–568 (2013)
5. Frydrychova Klimova, B.: Teaching Formal Written English, UHK, Gaudeamus (2012)
6. An example of a smart learning material layout. http://bboard.uhk.cz/webapps/portal/frameset. jsp?tab_tab_group_id=_2_1&url=%2Fwebapps%2Fblackboard%2Fexecute%2Flauncher% 3Ftype%3DCourse%26id%3D_1507_1%26url%3D (2014)
7. Bednarikova, I.: Didakticke aspekty uciva a e-learning (Didactic principles of learning materials and e-learning. In: Sedlacek, J. (ed.) Sbornik prispevku ze seminare a souteye e-learning 2003, pp. 164–171. Gaudeamus, Hradec Kralove (2003)
8. Dunleavy, P.: Authoring a PhD. Macmillan, Palgrave (2003)
9. Wright, R.C.: Criteria for evaluating the quality of online courses. http://elearning.typepad. com/thelearnedman/ID/evaluatingcourses.pdf (2012)
10. Branch, R.M., Dohun, K., Koenecke, L.: A Checklist for evaluating online materials. http:// www.libraryinstruction.com/evaluating.html (2011)
11. Simonova, I., Poulova, P.: Learning Style Reflection within Tertiary E-Education, WAMAK (2012)
12. Klimova, B., Poulova, P.: Surveying students' satisfaction with online study materials. In: Proceedings of the 39th Annual International Computer, Software and Applications Conference. Taichung, Taiwan (2015, in press)
13. Frydrychova Klimova, B., Poulova, P.: Impact of a form of online materials on the quality of education – a case study. Int. J. Digit. Inform. Wireless Commun. 3(1), 43–49 (2013)
14. Cechova, I., Zerzanova, D., Berankova, J.: Materials development in language training: online course of military English. In: European Conference on E-learning, pp. 80–89 (2012)
15. Hwang, N.Y., Wang, C.Y., Sharples, M.: A study of multimedia annotation of web-based materials. Comput. Educ. 48(4), 680–699 (2007)
16. Karuppan, M. C., Karuppan, M.: Empirically based guidelines for developing teaching materials on the web. Bus. Commun. Q. 62(3), 57–45 (1999)
17. O'Daniel, M.: Online versus printed materials, Computimes. vol. 1, Malaysia (2001)
18. Gerlich, R.N. Web-assisted courses: a case study of how on-campus students use online materials. In: Proceedings of the Allied Academies International Conference, Academy of Educational Leadership, pp. 3–7 (2002)
19. Baki, A., Guveli, E.: Evaluation of a web based mathematics teaching material on the subject of functions. Comput. Educ. 51(2), 854–863 (2008)

20. Jung, I., Choi, S., Lim, C., Leem, J.: Effects of different types of interaction on learning achievement, satisfaction and participation in web-based instruction. Innov. Educ. Teach. Int. **39**(2), 153–162 (2002)
21. Lindfors, J.: Children's Language and Learning. Englewood Cliffs, NJ, Prentice-Hall (1987)
22. Sperling, R.A., Seyedmonic, M., Aleksic, M., Meadows, G.: Animations as learning tools in authentic science materials. J. Instrumental Media **30**(2), 213–221 (2003)
23. Reid, E.: Multimedia used in development of intercultural competences (ICC). In: CALL and Foreign Language Education: E-Textbook for Foreign Language Teachers, Nitra, UKF, pp. 83–92 (2014)
24. Frydrychova Klimova, B.: Multimedia in the teaching of foreign languages. J. Lang. Cult. Educ. **1**(1), 112–121 (2013)
25. Mbarha, V., Bagarukayo, E., Shipps, B., Hingorami, V., Stokes, S. et al.: A multi-experimental study on the use of multimedia instructional materials to teach technical subjects 1. J. STEM Educ. Innov. Res. (suppl. Special Edition), pp. 24–37 (2010)
26. Mayer, R.E.: Multi-media aids to problem-solving transfer. Int. J. Educ. Res. **31**, 611–623 (1999)
27. Mayer, R.E.: The promise of multimedia learning: using the same instructional design methods across different media. Learn. Instr. **13**, 125–139 (2003)
28. Winters, N., Walker, K., Rousos, D.: Facilitating learning in an intelligent environment. In: The IEE International Workshop on Intelligent Environments, London, Institute of Electrical Engineers, pp. 74–79 (2005)
29. Yang, S.J.H.: Context aware ubiquitous learning environments for peer-to-peer collaborative learning. Educ. Technol. Soc. **9**(2), 188–201 (2006)
30. Mikulecky, P.: Smart environments for smart learning. In: Proceedings of the 9th International Scientific Conference on Distance learning in Applied Informatics, Nitra, UKF, pp. 213–222 (2012)
31. Yang, S.J.H., Okamoto, T., Tseng, S.S.: Context-aware and ubiquitous learning. Educ. Technol. Soc. **11**(2), 1–2 (2008)

Part II
Smart Educational Technology

Smart Study: Pen and Paper-Based E-Learning

Dieter Van Thienen, Pejman Sajjadi and Olga De Troyer

Abstract Smart Study is an educational platform allowing learners to continue using pen and paper for exercise solving, accomplished by using a digital pen and Anoto paper. Handwritten solutions are digitalized, automatically corrected, and animated feedback is provided by means of a tablet device. By using pen and paper, read-write learners can still use their preferred learning method while other types of learners can practice their handwriting. The use of a tablet for feedback and additional information makes it a lightweight and mobile platform, easy to use by children. The platform was evaluated by means of a case study involving a group of 15 children, showing that they were more motivated than usual and enjoyed making exercises using the Smart Study platform.

Keywords E-Learning · Digital pen · Interactive paper · Tablet · Read-write learners

1 Introduction

With the emergence of tablets, smartphones, and e-readers, one can question whether pen and paper still has some use. Information on paper is static, while being displayed on a screen it can become dynamic, allowing adaptation to the individual. Those devices also become increasingly affordable for a larger audience. A paperless environment seems to become a realistic scenario. Even school environments go

D. Van Thienen (✉) · P. Sajjadi · O. De Troyer
Vrije Universiteit Brussel, Department of Computer Science, WISE,
Pleinlaan 2, 1050 Brussels, Belgium
e-mail: dvthiene@vub.ac.be

P. Sajjadi
e-mail: ssajjadi@vub.ac.be

O. De Troyer
e-mail: Olga.DeTroyer@vub.ac.be

© Springer International Publishing Switzerland 2015   93
V.L. Uskov et al. (eds.), *Smart Education and Smart e-Learning*,
Smart Innovation, Systems and Technologies 41,
DOI 10.1007/978-3-319-19875-0_9

along with this evolution by using computers, tablets, and other electronic devices for teaching and learning.

But as this digital shift happens, the traditional way of writing with a pen on paper tends to disappear. Writing with a pen on paper is progressively being replaced by typing on a keyboard or writing with a stylus on a capacitive screen. Whether this is a desirable evolution from an educational point of view is an issue of discussion, with arguments that are for and against. However, rather than joining one of the two sides, we have investigated what could be done in order to go along with this digital evolution, while keeping children stimulated to keep practising their handwriting. Instead of replacing pen and paper by modern technology, we propose to combine them. This could preserve the advantages of reading and writing on paper, while at the same time providing the advantages of digital information. Digital pens, i.e., pens that can digitally capture what is written on paper, make this possible.

Smart Study is an educational platform using a digital pen and paper for writing and combining this with a tablet for providing digital feedback and information. Smart Study aims at supporting learners to practice, i.e., exercise solving, more autonomously while at the same time stimulating their handwriting. The digital pen writes like a normal pen, but can digitally capture everything written down on, so-called, interactive paper. Next, intelligent character recognition is used to make sense of what has been written. A tablet app automatically corrects the handwritten answers, providing feedback in an animated way on how to correctly solve the exercises. The app also provides an overview of the exercises made by the learner. All data is stored in a database, which can be consulted for generating reports. Since all answers are corrected automatically, the teacher's work is relieved as well. Currently, primary school children are targeted, but the approach is not limited to this group.

The paper is organized as follows: Section 2 presents related work. Section 3 discusses the importance of paper, Sect. 4 presents the Smart Study platform and Sect. 5 its evaluation. Section 6 provides conclusions and future work.

2 Related Work

There are plenty of tablet apps for educational purposes, see e.g., Apple Store or Google play. These will not be discussed. On the Livescribe store, there are also a few digital pen applications for educational purposes. These apps either use the display of the pen or the speakers to provide feedback to the learner. For instance, the SpellingMe app tests students on spelling words for students grades 1–5. The student has to spell a word, pronounced through the speaker, and the Smartpen provides aural feedback on the correctness of the spelling.

The LeapReader, provided by LeapFrog [1], is a digital pen for children age 4–8 that sounds out words, guides letter strokes, and helps build comprehension. In the past, the company produced the Fly digital pen, a pen similar to the Livescribe pen, but in 2009 the company discontinued both the manufacture and support of this pen. LeapFrog also has the LeapPad tablet, a learning tablet for children age

3–9. However, it is not clear whether there are apps that combine the digital pen and the tablet.

Although different researchers have proposed to augment paper by managing the link between handwritten and digital information (e.g., [2–4]), as far as we are aware, there are no educational apps combining pen/paper and a tablet. The only application using pen/paper interaction with feedback on a tablet that we came across is a game. Golem Arcana is a digitally enhanced board game [5], using a Tabletop Digital Interface Stylus which can read micro codes that are printed on the game pieces and over the terrain art of the board's regions. Players move their characters by tapping the appropriate game pieces with the stylus. The results are displayed on a tablet that is running the Golem Arcana app.

Some researchers investigated the use of a digital pen to improve learning. In [6], pen-based input in combination with groupware was used to facilitate interaction during computer science courses. However, no pen and paper but a tablet and a stylus were used for the writing. Miura et al. did a number of studies using a digital pen in a classroom [7–9]. In 2005, they developed AirTransNote [7], a digital pen learning system based on an ultrasonic digital pen (Inklink) and PDAs used to transmit the student's notes from Inklink to the teacher's PC via wireless LAN. In [9], an improved system was presented using an Anoto-based digital pen. The system was used to investigate the class activity and motivation of the children when adopting it to write calculations and draw diagrams. The results show that most children were more motivated, concentrated and enjoyed the lecture more.

3 Importance of Paper

The paperless office is a concept that has already been discussed for some time [10, 11] and it seems that this discussion has reached the classroom environment as well, questioning the need to practice handwriting. Recently [12], it was announced that from 2016 in Finnish schools, primary school children's typing skills would be stimulated instead of their handwriting. The proponents argue that handwriting has become useless in the digital age, as everybody will use computers and tablets in the future for communication. The opponents point out that handwriting helps children develop fine motor skills and brain functions [13], but proponents argue that this can also be achieved by handicrafts and drawing lessons. On the other hand, recent research [14] shows that students taking notes on laptops performed worse on conceptual questions than students who took notes longhand.

In 2003, Sellen and Harper [11] extensively discussed advantages and disadvantages of a paperless office. Although technology has evolved since then, some of their arguments are still valid: (1) *For a lot of people, paper is still the preferred way of reading and writing.* You simply pick up the paper and start reading, or pick up a pen and start writing. Electronic devices can simulate this, but writing on a capacitive screen with a stylus does not provide the same experience as writing with a pen on paper [15]; (2) *Electronic devices are dependent on a power source. Paper does not need this*. Although your pen could run out of ink, this is a very low-cost replacement; (3) *Physical paper is still a common method of information exchange*. Even today, although there is a computer on almost every office desk, a lot of paper is still used, often side-to-side with one or more screens; (4) *Clicking or scrolling to pages in a digital way will never be as fast as doing this with real paper*. Today, this still holds. However, an advantage of digital use is the ability to jump to a specific page, or searching for terms; (5) *The use of space*. In their case study, the subjects who used physical paper had no problems in spreading out the papers, cross-referencing to other papers, and were able to manipulate and rearrange physical documents easily. In the digital way, the subjects were frustrated by the limited ability of doing the same tasks. This is a problem that is hard to solve by simply using flat screens. Larger or multiple displays allow you to place a few documents next to each other, but manipulating and rearranging documents is still frustrating.

A similar discussion can be held for a school environment. In [16], Thayer describes the urge of schools to use e-readers as a replacement for printed textbooks, but also warns for the impact it can have on students. He questions the gain with e-readers, such as a possible decrease in costs, and what is lost when abandoning paper.

Besides the pros and contras that have been mentioned, we want to add an additional argument in favor of pen and paper in a learning environment. Different pedagogues have reported that individuals differ in how they learn (e.g., [17]), i.e., there are systematic differences in individuals' methods used to acquire and process information in learning situations. Different theories have been proposed that classify learning patterns into so-called learning styles. A commonly used categorization is the VARK model [18]. It categorizes learning into four major styles based on the preferred method in which stimuli are taken in. *Visual learners* have a preference for seeing (pictures, diagrams, etc.). *Auditory learners* have a preference for information that is heard or spoken (lectures, conversations, etc.) including talking to oneself. *Read-write learners* have a preference for written text and acquire and process information by writing it down. *Kinesthetic learners* learn best when carrying out a physical activity. In practice, learners may use a mixture of different learning styles.

According to this theory, the use of pen and paper is indispensable for people with a read-write learning style. One could argue that writing could be replaced by typing, but this may not be as effective for read-write learners. Recent research [14] has found that learning is more effective by writing than typing.

Learning style theories have been criticized by many researchers because they would lack a scientific basis and there would be a lack of evidence (e.g., [19, 20]). Whether this criticism is justified or not, it is a fact that read-write has always been, and still is, the dominant learning style applied in traditional education. To study, children and students are given textbooks and (paper) notebooks. Although the learning material is usually also explained aurally and sometimes also demonstrated, in the end, learners are supposed to use the textbooks for studying, are asked to write reports, essays, solve exercises, and are stimulated to write down what they need to learn as much as possible.

It is clear that there is no easy solution for schools to adapt to a modern environment in such a way that learning is improved and the budget is decreased. It looks like paper is something that cannot be easily replaced by electronic devices in a school environment. So instead of trying to replace paper by modern technology, why not combining both?

4 Smart Study Platform

Smart Study is an educational platform that allows children to practice (i.e., make exercises) more autonomously while at the same time stimulating their handwriting. The platform is composed of a digital pen, exercise sheets (Anoto paper), and a tablet. We first discuss the technology used to realize the platform; next we present the architecture; and subsequently the workings of the platform.

4.1 Technology Used

The digital pen used is the Livescribe Echo Smartpen [21]. This pen writes like a normal pen but can capture everything written on paper, using a tiny infra-red digital camera next to the pressure tip of the pen, which is activated every time you write something down on the paper. The paper contains an Anoto dotted pattern [22]. These are tiny printed dots (hardly noticeable) that have unique positions, so the camera in the digital pen can determine the exact location of the tip of the pen on the paper. The pen has a small display that can present information when certain actions have been done, a microphone to record audio, and a built-in speaker that can play back audio. The pen also has built-in memory that can be used to store what has been captured. It also has a processor and a built-in Intelligent Character Recognition (ICR) engine, which is used to recognize the handwriting. This pen has a micro-USB connector to connect it via a cable to a computer device. Newer Livescribe Smartpens have the ability of connecting to a computer device, tablet, or smartphone in a wireless way via Bluetooth or WiFi. However at the time of development, we did not have an SDK for these newer pens at our disposal. Therefore, we decided to develop a prototype with the Echo pen for which the SDK was at our disposal.

The digital pen and dot paper are used by the learner to solve exercises. An Android tablet app is used for providing feedback to the learner on the correctness of his or her solutions. Since tablet devices are light, mobile, and easy to use by providing a touch screen, this was preferred over the use of a laptop or a desktop computer. Android was deliberately chosen because of the wide variety of available tablets in different price ranges.

4.2 Livescribe Platform

In order to develop an application for the Echo pen, two components need to be developed: a Penlet and a Paper Design, supported by the Livescribe SDK. A Penlet is the application running on the pen that allows interactions, such as tapping or writing, and can detect specific regions on the dot paper, called active regions. These active regions are defined in a Paper Design, which is software deployed onto the pen to know the layout of the dot paper. Such a Paper Design can contain static regions of different shapes with specifically defined functions by linking them to a Penlet. A developer can define these regions, add images and text to the paper, generate the dotted pattern overlay, and finally produce Adobe PostScript files, which can then be printed by a qualitative Laser printer.

To use the application, the Penlet and the Paper Design need to be installed on the pen. The Penlet can be activated by navigating to it using the Smartpen's menu or by directly interacting with the printed dot paper.

4.3 Architecture

Figure 1 presents the main components of the platform. Because the SDK of the Echo pen does not allow direct communication with a tablet, a laptop was used as a communication bridge. Next, there is a SQL database (accessible over the Internet) in which correct solutions and feedback, as well as the data about the progress of the individual students, are stored.

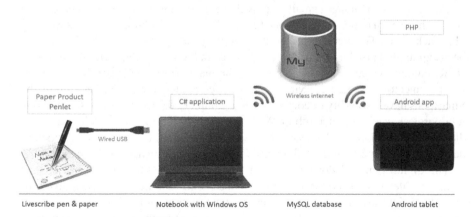

Fig. 1 Main components of the Smart Study platform

4.4 Workings of Smart Study

Before learners can start using the platform, the exercise sheets need to be prepared, i.e., creating the Paper Designs, the corresponding Penlet, and the tablet feedback. Currently, this must be done by software engineers. However, we aim at having an authoring tool that is easily usable by instructional designers, focusing on the content and generating the necessary software automatically. Such a tool is currently under development.

When exercise sheets are available (see e.g., Fig. 2), the learner starts by picking out a sheet and turning on the pen. The Smart Study Penlet is automatically activated when interacting with a recognized exercise sheet. The learner fills out the exercise page by writing the answers in the corresponding answer regions of the paper. Space outside these active answer regions is provided to allow the learner to work out the solution. Every time an answer is written in an answer region, it is recognized, stored, and displayed on the display of the pen. Exercises can be left open. The learner ends by tapping inside the area that displays the text *Ik ben klaar!* (Dutch for "I am ready!"). A text message (on the display of the pen), as well as a voice message is provided to inform the learner that the answers are saved successfully. An XML file containing the answers is automatically built and stored on the internal memory of the pen.

Next, the data captured on the pen has to be stored into the database. As mentioned before, this is done by connecting the pen (with a micro-USB cable) to a computer running the synchronisation program, which will automatically execute the entire process cycle, including the processing of the information, comparing the answers with the solutions, and storing the data into the database. The learner can now switch to the tablet to get feedback. In order to do this, the Smart Study app needs to be launched. An active Internet connection is required. After authentication, the home screen is presented, providing the choice between correcting answers and viewing an overview of all results obtained so far.

When the learner opts for correcting the handwritten answers, the app will load the data related to the sheets filled out by the learner. The screen will show an overview of all the exercises on the sheet: the questions, the recognized answers, and an indication whether the answers were correct or not. Correct answers are shown in green, incorrect answers in red along with the correct answer in blue. See Fig. 3 for an example screen. The learner can tap an exercise to obtain feedback on how to correctly solve the exercise (static text or animated feedback).

A demonstration video was created and uploaded to Youtube,[1] showcasing the final product.

---

[1] http://youtu.be/eoEacSLyfOY.

Fig. 2 Example of an
exercise sheet

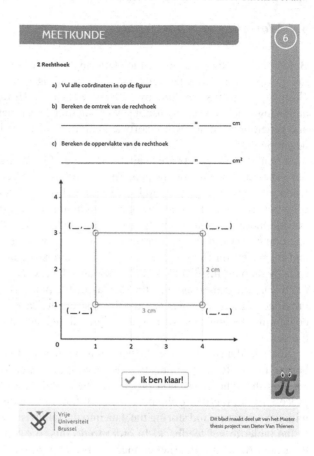

5 Evaluation

A pilot evaluation has been performed to obtain a first evaluation of the platform. The evaluation was conducted in three primary schools, having 5 children of the fifth grade participating per school (age 10 or 11): 7 boys (47 %) and 8 girls (53 %). Different schools were involved because of a possible difference in attitude towards the use of new technology and experience with educational software.

The evaluation was done individually, using math exercises. First, the child was asked for his or her experience with computers and tablets, and experience with educational software. Next, a brief explanation of the platform was given to the child, where after the child had to solve the math exercises using the platform. Subsequently, we did a structured interview with the child to elicit the child's opinion about the platform. In addition, and after all children of one school did the evaluation, an additional structured interview was done with the teacher of that class to obtain his or her opinion. The questionnaires used for the children and teachers were

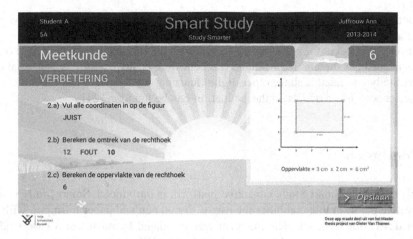

Fig. 3 Example of a solution screen on the tablet

based on the questions in [23]. The math knowledge of a child was obtained from the teacher (i.e., average of the child's math scores).

During the experiment, we observed the attitude of the child, as well as the accuracy of the handwriting recognition and the pen grip of the child. Because the handwriting of children is still developing, it was important to evaluate the accuracy of the built-in ICR. When an answer was recognized incorrectly, we asked the child to write it once more. This way, the child was not penalized for the misrecognition, and the overall accuracy of the ICR could be calculated based on the amount of retries. Because the pen is rather big and thick for a child, we also wanted to evaluate the child's pen grip and whether a bad grip would have an impact on the ICR.

The results of this pilot evaluation were positive. The ICR accuracy of the pen was very high (on average 97.39 %), keeping in mind that these are handwritten notes conducted by children. Most children had a good pen grip (11 out of 15) as well. For 11 children the handwriting recognition was 100 % accurate; 9 of them having a good pen grip and 2 having a bad pen grip but a neat and clean handwriting. For 3 children the handwriting recognition was below 90 % accuracy; 2 of them having a bad pen grip and a bad handwriting. The other child had a good pen grip, but wrote the number 1 in such a way that the ICR interpreted it as 7. All children were excited and enjoyed using the platform. They found it fun and easy to work with. Compared to their average performance in math, they performed better than usual. They were more motivated and were actively working with the tablet app, closely studying the feedback for an exercise they filled out wrong. However, because of the limited number of participants, it is not possible to draw general conclusions here.

The teachers were asked more reflective questions regarding the potential of the platform and how the platform could be improved. All teachers were satisfied with the current state of the platform. They also indicated that this platform would be ideal for children with learning problems. A common remark was the need to connect the pen to a computer using a USB cable. This can however be eliminated when using a wireless pen which is able to directly connect to the tablet. All teachers also questioned the cost of the platform in a regular class environment, as in principle one pen is needed per student. However, all teachers were positive about the potential of Smart Study and rated it above other digital learning platforms, already being used by the schools. More details on the evaluation can be found in [24].

6 Conclusion

In this paper, the Smart Study educative platform is introduced. Smart Study is an innovative platform that combines the use of digital pen and paper for input, with a tablet device for feedback. The platform was evaluated by means of a case study involving a group of 15 children, age 10–11. The ICR accuracy of the pen was very high (on average 97.39 %), while the children were more motivated than usual and enjoyed making exercises using the Smart Study platform. Their teachers were enthusiastic as well, but questioned the cost of its use in a classroom environment. However, the platform could also perform well in a remediation situation, at home, or for children that are hospitalized.

In future work, we will investigate the replacement of the Echo pen by a wireless digital pen allowing to have direct communication between the pen and the tablet. We are also considering creating a Web application for the feedback instead of (or next to) the native tablet app, so it could be used on a broader range of devices. Currently, we are working on an authoring tool to support the creation of new courses and exercise sheets. Such a tool will allow instructional designers to easily create the Paper Designs, Penlets, and feedback screens.

Acknowledgments We like to thank the children, teachers, and schools involved in the evaluation.

References

1. LeapFrog: Leapfrog. http://www.leapfrog.com
2. Mackay, W.E., Pothier, G., Letondal, C., Bøegh, K., Sørensen, H.E.: The missing link: augmenting biology laboratory notebooks. In: Proceedings of the 15th Annual ACM Symposium on User Interface Software and Technology, pp. 41–50. ACM (2002)
3. Arai, T., Aust, D., Hudson, S.E.: Paperlink: a technique for hyperlinking from real paper to electronic content. In: Proceedings of the ACM SIGCHI Conference on Human factors in computing systems, pp. 327–334. ACM (1997)
4. Luff, P., Heath, C., Norrie, M., Signer, B., Herdman, P.: Only touching the surface: creating affinities between digital content and paper. In: Proceedings of the 2004 ACM Conference on Computer Supported Cooperative Work, pp. 523–532. ACM (2004)
5. Harebrained Schemes LLC: Golem Arcana. http://www.golemarcana.com
6. Berque, D., Bonebright, T., Whitesell, M.: Using pen-based computers across the computer science curriculum. In: ACM SIGCSE Bulletin, vol. 36, pp. 61–65. ACM (2004)

7. Miura, M., Kunifuji, S., Shizuki, B., Tanaka, J.: Airtransnote: augmented class- rooms with digital pen devices and rfid tags. In: IEEE International Workshop on Wireless and Mobile Technologies in Education 2005 (WMTE 2005), pp. 56–58. IEEE (2005)
8. Sugihara, T., Miura, T., Miura, M., Kunifuji, S.: Examining the effects of the simultaneous display of students' responses using a digital pen system on class activity-a case study of an early elementary school in Japan. In: 2010 IEEE 10th International Conference on Advanced Learning Technologies (ICALT), pp. 294–296. IEEE (2010)
9. Miura, M., Sugihara, T., Kunifuji, S.: Improvement of digital pen learning system for daily use in classrooms. Educ. Technol. Res. **34**, 49–57 (2011)
10. Liu, Z., Stork, D.G.: Is paperless really more? Commun. ACM **43**(11), 94–97 (2000)
11. Sellen, A., Harper, R.: The Myth of the Paperless Office. MIT Press (2003)
12. BBC, Finland: Typing takes over as handwriting lessons end. http://www.bbc.com/news/blogs-news-from-elsewhere-30146160
13. Feder, K.P., Majnemer, A.: Handwriting development, competency, and intervention. Dev. Med. Child Neurol. **49**(4), 312–317 (2007)
14. Mueller, P.A., Oppenheimer, D.M.: The pen is mightier than the keyboard advantages of long-hand over laptop note taking. Psychological science, p. 0956797614524581 (2014)
15. Ozok, A.A., Benson, D., Chakraborty, J., Norcio, A.F.: A comparative study between tablet and laptop pcs: user satisfaction and preferences. Int. J. Hum.-Comput. Interact. **24**(3), 329–352 (2008)
16. Thayer, A.: The myth of the paperless school: replacing printed texts with E-readers. In: Proceedings of the Child Computer Interaction, pp. 18–21 (2011)
17. James, W.B., Gardner, D.L.: Learning styles: implications for distance learning. New Dir. Adult Continuing Educ. **1995**(67), 19–31 (1995)
18. Fleming, N., Baume, D.: Learning styles again: varking up the right tree!. Educ. Dev. **7**(4), 4 (2006)
19. Curry, L.: A critique of the research on learning styles. Educ. Leadersh. **48**(2) (1990)
20. Coffield, F., Moseley, D., Hall, E., Ecclestone, K., et al.: Learning styles and pedagogy in post-16 learning: a systematic and critical review (2004)
21. Livescribe: Echo smartpen. http://www.livescribe.com/en-us/smartpen/echo/
22. Anoto: The Pattern. http://www2.anoto.com/the-paper-3.aspx (2014)
23. Donker, A.: Human factors in educational software for young children. Phd thesis, Vrije Universiteit Amsterdam (2005)
24. Van Thienen, D.: Smart study: an educational platform using digital pen and paper. Master's thesis, Vrije Universiteit Brussel (2014)

Enhance Teleteaching Videos with Semantic Technologies

Matthias Bauer, Martin Malchow and Christoph Meinel

Abstract Teleteaching systems have existed for more than a decade now. During that time, thousands of lectures have been recorded. These recordings are usually made available as single or dual stream videos. In the case of dual stream one video shows the speaker and the other one the slides. However, the content of the slides and the speaker's talk are not accessible through automatic search functionality provided by the e-lecture platform hosting the videos. In order to change this situation optical character recognition (OCR), automatic speech recognition (ASR) and common methods of the semantic web are used to analyze the video content and its semantics. Nevertheless, the semantic information found has to be made accessible to the user. Therefore, a novel HTML5 video player is introduced to enable the students to utilize the found semantic information while watching the video. This will enhance and simplify their online learning or research process.

Keywords e-learning · Teleteaching · Distance learning · Education technology · Semantic web

M. Bauer (✉) · M. Malchow · C. Meinel
Hasso Plattner Institute, University of Potsdam, Potsdam, Germany
e-mail: matthias.bauer@hpi.de

M. Malchow
e-mail: martin.malchow@hpi.de

C. Meinel
e-mail: christoph.meinel@hpi.de

© Springer International Publishing Switzerland 2015   105
V.L. Uskov et al. (eds.), *Smart Education and Smart e-Learning*,
Smart Innovation, Systems and Technologies 41,
DOI 10.1007/978-3-319-19875-0_10

1 Introduction

1.1 The Lecture Video Portal Used for Illustration

The research project the authors of this paper are working on has existed since 2002 and has been dedicated to lecture recording, postproduction, analysis and distribution. The lecture video web portal[1] of the project contains around 5,600 lectures and presentations in more than 400 collections. The recordings are also available as video podcast episodes, cut at the chapter marks created during or after the recording. Every week, around 30 h of new lecture videos are added [1]. Lately, we found that it would be helpful for the students consuming our lecture recordings if they were able to use additional functionality that can enhance their learning efficiency. There have been studies regarding collaborative learning and social web features, but for learners who prefer learning by themselves, we realized that there had to be more assistance from the web platform.

The keywords we can extract from OCR and ASR analysis are just plan keywords without semantics. This is a limitation to the further usage of the extracted terms which shall be overcome with the work described in this paper.

1.2 Semantic Web

The Semantic Web refers to structured, linked information in the World Wide Web that can be processed by machines like a global linked database. It is a collaborative effort led by the World Wide Web Consortium (W3C), providing a common framework (mostly RFD[2]) for formalizing information. Much research has been conducted in the field of Semantic Web Technologies, also at the Hasso Plattner Institute. However, this research often features mostly academic examples. Having realized this, the W3C also provides some examples of use cases and case studies where Semantic Web technologies were employed.

It would have been feasible to build up a database of semantically linked information contained within this ecosystem, connecting it to other semantic data sources (e.g. DBpedia) and making it accessible publicly and queryable e.g. via SPARQL. However, this approach might not be the most useful one to the everyday user.

---

[1]https://www.tele-TASK.de/, accessed on 02 March, 2015.
[2]https://www.tele-TASK.de/, accessed on 02 March, 2015.

In our work we focused on applying semantic approaches to improve the direct experience of users of a lecture video web portal. The website our where we implemented and observed our ideas is a platform for watching lecture videos, mostly used by students to catch up on missed lectures and review for exams. As such, improving the core features, such as search and displaying videos to the user, seems worthwhile.

2 Related Work

There has been research in our group earlier with interesting results, but the works concentrated mostly on the technical aspects on how to gather data out of speech and text [2], on how to train the tools [3] or how to extract the relevant data [4]. In [5] it is shown how to combine the analysis results of the OCR and ASR analysis. But still the result of the analysis and improvement processes were keywords without semantics. Töpper [6] and Waitelonis [7] actually do automatic video segmentation, OCR and extraction of semantic entities from video lectures, but they do not concentrate on extraction and provision of Uniform Resource Locators (URL), Request for Comments (RFC), or International Standard Book Numbers (ISBN) on the lecture slides. Neither do Imran [8] in their proposal of a framework for semantic extraction and utilization of metadata. [9] do interesting work concentrating on highlights in sports videos and also consider close-caption text, but still our needs for learning-oriented resources is not satisfied. Also the interesting work of [10] who researched on visual summaries of lectures and enhance them with text captions is remarkable (especially considering it was done already in 1999) but it does not fit our requirements and idea entirely.

3 Approach

3.1 Video Search

Maybe the second most important function of lecture video web portals, besides the simple delivery of video content, is enabling the user to find the content he or she is looking for. When searching for information on the Internet, search engines (such as Google, Bing or DuckDuckGo) are undoubtedly the way to go. We found that the internal search function offered by the web portal, is mostly used to find specific videos and not for general research on a topic. In order to help improving the way search engines display and process the learning resources, they need to understand the content of the page and the video. One way of achieving this is to use semantic annotations in the HTML source of the lecture pages.

```
<!-- Dublin Core -->
<link rel="schema.DC"
  href="http://purl.org/dc/elements/1.1/" />
<link rel="schema.DCTERMS"
  href="http://purl.org/dc/terms/" />
<meta name="DC.title"
  content="Internet Security" />
<meta name="DC.creator"
  content="Teaching Team" />
<meta name="DC.subject"
  content="Weaknesses" />
<meta name="DC.language"
  content="en" />

<!-- schema.org -->
<div itemprop="video" itemscope
  itemtype="http://schema.org/VideoObject">
  <meta itemprop="duration"
    content="T1H5M48S" />
  <meta itemprop="description"
    content="Part of series: Internet Security" />
  <meta itemprop="thumbnailUrl"
    content="thumb.jpg" />
  <meta itemprop="uploadDate"
    content="2014-10-15T12:54:18" />
  <meta itemprop="inLanguage"
    content="en" />
  <!-- [...] -->
</div>
```

This has already been done in parts of the system. However, still the Dublin Core metadata set,[3] which was originally published in 1995, is used. Today, Google suggests using the schema.org metadata set[4] (released in 2011), which is supported by Google, Bing, and Yahoo. Using the schema.org VideoObject construct allows more meta information (such as video thumbnails, slide previews) to be added to the page and thus be utilized for better search and display possibilities. In the program code above we show the difference in code between Dublin Core and schema.org. One might argue why, in the context of a lecture video web portal, not the approaches specialized on e-learning IEEE LOM[5]

---

[3]RFC 5013: The Dublin Core Metadata Element Set.
[4]https://schema.org/, accessed on 19 January, 2015.
[5]IEEE Standard 1484.12.1-2002.

or ISO MLR[6] were utilized. In order to keep the project as general as possible we chose not to concentrate only on videos with metadata taken from the e-learning domain but rather use DublinCore which is a more universal approach.

3.2 Online Video Player

As of shortly, a video player implemented in Adobe Flash[7] was the most used display mode on the on the examined web platform (others are RealVideo[8] and direct downloads of MP4 files). Figure 1 shows a choice of available formats of the lecture "Similarity Measures". The former standard (Flash) player was developed using the OpenLaszlo[9] framework and streams videos using the Real-Time Messaging Protocol (RTMP[10]). Unfortunately, the latest stable release of OpenLaszlo was published in October 2010. The last nightly build was made publicly available in November 2012. So, today it can be considered outdated. For quite a lot of years, Adobe Flash was the best suiting technology to deliver video to users, especially with the requirement of playing to videos synchronously. So, when this video player was built in 2010, these technologies were a good choice considering.

Today, with the specification of the HTML5 standard[11] by the W3C and specifically the ability to natively deliver video content, this seems like the better approach. By switching to a video player that uses HTML5 and JavaScript we have improved on areas such as security, mobile experience and accessibility. Especially, we make it easier for developers to extend the player. It is possible to add features using well-known web technologies, removing the learning curve that ActionScript and the OpenLaszlo framework imposed. This is also an important factor when it comes to student projects or guests writing a bachelor or master thesis because it helps them to understand the software project faster and hence contribute earlier. The new HTML5 player, as shown in Fig. 2, supports all features the original Flash player did, as well as adding a few quality improvements. For example, it is now possible to adjust the playback speed and show or hide the chapter overview as needed.

An interesting side note is that due to the dynamic nature of JavaScript, changes are possible on the fly and in contrast to Adobe Flash, to completely change the user interface on the client side to suit individual needs. This is called augmented browsing and can for example be realized using browser add-ons like Greasemonkey or others.

---

[6]ISO/IEC Standard 19788.

[7]http://www.adobe.com/products/flashplayer.html, accessed on 19 January, 2015.

[8]http://real.com/, accessed on 19 January, 2015.

[9]http://www.openlaszlo.org/, accessed on 19 January, 2015.

[10]http://www.adobe.com/devnet/rtmp.html, accessed on 19 January, 2015.

[11]http://www.w3.org/TR/html5/, accessed on 19 January, 2015.

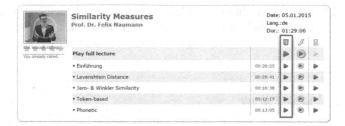

Fig. 1 Available video players for the lecture: HTML5 player, Flash player, Podcast download

3.3 Additional Features

Additional secondary features such as playlists, tags or the extracted slides and keywords of the lecture, reside in tabs below the video in the Flash player. Unfortunately, this means that they are usually below the fold, and are thus less likely to be discovered and used by the user. A study of the question which tools are most helpful for students doing online research in our web portal showed that some of the offered features are not that easy to find and hence not always used as intended [3].

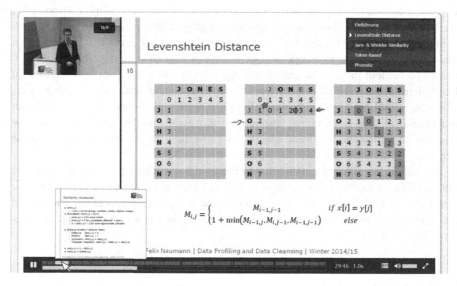

Fig. 2 Screenshot of the new HTML5 player including chapter overview (top right), slide preview and control elements for playback speed adjustment, full screen mode etc. (bottom)

Due to the easier extensibility of the HTML5 player, new information as well as that contained in the existing tabs can be displayed directly on top of or next to the video.

Links. In the current static display of slides in the lecture videos, following links that are displayed comes with a certain effort. The user has to copy the link manually into the address bar, which is especially difficult with long links. We have implemented a system that allows the user to simply click the link on the slide, opening a new browser tab with the requested page.

Our system searches the existing OCR data of the lecture slides for patterns of Internet links (e.g. text starting with "http"). As the characters in these links are often wrongly detected by the OCR software due to special characters and sometimes small font, we then try to reconstruct the correct URL as shown in Fig. 3. This is done using various heuristics trying to compensate for the most frequent detection errors (e.g. replacing spaces with dots or hyphens). In order to make sure that there is a valid resource behind the reconstructed link, we send a HTTP HEAD request to the URL. This allows checking for a valid resource without having to download the entire page.

Fig. 3 Process of extracting links from recorded lecture slides

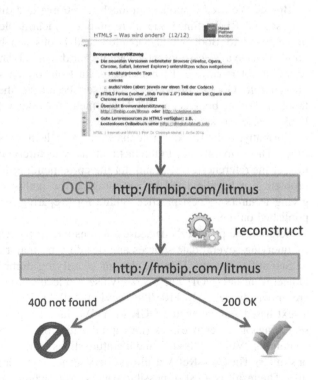

Challenges. Although we tried to minimize false positive errors with this approach, some challenges remained. One phenomenon that we encountered is typosquatting, where URLs of slightly misspelled domains are registered to serve

ads to people who input the incorrect URL into their browsers [11]. However, instead of the user, it is the OCR that introduces the mistakes. An example of this is the OCR result http://en wikipedia org which was reconstructed to http://en-wikipedia.org instead of http://en.wikipedia.org. One would assume this to return a HTTP status code of 404. However, the HTTP status code returned (after some redirects) is 200. The delivered page is full of advertisements and not the one that was desired. Another problem we frequently encountered was the apparent misconfiguration of webservers. Instead of delivering an HTTP 404 status code when the requested URL was not found, these servers returned a styled "page not found" page, but with a 200 status code. The solution here again was to try to filter these pages using heuristics that search for certain keywords on the page.

RFCs. Since the examined lecture archive as well as others contain a notable amount of lectures dealing with the Internet, there are many references to RFCs (Request for Comments). To enable the user to quickly look up details on these Internet Standards, the OCR data of the lectures is searched for RFC references. Using its identification number, the RFC metadata is retrieved from the RFC repository of the Internet Engineering Task Force. Users can then see the title of the RFC and follow a link to directly read it.

Books. We used a similar approach to retrieve and display book metadata that we used for RFCs. Since books are uniquely identifiable by their ISBN (International Standard Book Number), referenced books can be easily detected. ISBN numbers have a check digit which allows for direct validation. In order to prove the feasibility of retrieving the book metadata from online sources, we implemented querying the Google Books API. This service returns the title, author, year and a thumbnail of the book cover. While this method works well for correctly referenced book, it also has some drawbacks:

Currently, only 15 book references in 4951 lectures are detected by our algorithm. This is mainly due to the fact that many lecturers seem to mainly include the title of the referenced book (and not the corresponding ISBN) in their slides. This problem can be alleviated by encouraging the inclusion of ISBNs in slides by giving lecturers a "best practice" guide for preparing lectures to be recorded and published online.

Seldom, the provider's database contains wrong metadata. This could be solved by querying several data sources and deciding by majority vote.

Scientific Papers. There is also an identifying number for papers, the Digital Object Identifier (DOI), but it is rarely used on lecture slides, possibly because DOIs are unwieldy (e.g. 10.1103/PhysRevLett.110.228701). Therefore we employed a text-based search on the OCR data. We identify possible paper references by searching for conference titles, (i.e. capital letters followed by a year), journals (strings containing "Vol." or "Issue") and literature slides (with a title containing "Literature" or similar). The CrossRef API allows querying academic papers by their DOI and their title. The result is a list of possible papers, from which we filter the correct one by comparing the titles with a Levenshtein distance metric and by trying to match the author and the year in both references.

Related Articles. To help users continue their research on the topic discussed in the lecture they are viewing, related Wikipedia articles are displayed next to the slides. This is time sensitive, as the information is relevant especially to the slide that is shown at the moment. Thus, we decided to allow users to show the related articles as an overlay on top of the video. This updates automatically when the next slide is shown. We use DBpedia Spotlight[12] to extract relevant topics. This service relies on DBpedia,[13] a semantic version of Wikipedia, to extract the important words on a slide and to disambiguate their meanings based on the context as illustrated in Fig. 4.

Search Based on Semantic Resources. The internal search function for lectures of the examined lecture video portal so far matches the title, the description and the lecturer. Because semantic information about the content of the lecture and the mentioned books, RFCs and papers is now available, it is possible to extend the search to include this metadata. As a prototypical implementation we implemented and extended the search functionality for book titles. The entered search pattern will now be searched in all book titles as well. If a lecture is found because of a book mentioned in that lecture, this result will be marked with a "book found" label. Hovering over this label will display the book's title.

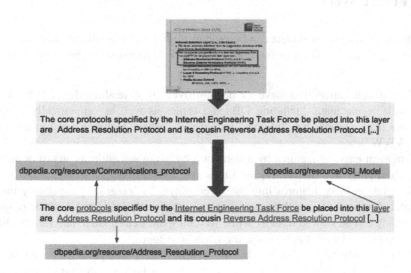

Fig. 4 Using DBpedia to find related articles

---

[12]https://github.com/dbpedia-spotlight/dbpedia-spotlight/wiki, accessed on 19 January, 2015.
[13]http://dbpedia.org/About, accessed on 19 January, 2015.

4   Results

In this work we explored the possibilities that techniques and approaches from the Semantic Web offer in real life scenarios in the context of a lecture video distribution platform. By using the textual data extracted from the lecture slides we were able to construct metadata entities. We were able to extract more than 70 paper references, 130 URLs, 330 RFCs as well as more than 160,000 DBpedia resources. The focus was on using these to supply additional features to the end user, thereby improving the user experience. We updated the technologies used in the platform to allow easier integration of these new features and decrease the learning curve for future developers.

All of the shown ideas and approaches are also applicable in other contexts, not only in the e-learning domain working with lecture videos. As long as there is a video and a method of extracting metadata or adding it manually, it is possible to enhance the video watching experience as described in this work to give the user a bigger choice of further research or related topics.

5   Future Work

We have shown how metadata can be used to improve a video distribution website. However, there is still room for improvement. Future work can be done in these areas:

- The search capabilities have to be improved by extending the search to all available metadata, including DBpedia topics as well as semantically related ones (e.g. a search for "Internet" would also retrieve results having to do with "WWW").
- Augment existing features by automation by using semantic means to populate the currently few (manually) used tags and annotations.
- Integrate existing features into the video player by integrating especially time-related information, such as user annotations directly into the UI of the player, so they can be viewed directly while watching the lecture video and no scrolling is necessary which might make the users lose focus.
- Refine newly added features by monitoring user participation. A user study should be conducted to ascertain exactly which features are most useful and should be focused on in future work.
- Involve the user by allowing improvement of the data. As mentioned in Sect. 3.3, some extracted URLs might be incorrect. A possible solution would be to allow users to correct the URL using a textbox where they can input the corrected data and additionally allow them to rate the relevance and correctness of the results.

References

1. Kandzia, P., Linckels, S., Ottmann, T., Trahasch, S.: Lecture Recording—A Success Story. it 3/2013 vol. 55, pp. 115–122 (2013)
2. Yang, H., Siebert, M., Lune, P., Sack, H., Meinel, C.: Automatic lecture video indexing using video OCR technology. In: Proceedings—2011 IEEE International Symposium on Multimedia, ISM 2011, pp. 111–116 (2011)
3. Repp, S., Meinel, C.: Semantic indexing for recorded educational lecture videos. In: Proceedings—Fourth Annual IEEE International Conference on Pervasive Computing and Communications Workshops, PerCom Workshops 2006, pp. 240–244 (2006)
4. Repp, S., Meinel, C.: Automatic extraction of semantic descriptions from the lecturer's speech. IEEE International Conference on Semantic Computing, pp. 513–520 (2009)
5. Yang, H., Grünewald, F., Bauer, M., Meinel, C.: Lecture video browsing using multimodal information resources. In: Lecture Notes in Computer Science (Lecture Notes in Artificial Intelligence and Lecture Notes in Bioinformatics), LNCS, vol. 8167, pp. 204–213 (2013)
6. Töpper, G., Knuth, M., Sack, H.: DBpedia ontology enrichment for inconsistency detection. In: Sack, H., Pellegrini, T. (eds.) Proceedings of the 8th International Conference on Semantic Systems (I-SEMANTICS '12), pp. 33–40. ACM, New York, NY, USA (2012)
7. Waitelonis, J., Sack, H.: Towards exploratory video search using linked data. Multimedia Tools Appl. **59**, 645–672 (2012)
8. Imran, A.S., Rahadianti, L., Cheikh, F.A., Yayilgan, S.Y.: Semantic tags for lecture videos. In: Proceedings—IEEE 6th International Conference on Semantic Computing, ICSC 2012, pp. 117–120 (2012)
9. Babaguchi, N., Nitta, N.: Intermodal collaboration: a strategy for semantic content analysis for broadcasted sports video. In: Proceedings of the IEEE International Conference on Image Processing (ICIP), vol. 1 (2003)
10. Uchihashi, S., et al.: Video manga: generating semantically meaningful video summaries. In: Proceedings of the Seventh ACM International Conference on Multimedia (Part 1). ACM (1999)
11. Moore, T., Edelman, B.: Measuring the perpetrators and funders of typosquatting. In: Lecture Notes in Computer Science (Lecture Notes in Artificial Intelligence and Lecture Notes in Bioinformatics), LNCS, vol. 6052, pp. 175–191 (2010)

Tests Generation Oriented Web-Based Automatic Assessment of Programming Assignments

Yann Le Ru, Michaël Aron, Jean-Pierre Gerval
and Thibault Napoleon

Abstract This paper describes a Web-based system that is aimed to help both academic staff and students in the task of automatic assessment of programming assignments. The server implements LAMP (Linux, Apache, MySQL and PHP) architecture powered by the Laravel Framework and designed by Twitter bootstrap. The system focuses on adaptability, scalability and simplicity. It also enables teachers with few skills in the field of software testing to use such a tool. This system implements various types of technologies: databases, Web technologies, object oriented programming, responsive Web design, design patterns, and others.

Keywords Web-based system · Automatic assessment · Programming · Tests

1 Introduction

In this paper, we propose a Web-Based System that is aimed to help both academic staff and students in the task of automatic assessment of programming assignments.

The benefits of using such a tool are reviewed in [1]. Automatic assessment of programming assignments is of interest for all educational organisations to improve the quality in teaching and involve students in the learning process. Moreover,

Y. Le Ru (✉) · M. Aron · J.-P. Gerval · T. Napoleon
Institut Supérieur de l'Electronique et du Numérique, Brest 20 Rue Cuirassé Bretagne,
CS 42807 - 29228 Brest Cedex 2, France
e-mail: yann.le-ru@isen-bretagne.fr

M. Aron
e-mail: michael.aron@isen-bretagne.fr

J.-P. Gerval
e-mail: jean-pierre.gerval@isen-bretagne.fr

T. Napoleon
e-mail: thibault.napoleon@isen-bretagne.fr

© Springer International Publishing Switzerland 2015
V.L. Uskov et al. (eds.), *Smart Education and Smart e-Learning*,
Smart Innovation, Systems and Technologies 41,
DOI 10.1007/978-3-319-19875-0_11

117

assessing students is more objective as the teacher is not involved in the grading process. This is also faster because a repeating task is more adapted to computers.

The features to be assessed include but are not limited to functionalities, efficiency, testing-skills, robustness, quality by means of code analyses, etc. [2]. Sometime tools also provide anti-plagiarism features [3].

The goal of this paper is not to discuss about the utility but to deal with the tools and especially on the functionalities part of the assessment.

In the next sections, we will provide an overview of the existing tools' functionalities and explain the motivation of developing a new tool. Then, we will present an overview of developed Web-based system and details about our Automatic Assessment of Programming Assignments implementation.

2 An Overview of Tool's Functionalities

2.1 Web-Based Systems

First, most of the existing solutions in automatic assessment of programming assignments are web-based [4, 5]. Indeed, it provides an autonomous way for students to assess their own program, and an easy way for teachers to supervise. Usually the software includes two different interfaces. From student's interface, the user uploads his work, it runs on the server side and provides a feedback. From the teacher's interface, the user access to a dashboard to supervise all the works.

2.2 Programming Languages

Tools exists for many languages such as C, C++, Python, Pascal, shell, Assembler, etc. and Java is the most represented [2]. Some tools are specialized for one of programming languages; others implement few of them by the mean of plug-ins.

2.3 Tests Implementation

To assess the functionality part of a programming work, a majority of the systems use xUnit tests frameworks. Tools such as Web-Cat [6] focus on assessing the student's performance at testing his or her own code by writing their own xUnit tests. The idea is to make the student aware of the importance of testing. Others grade only the skills of programming, the teacher write the xUnit test for the specific exercise. In both cases, the tool requires skills in writing tests for teachers and eventually for students.

3 System Overview

3.1 Motivations

Several ideas motivated the creation of a new software.

The first one is to simplify the process of programming assignment creation. During this step, the teacher must define program functionalities he or she wants to assess in the case tests are not students' charge. Usually, it is done by writing xUnit tests that require specific skills. Teachers in charge of basic programming do not all master this domain and give up the tool. To avoid this fact, the idea is to provide a test generator requiring just a limited knowledge in testing.

The second is to make the tool smart. Thanks to the web-based scheme, compiling, executing and testing program take place on the server side and therefore is transparent for the end user. There is no need for special tool on the client side, only a browser and an editor to write the code. The student can test his program from everywhere and every device.

Last, an application is used on a mobile device only if it is designed especially for these kinds of target, time saving is essential for mobile users. This requires a special attention on software design, namely ergonomics and performances.

The software is aimed to address three criteria: adaptability, scalability and simplicity.

3.2 Technical Overview

An overview of the system (Fig. 1) is presented hereafter. It points out the main components: server, database and roles: teacher and student.

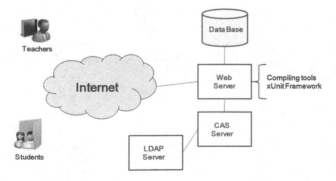

Fig. 1 System overview

Server and database implement Linux, Apache, MySQL and PHP. This architecture has come to be known simply as LAMP. The application is powered by Laravel Framework [7]. Status and information about users come from CAS server connected to a LDAP server.

3.3 Conception

We have two different roles: teacher and student.

3.3.1 Teachers

Academic staff module's main functionalities (Fig. 2) are project's management and project's life.

The first one consists of project creation and management including a user-friendly xUnit-tests generator (more information in Sect. 2.4). Once a project is created, only owners can access the project and his management.

From the projects life module, teachers are able to launch tests for a student and have access to the results pages. Different views are available for the results depending on the information needed. Student result view shows detail information for every tests. Group view shows only the project marks for all students, details are accessible by clicking on the student name.

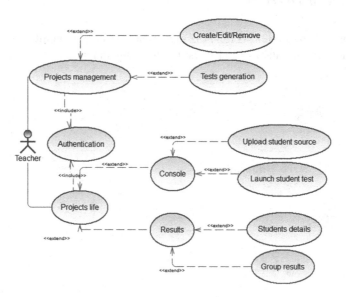

Fig. 2 Academic staff user case

3.3.2 Student

Student module (Fig. 3) implements only projects life with a different level of information in the results view.

Students can test the projects corresponding to their class and have access only to their own results without details. They cannot see the tests details but only if the functions tested work or not.

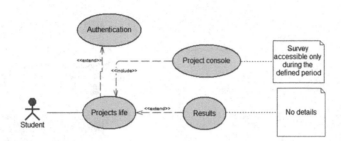

Fig. 3 Student user case

3.3.3 Authentication

The software must be adaptable to every university academic organisation requirements. To respect this criterion, there is no link with our information system apart from the CAS server, which is easy to adapt to any kind of authentication.

3.3.4 Responsive Web Design

The emergence of the "Bring Your Own Device" is a wide phenomenon and is even stronger in student population [8]. Academic organisations have to adapt their tools for this new trend and therefore must be compatible with any kind of devices (computer, smart phones, tablets and certainly in the future other smart machines, etc).

The software is responsive design (Fig. 4), it uses the twitter bootstrap framework version 3.

3.3.5 Adaptability

The software is adaptable to any programming language by means of modules. Currently, it is efficient for C and Java languages.

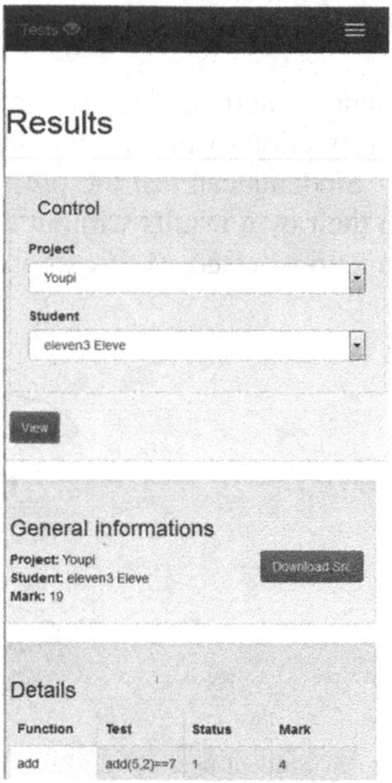

Fig. 4 Results mobile design

3.4 Tests Design

3.4.1 Considerations

The tool focuses only on unit tests and especially on limit values or exceptions, for instance the test of the division by zero. The pedagogical purpose is to make students aware of these problems.

3.4.2 Tests Conception

A project contains functions to test (Fig. 5). Each function contains several tests. A test consists of verifying an assertion for the specific function. For instance, if the function is the division, the tests could be for the first one the verification of the operation 8/4, the second 9/4 and the third 9/0.

The implementation uses weak relationships to improve the database queries performances.

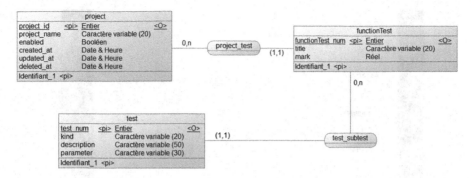

Fig. 5 Tests database

3.4.3 Test Generator

A test generator limits the xUnit skills required as explained in Sect. 2.2. The tool proposes a limiting set of xUnit function to the user, depending on the language chosen for the project (Fig. 6).

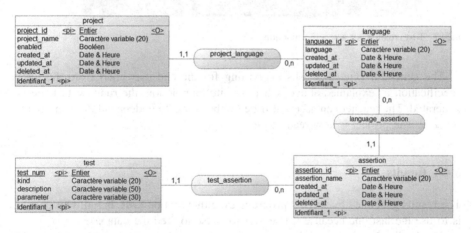

Fig. 6 xUnit assertions and language database

The knowledge required to create a test suite is just the parameters to pass to the assertion function as shown in Fig. 7.

A structure of a xUnit program includes:

- a list of tests to run
- parameters to indicate the form of the results
- a runner

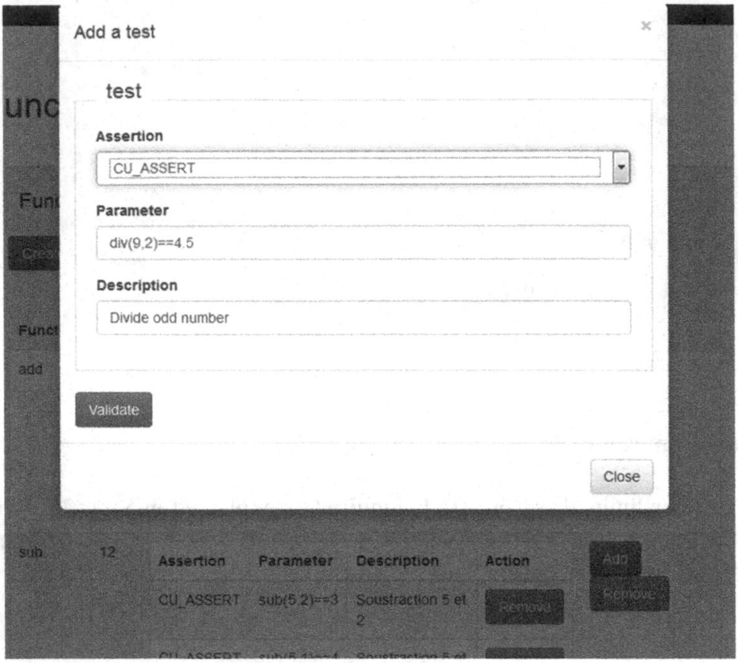

Fig. 7 Test creation for a specific function

The generator (Fig. 8) masks everything for the end user apart from the tests specification as explained above. On test suite completion, the running test file is generated. The teacher can access it if he or she wants to understand the generation process. The project is now ready to use.

3.4.4 Results

The results generated by a xUnit program are either text or XML based. Our choice is to use the last one because it is easier to parse to feed the database.

The feedback depends on the status of the user (Fig. 9): student on the left side and teacher on the right side.

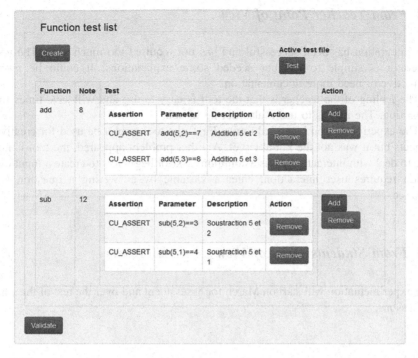

Fig. 8 Test generator

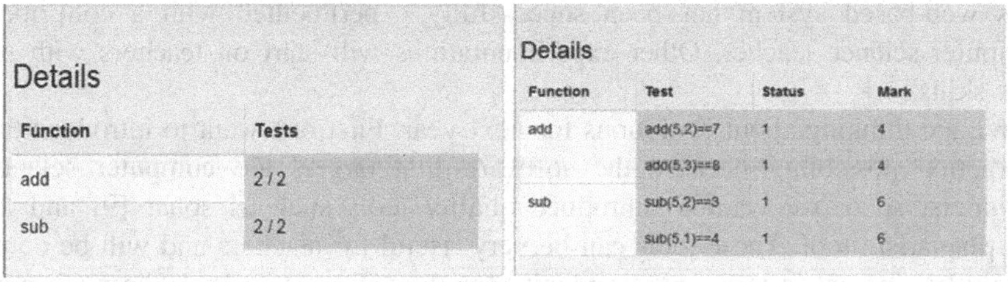

Fig. 9 Results from student and teacher's view

4 Experimentation

4.1 General Considerations

The software was efficient at the end of November 2014. It is currently testing by a teacher with xUnit skills. The plan is to use the tool on March for a programming assessment with first year students who are beginners in computer science.

The experiment was to adapt an existing assessment to integrate it into the tool.

4.2 From Teacher Point of View

The integration has been successful and has not required too much time. The test generator is simple to use but needed some explanations. It could be more instinctive or need more documentation.

The feeling of the teacher is that the tool is easy to use and will save times for correction. The idea is to use it also during lessons.

The experimentation identified limits. First, the tool cannot be used for creative projects but it was not the initial target. Another problem appeared, the tool is not able to deal with interactions. It is a common task for beginners to create a function, which requires user interaction, enter a variable by answering a question for instance.

4.3 From Students Point of View

The experimentation will start on March for assessment and over the rest of the year for lessons.

5   Conclusions and Future Work

This web-based system has been successfully experimented with a confirmed computer science teacher. Other experimentations will start on teachers with no tests skills.

We are thinking about evolutions for next year. First, we want to introduce the interaction possibility because the software first targets are computer science beginners. Also, we want to introduce quality tools such as sonar [9] and an anti-plagiarism tool. These tools can be very useful for teachers and will be completely transparent in terms of complexity for the end-user. Indeed, the main purpose is to make a tool easy to use.

References

1. Ihantola, P., Ahoniemi, T., Karavirta, T., Seppala, O.: A survey of automated assessment approaches for programming assignments. Comput. Sci. Educ. **15**(2), 83–102 (2005)
2. Ihantola, P., Ahoniemi, T., Karavirta, T., Seppala, O.: Review recent systems for automatic assessment of programming assignments. Comput. Sci. Educ. (2010)
3. Subba, L.T.: An anti-plagiarism add-on for web-CAT. Masters thesis, National University of Ireland Maynooth (2013)

4. Shah, A.: Web-CAT : A web-based center for automated testing. Master thesis, Faculty of Virginia Polytechnic Institute and State University (2003)
5. Spacco, J., Hovemeyer, D., Pugh, W., Hollingsworth, J., Padua-Perez, N., Emad, F.: Experiences with marmoset: Designing and using an advanced submission and testing system for programming courses. In: ItiCSE '06: Proceedings the 11th Annual Conference on Innovation and Technology in Computer Science Education. ACM Press, New York (2006)
6. http://marmoset.cs.umd.edu
7. http://laravel.com/
8. Emery, S.: Factors for consideration when developing a bring your own device in higher education. Master thesis, University of Oregon (2012)
9. http://www.sonarqube.org/

"Living Books" and the Advanced Network Technologies for Education in a Technical University

Yulia Antokhina, Nikolay Balonin and Mikhail Sergeev

Abstract This article presents an analysis of the future of technical books. The process of formation of network technologies that change the very concept of books, -in particular, technical books - is considered. A new image of a modern technical book, containing algorithms executable from pages and linked to described objects, is suggested. An overview of implementation technologies, including the Java-MatLab language developed by the authors of this article and the new FB + format for representation of information, is presented.

Keywords Technical book · Mathematical network · Internet · Network robotics · Living books · Executable algorithms · Data storage formats · CDF · Java-MatLab · FB · FB+

1 Introduction

The rapid development of network technology today has led to the emergence of new forms of distance learning and knowledge control [1–5]. Today, we can say that communication technologies have developed so much that, for effective learning, student groups can be formed of students from around the world, and not only those in a single city. The physical migration of students has begun to give way to migration of virtual departments and universities. Using virtual technologies

Y. Antokhina (✉) · N. Balonin · M. Sergeev
Saint-Petersburg State University of Aerospace Instrumentation,
67 Bolshaya Morskaya Street, Sankt-Petersburg 190000, Russian Federation
e-mail: antoxina.j@yandex.ru

N. Balonin
e-mail: korbendfs@mail.ru

M. Sergeev
e-mail: mbse@mail.ru

© Springer International Publishing Switzerland 2015   129
V.L. Uskov et al. (eds.), *Smart Education and Smart e-Learning*,
Smart Innovation, Systems and Technologies 41,
DOI 10.1007/978-3-319-19875-0_12

today, you can even perform operations in real laboratories with expensive and rare equipment.

This all has been made possible, but the mechanisms of creation of network resources for shared use largely remain the same, and are not relevant for the near future or sometimes even the present.

This paper discusses the principles of organization of training within a highly productive mathematical network, based on "living book" technology [6].

2 Book Development, Inevitability of Renewal of Design

First of all, progress in the field of education must affect the book and change its appearance. We're talking about real changes, not just the conversion of books to electronic formats for viewing on computers, tablets, smartphones, etc. The book has crossed centuries and millennia, and through its evolution, its appearance, method of manufacture, and capabilities of presentation of information changed. Books in the times of Pushkin were different from books in the times of Tolstoy, even in terms of the way they were manufactured. In Tolstoy's time, industrial manufacture of books had already begun, which provided, it is important to mention, funds to support a new way of life and changes in society. However, it should be stated that, before the end of the twentieth century, the technology of sharing of knowledge through books had been developing very slowly.

Today, we can think as follows: books are not even the computers, but the Internet of the past, though of course this term is not seen in books from the past because no one knew such a word. The whole period of the existence of books is a preparation period for our luxurious century of computer innovations, wherein the means of transfer of information from one generation to another cannot stay the same.

The changes have already affected books and textbooks, which are the basis of education. Paper books are gradually being replaced by books in electronic formats like PDF, FB2 and many others. They eliminate the difference between Internet products, consumed from the Internet via computers, tablets, smartphones, etc., and paper books.

But the important thing is that all this is not adequate for the new possibilities of modern technology, especially in terms of training of technical specialists.

Mathematics is the basis of any science and technology education. One can argue about its value, but pay attention to one important detail – the carriers of today's books are computers. Separation of information into text and a program, which is to be executed in some software environment, is no more than just a technical confusion, very pathetic and impractical.

Why would one use images to describe Euclid's algorithm in a book of the future?? Why use a textual description of an algorithm in an electronic book, viewed on a computer with a powerful processor? Even if a computer for reading texts is made with the use of energy saving technologies, and making calculations on it is physically difficult, information about user needs and requests can still be delivered via the Internet to a processing core of any capacity.

Obviously, the Internet is a necessary stage, the logical outcome of the current pattern of development. Printed books give way to e-books, downloaded from the Internet to a tablet, bookreader, etc. Algorithms, executed within a book, are a quite obvious feature of computer books of the near future. Executable algorithms cannot be used in the pages of this article, because it is written using an old technology, but it already has an alternative – text on the Internet.

In today's technical textbooks, the emphasis should be placed on the algorithms executed from pages and on presentation of mathematical (computed) animations and graphs. This, certainly, is modern and timely. There already are online services, including google.com, that are experimenting, for example, with online text editing. Our proposal is the creation of books based on "living book" technology [5, 6], which is a combination of an online text editor and executable algorithms, linked to a means of visualization and providing bilateral exchange of data with the objects of study.

3 Executable Algorithms

New objectives for mathematical systems have been clearly defined by the leader in development of such technologies, Stephen Wolfram [7]. He noted that the PDF electronic document format, developed in the last century, has a major drawback – it does not allow for the inclusion of executable algorithms. This solution was addressed by the MatLab system by its alignment with the Word text editor [8]. According to the Mathematica's author, the CDF format (Computable Document Format) [9] can bring elements of mathematical calculations to the document, which otherwise is very similar to PDF. Thanks to the new format, users of technical books with mathematical formulas can make variations to the original sample data, see more solutions and corresponding graphs, use it like a standard template to change it, and solve their own tasks. These ideas, which are still under discussion, have long been on the Internet, but they have not been brought to their final stage of development. The mentioned little-known website with the MatLab system, for example, has been closed due to the complexity of its support.

Internet-based technologies can support more than just mathematical calculations like x = A\b. Apart from mathematics, engineering training has always been associated with the experience of dealing with hardware: with educational and scientific apparatus, the launch of motors, control of manipulators, and robots. If, for example, there is an automatic telescope that operates continuously and is connected to the Internet, it is better to obtain data for processing by an executable algorithm in a book not from the text of CDF document, but by "reaching" the telescope and getting data from it directly. It can be obtained using a simple command, just like the one that launches the solution of linear algebraic equations. It is this kind of CDF format that suggests itself for executable documents on the Internet.

Apart from CDF, there are other ideas for the development of electronic books. For example, the Fiction Book (FB2) format, used in bookreaders, has become widespread [10]. This format does not yet include means allowing for displaying, for example, not only book chapters, but matrices, for example. However, this format is quite flexible and allows for development in the discussed direction.

Such a format, developed by the authors, has been conditionally named FB2 + [11, 12].

4 Matrix Operations on the Internet

Matrix operations on data arrays were not originally included in the standard Javascript language [13], although it has a mathematical part devoted to the calculation of mathematical functions. Javascript also has almost no graphics.

The name of the developed mathematical Internet system "Living Book" is derived by the fact that it allows for the creation of texts with executable algorithms on the Internet. Apart from algorithms, you can also place graphs generated by this system and other graphical objects in the text. Figure 1 depicts three-dimensional portraits of objects. "Living Books" can take data from their own pages, from the Internet, or from different educational laboratory stands.

Processing of information from laboratory set-ups is carried out on the network using the standard MatLab vector and matrix operations. Since Javascript, having some of the features of mathematical calculations, does not handle matrix expressions, the authors have proposed to use double curly brackets {{...}} for the isolation of matrix structures.

Fig. 1 Three-dimensional portraits of objects

An executable algorithm within the online text should be placed between the tags:

${{matrix operations}}$.

With the help of the precompiler, the contents of brackets gets translated from the language of vector-matrix calculus to the standard Javascript operations format. This configuration was named Java-MatLab [14] by the authors.

Such a system allows for all main matrix operations: transposition $\{\{X = A'\}\}$, algebraic addition $\{\{X = A + B\}\}$, multiplication $\{\{X = A * B\}\}$, left $\{\{A = A \backslash B\}\}$ and right $\{\{X = B/A\}\}$ multiplying by an inverse matrix, and the point-wise operations of Hadamard multiplication and division $\{\{X = A. * B; X = A./B\}\}$. Formulas can be written in brackets separately, or one after another.

Having added a library of common service functions (for example, solution of linear differential equations) and graphics for curve construction, the authors brought online everything that previously was achieved only in universities using comprehensive software packages.

However, in addition to these improvements, the authors have actively used the advantages of the Internet. Many developments have been performed and a friendly exchange of links is still maintained with the oldest university in Siena (Italy), where students have had access to laboratory work via the Internet since 2005. You can pour water in a jar, start an engine, apply a PID controller, etc. The system sends data on transition processes to students in MatLab format. But later, students had to buy the system, install it on their computers, and apply it to acquired data to solve educational tasks.

The last stage is excluded in Java-MatLab because the executable algorithms not only allow for processing the data received from the apparatus, but also for publication of reports on laboratory work and scientific research studies on the Internet. Our system has been developing since 2005 and is now brought to practical use in the educational process as a "living laboratory" at the livelab.spb.ru website [15]. It was launched before the Mathematica system and before the announcement of the CDF format as an educational website artspb.com [16] to support the teaching of mathematics, cybernetics, and programming. And this is understandable – the described innovations and ideas have long been expected to emerge.

5 Overview of Features of Executable Internet Algorithms

While developing Java-MatLab, we had to decide which loop construction to give preference to – the one typical for Javascript or the one typical for MatLab. We have implemented both variants. The last one was named iMatLab (Internet-MatLab) [14].

The selection of loop construction in the Javascript standard and indexation of matrix elements starting from zero (not from 1, as it is in MatLab), led to fewer mistakes among students. This solution is selected as the main one, because this type of notation is the most common one in programming languages. The proposal to use the established habits of programmers instead of breaking them allowed for a significant increase in the efficiency of use of the tool. A synthetic language was created which has incorporated both elements: syntaxes of the popular vector-matrix calculus toolkit and a widespread Internet programming language, the syntax of which originated in Java. This synthesis, in addition to the above, is reflected in the paired name Java-MatLab.

In the described version of the language, the matrix formulas are written tightly, without spaces. The functions are placed outside the brackets. For example, the Kronecker matrix product looks like A = kron(B,C). A total of 100 functions are implemented as of today [14].

Java-MatLab enables students to solve standard problems in linear algebra, including solving systems of linear algebraic equations and algebraic problems with eigenvalues. This is natural, because these are the most common problems which other ones can be reduced to. It is possible to deploy procedures for analysis and synthesis in linear dynamic systems, implement the Runge-Kutta method, analyse frequency of signals and systems [17], perform unerring solution of integer systems of equations [18], etc.

The current version of the Java-MatLab system provides the ability to write and add a new custom toolbox.

Modelling of dynamic systems. According to a tradition that goes back to MatLab, a "living laboratory" includes means for the modeling of linear dynamic systems, described by matrices of the state space model or the coefficients of numerator N and denominator D of the transfer function $Q(p) = N(p)/D(p)$.

The typical scenario is as follows:

$t = time(10); u = one(t); N = 1; D = [1,2,1]; y = lsim(N,D,u,t); plot(t,u,y);$.

In this case the t = time(10) operator defines a vector of time counts from 0 to 10 (one hundred points; this number is affected by the second argument of the time generation function), u = one(t) calculates the input single step signal, vectors N = 1 and D = [1, 2, 1] contain the numerator and denominator of the transfer function of the dynamical system, function y = lsim(N,D,u,t) calculates the output signal of the system using the input signal (there is a similar function in MatLab), and plot(t,u,y) outputs process graphs.

This is largely important in the study of automatic control theory, theory of electrical engineering, modeling of dynamic systems, and other similar disciplines.

Basic linear algebra computations. Solving systems of Ax = b linear algebraic equations was discussed earlier. Together with setting the initial data, the problem is solved as follows:

$A=[[1,2,1],[2,1,2],[1,2,1]]; b=[4,5,4]; {{x=A\b}} puts('решение:'+x);$.

Here A=[[1,2,1],[2,1,2],[1,2,1]] describes, line by line, the elements of the system of equations, b=[4,5,4] defining its right part. This is followed by the solution and output of data to a line with a comment.

Row and column dimensions of a matrix are returned by functions $n = $ rows(A) and $m = $ cols(A). It is convenient to organize typical cycles, where n or m is specified as an upper limit. In the first place in A[i][j] there is, as it should be, the line index, although the numbering of elements starts with zero.

Real Jordan form D of matrix A = VD/V and eigenvectors of the matrix are returned by subprogram D = eig(A), V = eigv(A). They both can be found by M = eigs(A); D = M[0]; V = M[1]. The diagonal of eigenvalues highlights D = diag(D) or D = diags(D) – when searching for complex values: then D contains columns of the real and imaginary components. The second use of D = diag(D) is again diagonalizing the matrix, which is common for MatLab.

The matrix rank rank(A), determinant det(A), condition number cond(A), Cholesky decomposition chol(A), Gram-Schmidt orthogonalization orth(A), QR-decomposition of the matrix qr(A) and many more are implemented in the language.

Graphics and animation. Graphics capabilities of Javascript, PHP and others are inherited by Java-MatLab. However, the system is even richer than the previous one – you can easily visualize a mathematical experiment, when controlled objects are represented with movable gif-animations. There were no such capabilities earlier.

6 Connection of Java-MatLab Language with Robots Through the Internet

Mathematical calculations, performed online using the Java-MatLab language, can evolve into the creation of informative "living" illustrations in "living books" [6, 11, 12, 14] – documents in CDF format, published on the Internet, and online electronic journals with executable algorithms [19]. The time has come for these technologies.

In the performance of laboratory work, it is quite important to have the ability to generate an interface with control buttons and windows for displaying video-streams from web-cameras in training facilities [15]. An example of remote control and monitoring of an object in a laboratory is shown in Fig. 2. All of this has been developed within the traditional Internet technology and has reached the stage of mature solutions. A "living book" published on the Internet is not only able to send a signal to an object, but also to control transient processes. All this is not a goal in itself, but for the work and research studies with the use of real remote objects it is an undoubtedly positive quality of the technology.

The hardware access separation system was developed at the University of Siena. Its foundations were built into our implementation of Java-MatLab for the design of such interfaces.

Fig. 2 The interface with a control object

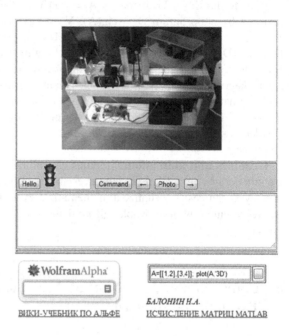

Network technologies today go further than direct connection of sensors and effectors though some interfaces on the Internet. That's why many other data transmission standards have been developed: Wi-Fi, Wi-MAX, Bluetooth, Wireless USB, ZigBee, Home RF, etc. [15–17]. Wireless technology with sensory ends experienced a real boom in its development, which was most clearly reflected by the development of WPAN personal networks based on ZigBee [20]. The peculiarity here is in the fact that just as with matrix operations, the operations with a coordinator, router, end nodes, and sensor nodes of such networks are executed as an exchange of text messages.

A ZigBee sensor network in a training laboratory is formed and reconfigured by a "living book" user remotely. Connection of robots through the network and increasing their number is performed in practically the same way, as the matrix operations are performed with data, received from the network.

The capabilities of the "living book" allow for the creation of an interface, or to regulate the entire data path from remote sensors. All end nodes of a ZigBee network are equipped with analog-to-digital convertors and gauges of input and output of binary signals, which allows not only for gathering information, but also for controlling remote objects.

Thus, the user of such a "living book" is not only enabled to consumer information from remote robotic sources, but also to perform engineering work or conduct software engineering.

7　Development of XML-Based FB2 Format

A significant role in the development of convenient document formats is played by the XML-tags, a quickly spreading set of rules for document layout with inclusion of additional information. XML is used in FB2 format. It is widely used on the Internet, and it is a common form of information layout. In connection with the need for the creation of documents that contain formulas and matrices, we would like to note that the currently existing table design tags are quite suitable for transfer of matrices (worse than formulas), but that they are excessively complex.

Therefore, it is reasonable, following the example of what happened to the book readers, to suggest tags for the generation of portraits of matrices:

<m>A = [[1, 2],[3, 4]]:option </m>.

These tags were implemented in the Java-MatLab system, which allowed for the carrying out of full-scale scientific research studies using illustrations generated by the network in articles submitted to the editorial boards of scientific journals. Such features were also used back in MatLab, but the old system, as we can ascertain, does not generate documents in the form that is required for articles. It is very hard to make the system work in this way.

The analysis of features of color volumetric and flat portraits of Hadamard, Mersenne, Euler, and Fermat matrices allowed for significant development of the theory of minimax orthogonal matrices, bringing the results of research to real use in the practice of processing and conversion of images and videos [21–24].

Writing and presentation of mathematical formulas on the Internet is a solvable problem. Computers gradually change the style of writing of mathematical formulas, which also did not appear accidentally. It was preordained by the convenience and ease of writing of marks and signs on paper. As we move away from paper technology, the former writing technology becomes unwieldy. Using the keyboard it is easier to define the norms of vectors in space R^n in the following way: $\| x \|_2 = (\Sigma_{i=1:n} x_i^2)^{1/2}$, specifying the index borders of sums (and integrals, if necessary) below, in line. The ten-year practice of posting documents on the Internet has not given any preference to alternatives in the form of gif-images (which are hard to format), or images, generated by browser plug-ins (you need to load these plug-ins). So it is better to stay conservative here.

8 Conclusion

The illustrative power of mathematical networks based on the "living book" technology is extremely high. It is convenient and easy to implement the principle of development of remote education in technical universities. The authors of this paper started with what they like and what was easiest for them – the design of a modern educational laboratory for performance of work in the "modeling" discipline.

It is obvious that in their first decade, the network robots have learned the material of university problems of the last century, and their "level of knowledge" is noticeably higher than that of an average student. The advent of machines that send analytical solutions to students' gadgets will inevitably change the requirements to what a student has to remember (learn by heart).

Mathematics is the most in-demand discipline for technical specialties, as well as for textbooks and books used in the educational process. For online projects that develop the "living book" technology, the use of mathematical services is based on their needs. This is the concept of the mathematical network, which was launched by the authors of this article and is already growing, existing in several varieties and with several addresses on the Internet: http://livelab.spb.ru, http://artspb.com, http://mathscinet.ru.

References

1. Gerval, J.-P., Le Ru, Y., Ghiron, S.: Autoring tool for SmartPhone pedagogical content development. In: Smart Digital Futures 2014, IOS Press, 2014, pp. 679–696. doi:10.3233/978-1-61499-405-3-679
2. Keenaghan, G., Horváth, I.: Using Game engine technologies for increasing cognitive stimulation and perceptive immersion. In: Smart Digital Futures 2014, IOS Press, 2014, pp. 657–667. doi:10.3233/978-1-61499-405-3-657
3. Le Ru, Y., Gerval, J.-P., et al.: Web-Based Course and teaching evaluation system including dashboards and evolutions reporting. In: Smart Digital Futures 2014, IOS Press, 2014, pp. 641–651. doi:10.3233/978-1-61499-405-3-641
4. Antokhina, Y.A.: Management of education quality at an Autonomous Technical University. Informatsionno-upravliaiushchie sistemy [6], 99–101 (2013) (in Russian)
5. Balonin, N.A., Sergeev, M.B.: Tachnical "Live Book": request for discussion. High. Educ. Russia 1(7), 141–144 (2013) (in Russian)
6. Sergeev, M.B., Balonin, N.A., Balonin, Y.N.: Certificate of state registration of the software "Live Book" . [1]2012661277, 11 Dec 2012 (in Russian)
7. http://blog.wolfram.com/?source=nav
8. http://www.mathworks.com/products/matlab/
9. http://www.wolfram.com/cdf
10. http://en.wikipedia.org/wiki/FictionBook
11. Balonin, N.A., Sergeev, M.B.: Modern network technologies in scientific researches. In: Informacionnye Tehnologii, vol. [1]2, pp. 23–26 (2014) (in Russian)

12. Balonin, N.A., Marley, V.E., Sergeev, M.B.: New opportunities of a mathematical network for collective research and modeling in the internet. Informatsionno-upravliaiushchie sistemy, [1]3 (70), 40–46 (2014) (in Russian)
13. https://www.codeschool.com/paths/javascript
14. Balonin, N.A., Sergeev, M.B.: Java-MatLab language release to distributed internet resources. Softw. Eng. 1(4), 16–18 (2014) (in Russian)
15. http://livelab.spb.ru
16. http://artspb.com
17. http://mathscinet.ru
18. Sergeev, M.B.: Exact solution of linear systems of equations with integer coefficients. Eng. Simul. 11(1), 151–155 (1993)
19. Balonin, N.A., Sergeev, M.B.: The concept of electronic magazine with executable algorithms. In: Fundamental'nye issledovanija, vol. [1]4–4, pp. 791–795 (2013) (in Russian)
20. http://www.zigbee.org
21. Balonin, N., Sergeev, M.: Expansion of the orthogonal basis in video compression. In: Frontiers in Artificial Intelligence and Applications, vol. 262: Smart Digital Futures 2014, pp. 468–474. doi:10.3233/978-1-61499-405-3-468
22. Balonin, Y., Sergeev, M., Vostrikov, A.: Software for Finding M-matrices. In: Frontiers in Artificial Intelligence and Applications, vol. 262: Smart Digital Futures 2014, pp. 475–480. doi:10.3233/978-1-61499-405-3-475
23. Balonin, N., Sergeev, M.: Construction of transformation basis for video image masking procedures. In: Frontiers in Artificial Intelligence and Applications, vol. 262: Smart Digital Futures 2014, pp. 462–467. doi:10.3233/978-1-61499-405-3-462
24. Vostrikov, A., Chernyshev, S.: Implementation of Novel Quasi-orthogonal matrices for simultaneous images compression and protection. In: Frontiers in Artificial Intelligence and Applications, vol. 262: Smart Digital Futures 2014, pp. 451–461. doi:10.3233/978-1-61499-405-3-451

Just Give Me a Hint! An Alternative Testing Approach for Simultaneous Assessment and Learning

Jerry Schnepp and Christian Rogers

Abstract The Millennial generation has grown up using computers as an essential part of their daily activities. Educators have successfully adapted formative assessments by incorporating Internet resources, gamification and pervasive technology to meet the needs of these students. However, summative assessments have changed very little. Point Barter is an online testing system that incorporates game-like features to equitably assist students in answering test questions. It allows students to dynamically trade test points for hints about the correct answer. This paper presents the results of a study that evaluated Point Barter with two sections of the same course. One section used Point Barter throughout a semester while the other used the standard testing system. Upon completion of the course, all participants took a standardized cumulative final exam. The experimental group scored significantly higher than the control group, indicating that using Point Barter throughout the semester may have led to increased learning.

Keywords Summative assessment · Hints · Crib sheets

1 Introduction

Most current college students belong to the millennial generation: the first to grow up with computers in their homes and convenient Internet access [1]. Millenials have a depth and duration of experience with technology that differs vastly from previous generations. Consequently, their workflow is different. Particularly, Millennials are accustomed to conducting tasks with the support of technology [2].

J. Schnepp (✉)
Visual Communication Technology, Bowling Green State University, Bowling Green, OH, USA
e-mail: jschnepp@bgsu.edu

C. Rogers
Indiana University – Purdue University Indianapolis, Indianapolis, IN, USA
e-mail: rogerscb@iupui.edu

© Springer International Publishing Switzerland 2015   141
V.L. Uskov et al. (eds.), *Smart Education and Smart e-Learning*,
Smart Innovation, Systems and Technologies 41,
DOI 10.1007/978-3-319-19875-0_13

Regardless of generation, the ubiquitous incorporation of computer technology is pervasive across industries. In fact, few jobs require that employees store all-important knowledge "in their heads". Instead, they use computers and mobile devices to supplement expertise. While the acceptance of, and reliance on, technology spans generations; it is especially emblematic of Millennials.

Educators have responded to this paradigm shift by incorporating multiple modalities into curricula. Instead of traditional lectures, flipped classrooms [3] and gamification techniques, such as ask-hint strategy, badging and reverse points [4] are commonly used to accommodate students who have become accustomed to a different style of learning. Students respond well to these approaches as evidenced by studies conducted by Enfield and Strayer [5, 6]. It seems teachers are developing new ways to engage Millennials in the spirit of Ignacio Estrada, who famously said: "If a child can't learn the way we teach, maybe we should teach the way they learn."

While classroom instruction and assignments undergo a technological renaissance, little has changed regarding summative assessments such as quizzes and exams. Should the assessment of Millennials follow the same structure as those of previous generations? Harris and Hodges define assessment as "the act or process of gathering data to better understand the strengths and weaknesses of student learning" [7]. Since millennial students are exposed to different ways of learning, modern educators should adjust their evaluative approach as well as their instructional approach. Rawson and Dunlowski further suggests that the act of testing is more than a means to evaluate learning [8]. Testing can be used to improve learning, specifically when students are provided feedback.

This paper describes a study that evaluated a novel approach to student assessment called Point Barter [9]. Point Barter is an online quiz system that allows students to sacrifice partial credit in exchange for hints about the correct answer. The purpose of this approach is to present students with alternate strategies to earn quiz points. Our goal in this line of inquiry is to identify a combined learning and evaluative activity that will lead to more effective knowledge acquisition.

2 The Point Barter System

Point Barter is an online quiz system that allows students to trade points for hints. For each question, the student has the option to barter a pre-determined point value. If the student does not know the correct answer and chooses to barter, points are automatically deducted from the overall point value of the question and the hint is displayed. Each question can have multiple hints. After the bartering transaction, the student answers the question. If answered correctly, the student is awarded the remaining points (Fig. 1).

Point Barter's key feature is similar to an important element of strategy games. It gives students the option to trade something of value - quiz points - for information that leads to the achievement of a goal - a correct answer. More importantly, it reinforces that there are multiple paths to success. A testing system that allows students to draw on outside information promotes ingenuity and strategy.

We conducted a pilot study of the Point Barter system to evaluate usability and student perception [10]. After using Point Barter to take an exam, participants overwhelmingly reported a positive experience. Many identified a reduction in test anxiety. Ninety-five percent of participants would recommend that Point Barter be used in other classes.

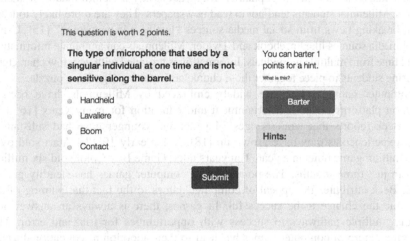

Fig. 1 The point barter interface

This interactive testing activity engages students using a game-like format for assessment that provides continual feedback as the student participates. Students can gain insight on the subject while also being evaluated on that knowledge. The exchange of resources - points for hints - is familiar to many Millennials because of its commonality with strategy games in which a character trades currency (points, coins, etc.) for weapons, powers, or even clues to advance their mission. Success in such games is achieved through careful compromises and an exploration of options. In games, there is *always* a way to achieve success, and usually there are multiple ways to that accomplish that goal.

3 Literature Review

The millennial generation poses unique challenges to educators. Because they were raised with consistent access to the Internet, millennial students tend to receive information and interpret that information differently than generation X or the baby boomer generation [11, 12]. They are used to accessing information quickly and readily in a just-in-time fashion. Because of their experience with the way information is displayed on the Internet, Millennials expect to acquire information in small chunks and to have the option to customize many aspects of their experience [13].

As of January 2014, 97 % of Americans between the ages of 18 and 29 used the internet [14]. Students from the millennial generation have had access to technology throughout childhood, growing up with personal computers, and consequently relying heavily on them for productive work. They are confident in their abilities to use technology and find information on the world wide web.

Access to technology has affected how students send and receive information. For previous generations, television and newspapers were the primary sources of news. As social media outlets increase in popularity, young adults now seek information in smaller chunks. Millennial students tend not to read newspapers. They are more likely to learn about breaking news from social media sources "just as it is happening" [15]. Using social media sources like Facebook and Twitter, Millennials can aggregate information in real time from multiple individuals. Information is provided with very few characters, requiring students to piece together these chunks as if putting together a puzzle.

Computer games, which are widely embraced by Millennials, have been a common platform for both entertainment and education for many years [16, 17]. John Beck reports that workers ages 34 years and younger have had substantial game experience as they have grown up [18]. In the early 1980's, Atari sold over three million game units in a year. Ten years later, Game Boy Color sold six million units in just three months. The popularity of computer games has steadily grown since. Beck attributes the appeal of computer games to the fact that gaming offers everyone the chance to be successful. In games, there is always an answer and usually multiple pathways to success with opportunities for trial and error. The engaging nature of computer games has lead to their adoption as educational tools.

According to Squire [19], educational games create *experiences* in which learners use tools and resources to solve complex problems. Learners are immersed in an alternative environment that engages students in an entirely different way than traditional learning. By developing curricula with a game as the vehicle for content, educators appeal to the "gamer generation" directly, motivating the students to actively participate [20].

While research has been conducted on game-based learning in formative assessments, very little research has been conducted in the area of gaming with summative assessments. GAM-WATA [21] is one such system that incorporates "Ask-Hint Strategy" into a quiz structure. Students were provided immediate online hints that led to the correct answer, which they found empowering.

4 Methodology

We developed Point Barter specifically to evaluate how students would use an online testing system that incorporates an automated hints-for-points feature. An initial pilot study assessed the usability and acceptability of the software [10].

Students found the system easy to use and indicated a preference for it over their usual testing environment. A review of student comments prompted the question "If a student barters for hints on an exam, does he learn while taking the exam?" The current study aims to answer this question.

During the spring semester of 2014, two sections of the same Visual Communications Technology course participated in a study at Bowling Green State University. The same professor taught both sections. One section used Point Barter ($n = 18$) for quizzes and the midterm exam, while the other used Canvas ($n = 14$), the standard testing system used in most courses at the University. Other than the testing system, both sections were identical both in content and assessments, which consisted of twelve assignments, four quizzes and a midterm exam. Both sections took identical final exams using Canvas, allowing us to compare the scores between groups.

Students completed quizzes and exams during class sessions on lab computers. The format of the questions was either multiple choice, true/false, fill in the blank, short answer, or essay. Students viewed each question individually.

For each quiz or test question, students using Point Barter had the option to answer directly or use the barter feature. If a student chose to barter, the software subtracted the specified point value from the question's total point value and displayed the hint. Students could barter multiple times - up to four, depending on the question. If the student answered correctly, he received the remaining point value.

5 Results

The experimental group used Point Barter throughout the semester while the control group used Canvas, the standard testing system. However, for the final exam, both groups used Canvas so that we could objectively assess whether using Point Barter throughout the semester helped students learn. On the final exam, the experimental group scored an average grade of 80.33 % while the control group scored an average grade of 73.71 %, which is statistically significant ($p = 0.03$).

While the two groups used different testing systems for the midterm exam, the questions were identical. Interestingly, both groups scored similarly on the midterm exam. The experimental group earned an average grade of 77.2 % and the control group earned an average grade of 77.7 % ($p = 0.85$) (Fig. 2).

The course averages of both sections were similar. The experimental group course average was 82.77 % while the control group course average was 81.61 % ($p = 0.64$) (Fig. 3).

Figure 4 illustrates each instance of an assessment for the experimental group, consisting of four quizzes and the midterm exam. The graph compares the grade of the instance to the amount of barters for that instance for which there is a moderate correlation ($r = -0.406$). The negative slope of the regression line indicates that higher scores correlate to fewer barters.

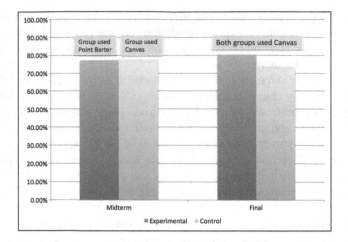

Fig. 2 Grade difference between class sections. The experimental group used point barter throughout the semester. The control group used canvas

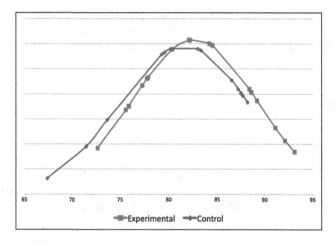

Fig. 3 Course average grade distribution of both class sections

6 Discussion

The group that used Point Barter throughout the semester scored significantly higher than the control group on the standardized and cumulative final exam. Since the course lectures, assignments and projects were identical for both groups, we assert that using Point Barter throughout the semester promoted additional learning. If students were unsure about an answer, they could barter for a hint, thus obtaining new knowledge. These students knew more about the subject after the quiz than they did before. They learned during assessment.

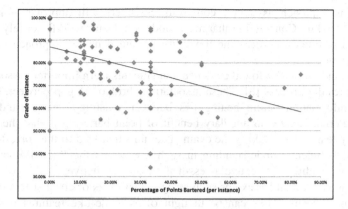

Fig. 4 Correlation between percentages of points bartered and grades for four quizzes and the midterm exam

Through discussion of the Point Barter system with fellow educators, we encounter a common initial opinion. Many assume that using this system will make it easy for weak students to do well on exams. After all, even those who do not study can obtain hints and potentially answer correctly.

While it is true that students who use the barter feature typically arrive at the correct answer, they do not necessarily receive high scores, since they must sacrifice points to obtain the hint. In fact, the strong students who *do not use the barter feature* tend to score highest. This is likely because they are well prepared for the exam and do not need help answering correctly.

The fact that the overall grade averages for both course sections were not significantly different from each other indicates that using Point Barter does not provide an unfair score advantage over traditional testing methods. Further, the grade distribution within the group that used Point Barter was similar to the control group. This indicates that the exchange of points for hints is equitable - weak students receive low scores and strong students score higher. In other words, scores remain consistent with preparedness, indicating that Point Barter is an equitable means of assessment.

We recognize that because the two sections were taught separately, there may exist confounding variables related to time of day and potential inconsistencies between lectures. However, we believe that these pose minimal impact. Still, we intend to address these issues in a future study.

7 Conclusion

The successful use of point bartering involves evaluation, deliberation, and prudence. By using the barter feature, the student has decided that instead of answering incorrectly, he will improve his score by choosing an alternate path. Some may criticize this approach as a lazy compromise. Others will approve of it as an important life skill. Considering that most modern industry workflows rely heavily on external resources, perhaps the flexibility ingrained in this approach is particularly valuable.

Rawson and Dunlosky found evidence that students learn important associations during the testing process [8]. These associations bridge the gap between familiar and new information, strengthening ties between existing concepts and newly learned material. Hence, an ancillary benefit of point bartering is that the student intrinsically learns while taking the exam. The hints that lead to the correct answer reinforce the student's understanding in such a way that the Point Barter system may be more useful in formative assessments than summative.

Adscititious material for exams such as open textbooks or cheat sheets has been used for many years. They can be thought of as game-like features in that the student must strategize in order to maximize their usefulness. Point Barter leverages this same utility, while rewarding the students who do not rely on hints with higher scores.

Of course, the way students will use Point Barter is highly dependent on the difficulty of the question, the usefulness of the hints, and the point value of the hints. This ambiguity makes an objective evaluation of the system difficult, and its implications beget many unanswered questions.

Educators should not ignore the profound effect technology has had on modern students. Testing environments should adapt to the way students learn, process, and use information. Traditional testing techniques are certainly valuable. However, explorations of new methods may lead to yet unidentified potential.

8 Future Work

While the size of this study provided significant data, it was not large enough to glean deep insight into the way students might learn through a testing system like Point Barter. However, small-scale experiments such as this are important stepping stones toward large-scale innovation. Having found evidence that Point Barter facilitates simultaneous learning and assessment, we can now focus on a more directed study with a larger population and fewer confounding variables.

In a future study, we intend to evaluate Point Barter within a closed environment. We will conduct a one-day workshop in which forty participants will learn about an unfamiliar subject. Participants will be randomly assigned to one of two groups. Both groups will participate in the workshop simultaneously with the same instructor. At the beginning of the workshop, both groups will participate in a pre-test that evaluates their existing knowledge of the topic. Throughout the workshop, participants will take four quizzes. The experimental group will use the Point Barter system while the control group will use an identical testing system that does not include the barter feature. At the completion of the workshop, both groups will complete a post-test, identical to the pre-test, without access to the barter feature. By reducing the number of confounding variables, we expect to obtain more accurate data upon which to base conclusions.

By using the barter feature, the student has decided that instead of answering incorrectly, he will improve his score by choosing an alternate path. Some may criticize this approach as a lazy compromise. Others will approve of it as an important life skill. Considering that most modern industry workflows rely heavily on external resources, perhaps the flexibility ingrained in this approach is particularly valuable.

References

1. Smola, K.W., Sutton, C.D.: Generational differences: revisiting generational work values for the new millennium. J. Organ. Behav. **23**, 363–382 (2002)
2. Blackburn, K., LeFebvre, L., Richardson, E.: Technological task interruptions in the classroom. Florida Commun. J. **41**(2), 107–116 (2013)
3. Mull, B.: Flipped learning: a response to five common criticisms. November Learning. http://novemberlearning.com/resources/articles/flippedlearning-a-response-to-five-common-criticismsarticle (March 29, 2012)
4. Apostol, S., Zaharescu, L., Alexe, I.: Gamification of learning and educational games. Elearning Softw. Educ. **2**, 67–72 (2013). doi:10.12753/2066-026X-13-118
5. Enfield, J.: Looking at the impact of the flipped classroom model of instruction on undergraduate multimedia students at CSUN. Techtrends: Linking Research & Practice To Improve. Learning **57**(6), 14–27 (2013). doi:10.1007/s11528-013-0698-1
6. Strayer, J.: How learning in an inverted classroom influences cooperation, innovation and task orientation. Learning Environ. Res. **15**(2), 171–193 (2012). doi:10.1007/s10984-012-9108-4
7. Harris, T.L., Hodges, R.E., International Reading Association.: The literacy dictionary: The vocabulary of reading and writing. International Reading Association, Newark, Del (1995)
8. Rawson, K.A., Dunlosky, J.: When is practice testing most effective for improving the durability and efficiency of student learning? Educ. Psychol. Rev. **24**, 419–435 (2012)
9. Schnepp, J.: US Patent No. 61,862,126. U.S. Patent and Trademark Office, Washington, DC (2013)
10. Rogers, C., Schnepp, J.: Students perceptions of an alternative testing method: hints as an option for exam questions. Manuscript submitted for publication (2014)
11. Gorman, P., Nelson, T., Glassman, A.: The millennial generation: a strategic opportunity. Org. Anal. **12**(3), 255–270 (2004)
12. Howe, N., Strauss, W.: Millennials go to College: Strategies for a New Generation on Campus: Recruiting and Admissions, Campus Life, and the Classroom. AACRAO, Washington (2003)
13. Tapscott, D.: Growing up Digital: The Rise of the Net Generation. McGraw-Hill, New York (1998)
14. Pew research internet project. Internet user demographic. http://www.pewinternet.org/data-trend/internet-use/latest-stats/ (2014)
15. Sweeney, R.: Millennial behaviors and demographics. New Jersey Institute of Technology, Newark **12**(3), (2006)
16. Gros, B.: Digital games in education: the design of games-based learning environments. J. Res. Technol. Educ. **40**(1) (2007)
17. Prensky, M.: Digital Game Based Learning. McGraw Hill Press, New York (2001)
18. Beck, J.C.: Got Game: How the Gamer Generation is Reshaping Business Forever. Harvard Business Press (2004)

19. Squire, K.: Game-based learning: present and future state of the field. University of Wisconsin-Madison Press, Madison (2005)
20. Kumar, R., Sarukesi, K., Uma, G.V.: OAMS—a game based online formative knowledge assessment system using concept map. J. Theor. Appl. Inf. Technol. **53**(1), 150–156 (2013)
21. Wang, T.-H.: Web-based quiz-game-like formative assessment: development and evaluation. Comput. Educ. **51**(3), 1247–1263 (2008). doi:10.1016/j.compedu.2007.11.011

M-MODEL: Design and Implementation of an On-Line Homework System for Engineering Mechanics

Edward E. Anderson

Abstract Students solve problems by developing mental models of the problem. Although these models are many and diverse, a common one used in engineering mechanics education consists of identifying the known and unknown variables, construction of a graphical problem representation, and developing a mathematical model derived from the preceding steps. This is also the case for courses in physics, mechanics, science and electrical circuits. M-MODEL is a computer-based implementation of this approach to problem-solving. It utilizes the visual graphic, variables identification and listing, and mathematical equation mental models as a construct for students to developing STEM problem solving skills. This paper describes how students use M-MODEL to solve problems and its coaching tools. This paper discusses the design objectives, intelligent tutor, and features of M-MODEL as applied to an engineering mechanics course. It also discusses how M-MODEL may be used to encourage students to develop mental model approaches to problem solving. An assessment of M-MODEL is also presented.

Keywords M-MODEL · Engineering mechanics · Problem solving · Assessment

1 Introduction

Students solve problems by constructing mental representations of the problem. These models take many forms such as graphical, mathematical, flow charts, process steps, and schematics to mention a few. As pointed out by Norman [1], these models can be contradictory, incomplete, superstitious, erroneous, and unstable, while varying in time. It is the task of the educator to help students learn how to form accurate and useful mental models and apply them to knowledge domain problems. M-MODEL is a computer-based tool that permits engineering and

E.E. Anderson (✉)
Department of Mechanical Engineering, Texas Tech University, Lubbock, USA
e-mail: Ed.Anderson@ttu.edu

© Springer International Publishing Switzerland 2015
V.L. Uskov et al. (eds.), *Smart Education and Smart e-Learning*,
Smart Innovation, Systems and Technologies 41,
DOI 10.1007/978-3-319-19875-0_14

science educators to develop problems using the principle mental models of the discipline in a consistent and flexible manner. This paper describes the user environment, the design philosophy behind M-MODEL, its intelligent tutor, and some of the embedded pedagogues.

Several engineering problem solving models or schema have been reported recently. These include the Wankat and Oreovicz [2] problem solving strategy, McMaster problem solving program of Woods [3] and Woods et al. [4], Gray and Costonzo [5] structured approach to problem solving, Mettes and associates [6] Systematic Approach to Solving Problems, and Litzenger et al. [7] Integrated Problem Solving Model. The Wankat and Oreovicz strategy divides problem solving into definite steps including motivation, exploration, and reflection as well as the more common define, plan, execute and check steps. The McMaster problem solving program uses a structure similar to that of Wankat and Oreovicz and implements it across entire curricula. Gray's structured approach emphasizes pattern-matching that starts with a small number of general equations that students reduce to fit a given situation. The Mettes problem solving schema is based upon a flow chart of problem solving steps and a constructionist approach to learning. Litzinger's integrated model emphasizes problem representation and the conversion from one representation (say problem statement) to another (say graphical).

The define, plan, and execute steps are the common thread among these various models. In engineering mechanics disciplines, these steps take the form of free-body diagram (FBD) development, listing of the given (known) and identifying (unknown) variables, creating a mathematical model consistent with the FBD and variables list, and final answer production. As pointed out by Gray and Costonzo, the current trend in engineering mechanics is to deemphasize the final answer production step and leave this to computational software. Four mental models are commonly used to build problem solutions. These are: problem statement, graphical representation such as a FBD, given/find representation typically in the form of lists, and a system of equations that will produce the final answers. These four mental models represent the core of science and engineering mechanics problem solving, have a long standing tradition in technical education, and are consistent with current trends in engineering and science education.

The newest trend in engineering mechanics education is the application of computer technology to teach students, engage them in the learning process, and to help them understand mechanics concepts and principles. These are many and varied. They are perhaps best illustrated by the works of Gramoll [8, 9], Dollar and Steif [10], Steif and Dollar [11, 12], Philpot [13, 14], Stanley [15], and Gray and Costonzo [16] to list a few. Many of these are similar to traditional textbook presentations with exceptions such as interactive examples, audio/video lectures, homework sets with immediate feedback, virtual experiments, and interactive animations to develop conceptual understanding. This approach to learning problem solving is based upon examples and homework problem sets and is fairly traditional. Problem interactivity has been added to keep the student engaged with the problem. Hints, intelligent coaching, instantaneous feedback, and intelligent correcting have been incorporated by many of these authors. But, they rely upon click-on-object, drop and drag, pair matching, multi-choice answers, and short answers (usually numerical) for user input and traversing the basic problem structure. They tend to be somewhat inflexible in that users must use notations, axis systems, vector directions, equation ordering and etc. as prescribed by the problem designer rather than allowing students to make choices and decisions on their own.

M-MODEL utilizes the four core mental models as is common amongst the various problem-solving research findings and requires users to fully develop their graphical, variables, and mathematical representations from the problem statement. Research finds that most users will develop their representations in that order, but it is not required and students can proceed however they deem appropriate as they develop their problem-solving skills. But, all user representations must be completed before a correct solution is possible. Users have complete freedom in naming their variables, orienting their FBD vectors and coordinate systems, selecting their units, and etc. as they set up their solution. These choices are graded against the problem designer's expectations and final answers. M-MODEL is therefore an extremely versatile system that gives students considerable freedom in developing their problem solution and encourages them to utilize in-depth problem solving skills and high-order cognitions.

The objective of this paper is to present the design philosophy and implementation of M-Model in a typical introductory engineering course. The results of an assessment of M-MODEL and antidotal findings are also reported.

2 Related Work

The Andes problem solving system for classical physics developed by Vanlehn et al. [18] and implemented at the United States Naval Academy by Schultz et al. [17] is based upon a Bayesian network representation of a problem. This system allows considerable flexibility and generates solutions, immediate feedback, and help comments based upon the path traversed by the user through this network. Hence, the user who elects to use one set of notations will be coached through the problem and produce a correct answer for that notation set just as the user who chooses to use some other notation system. This approach encourages students to think through the solution, plan their approach, and develop in-depth problem solving skills rather than charging directly and often blindly into and through the problem solution. Andes utilizes four mental models; problem statement, graphic representation, variables lists, and mathematical model, and requires users to develop each of these mental models (graphical representation is optional). Andes includes an equation solving tool although users can also solve the equations off-line. A research project conducted on some 330 students approximately one-half of whom were in a control group, resulted in a 3 % (1/3 letter grade) student performance improvement on departmental pencil and paper examinations by students who did Andes homework rather than traditional homework. Anecdotal results from Andes users [17] indicate that students are initially reluctant to carefully define their variables, some students ask for help on almost every step of a problem solution, giving effective hints and help is very difficult, and analysis of action logs reveal that students do not understand physics as well as might be thought.

3   M-MODEL Design Philosophy Objectives

The objectives for the design of M-MODEL were:

- Provide an interface that more accurately simulates problem solving processes which are known to effectively promote student learning.
- Allow for multiple solution paths to accommodate various learning styles.
- Provide intelligent tutoring, criticism, and feedback.
- Encourage and develop student critical thinking skills.

M-MODEL was conceived as a tool students can use to practice and develop their problem solving skills as well as to allow sufficient flexibility that varied, but correct, problem solution paths are possible. This latter objective is important in that engineering mechanics courses are typically those courses that begin the transition from well-framed problems to the more ill-defined engineering and design problems. It is also important that students learn how to formulate problems and that correct, but possibly different, answers depend upon that formulation.

M-MODEL was also designed to require users to use all four mental models common to the problem solving practices in the engineering educational literature. These mental models have a long standing tradition in engineering mechanics education and are familiar to engineering educators. M-MODEL also focuses on the model building process and leaves the computational details to an optional embedded equation solver. This feature is intended to channel students away from the rush to simply write equations, substitute values, and produce answers as is commonly the case. Rather, students must carefully build complete models before a final solution is possible. Focusing on model building is also the current trend in engineering and science education.

M-MODEL is also intended to give problem developers tools that they can use to develop their own problems and homework sets. It is also designed to reduce the task of grading student solutions. It evaluates many student mistakes and misconceptions, assigns a grade based upon mistakes and misconceptions, and records these grades when implemented on a database and active page server. The software also records user activities, sequences, and time-on-task for research and verification purposes.

4 M-MODEL User Interface

The initial user screen is shown in Fig. 1. This screen is divided into 6 different areas: problem statement, video tutor, graphical representation, variables listing, pre- and post-calculations, and equation system areas. The problem statement is presented in the upper, left portion of the screen. Problems can contain up to two random parameters. This example contains two, the force W and length 12. Users select a free-body from the problem statement diagram by clicking on the appropriate object. Users are not required to do this at the beginning, but ultimately they must select a free-body and develop a FBD. Points are deducted from a student's score and an error message appears if an incorrect free-body is selected.

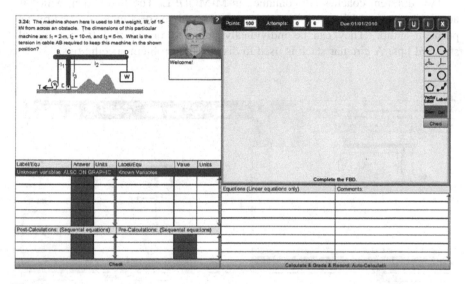

Fig. 1 Initial user interface

Once a user selects the correct free-body, it appears in the graphics construction window of the upper, right-hand screen where the user can use the tools in the toolbox to complete the FBD. These tools include: lines, arrows, clockwise and counter-clockwise moments, two- and three-dimension axis, points, circles, vector (bold) labels, plain labels, screen clearing, and object deleting. Any time a tool is selected by the user, a pop-up screen with user instructions appears. Some of the objects produced by the tools (e.g., axis and moments) can also be rotated as appropriate for the problem. Users click the "Check" button below the tool box to determine if their FBD is complete. An example of a completed FBD is shown in Fig. 2.

Problem designers set the minimum number of graphic objects users must add to the FBD. This determines the detail that is expected for correctly completing a

FBD. This may be as simple as 4-arrows, 4-bold labels, 3-labels, and 1-two-dimensional axis for the example of Fig. 2.

The graphical mental model window includes a grade display, attempts counter, and three help buttons. Users begin with 100 points. This number is reduced whenever the user makes a mistake. This reduction depends upon the significance of the mistake. This deduction is set to be higher for a major error like selecting an incorrect free-body and less for a minor mistake like indistinct labels. The magnitude of the deducted points is at the discretion of the problem designer. The "U" button in the upper right-hand of this window activates a pop-up list of the unit symbols in M-MODEL. The "I" button activates another pop-up with an abbreviated set of instructions as a quick reference for users. Full user instructions are available in a separate file.

Two different coaches are contained in M-MODEL. The first coach, which is accessed with the "T" toolbar button, is a set of problem specific tips created by the problem author. They can be individually authored or selected from a file of standard tips. A circular stack is used to display the problem specific tips.

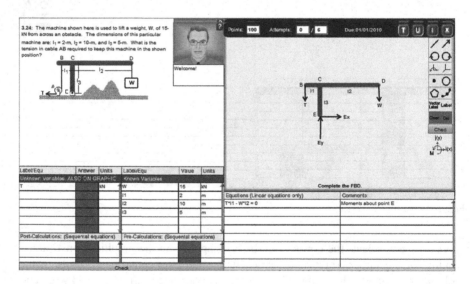

Fig. 2 User screen prior to completing calculations

The second coach or tutor, is accessed with the "?" button in the video tutor window. This tutor is an intelligent tutor based upon a prioritized list of procedural tasks as identified by content experts. This stack consists of major tasks and sub-tasks. For example, a major task is "complete the graphic object" with sub-tasks like "draw all arrows", "label all arrows", and "label all dimensions". When a user requests help from the intelligent tutor, the tutor identifies where the user is in the prioritized procedure list. If the user is working on a sub-task he or she completes that sub-task. Once that sub-task has been completed, the tutor suggests that the user proceed with the next sub-task in the prioritized list until all sub-tasks have been completed. Once all the required sub-tasks have been completed, the tutor suggests that the user proceed with the highest priority undone major task in the list. For example, major task graphic model construction has a higher priority than the listing major task. The listing task includes the unknown listing and known listing sub-tasks. So, the user who starts at the lower priority listing major task before completing the graphic construction major task will be tutored to complete the known and unknown listing sub-tasks. Once these sub-tasks are completed, the user is tutored to proceed with the graphics construction major task.

The left-hand side of the user interface also contains a variables list, pre-calculation, and post-calculation section in the lower quadrant as shown in Fig. 2. The variables list is divided into lists of unknown variables and known variables. These are completed by entering the labels and units for the problem known and unknown variables. The value of the known variables must also be entered by the user. The calculated value of the unknown variables is displayed once the user completes her model and clicks the "Calculate" button to the lower right. Once this "Check" button is clicked, all labels are checked to insure they are distinctly named, and all units are checked for internal consistency. Error messages are displayed and points deducted if any of these errors occur. These error messages provide immediate feedback and give the user the opportunity to make corrections prior to completing the full solution.

Pre-calculation equations and units of the resulting pre-calculation variables are entered in the "Pre-Calculations" list. Only constants and known variables can be used to build the pre-calculation equations. These are entered in equation form with the pre-calculation variable label on the left-hand side of the equation. Typical examples of pre-calculations are items like moments of inertia and areas. The pre-calculation equations are also sequentially evaluated. That is, any equation can use the known values or variables of any pre-calculation equations above it. The pre-calculation equations are the first to be evaluated when the "Calculate" or "Check" button is clicked. The results of these equations are displayed at this time and the pre-calculation variables become available for the subsequent evaluation of the equation system in the lower, right-hand section of the user interface and the post-calculation equations. The final user screen illustrating the result of these calculations is shown in Fig. 3.

The "Post-Calculations" section is organized in the same manner as the "Pre-Calculations" section. The only difference is that post-calculations are done once the unknown variables have been determined by solving the pre-calculation and equation system equations. At this point all variables are available for the post-calculation equations and they may be evaluated. These equations are also sequential. A typical example of a post-calculation is an axial stress calculated from an unknown internal force (from the "Equations" section) and section area (from the "Pre-Calculation" section).

The last section to be completed by the user prior to calculating final answers is the "Equations" section in the lower, right-hand quadrant of Fig. 2. This system of

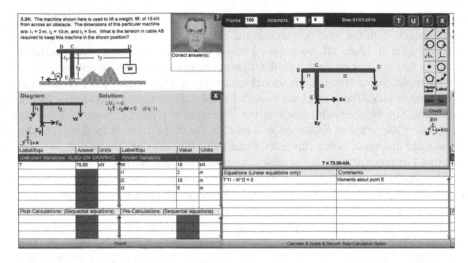

Fig. 3 Final user screen

equations is commonly, but not necessarily, the equations of force and moment equilibrium. This is divided into two lists: equations and comments. Comments are optional. Equation entry is intuitive and subject to very few rules.

Terms involving only constants, known variables, and pre-calculation variables are entered into these equations as appropriate. The equation solver includes the ^, *, /, +, and − operators. The equation solver also includes the following functions: sin, cos, tan, asin, acos, atan, pow (x^y), and ln (natural logarithm). These operators and functions are used to build the terms and coefficients of the pre-calculation, post-calculation, and equation system equations.

Equations entered in this section must be independent, linear equations with the number of equations matching the number of unknown variables. Any non-linear calculations (e.g., diameter of a circle given the area) must be done in the "Pre- or Post-Calculation" sections. When the "Calculate" button is clicked, several equation format checks are done, error messages displayed, and points deducted if appropriate. The system of equations is also checked at this time to insure that they are independent. The system of equations and post-calculation equations are then solved. Next, the values of the unknown variables and post-calculation variables are compared against the answers produced from a set of correct answer equations provided by the problem designer. If the problem statement includes random parameters, the problem designer must provide correct answer equations for the unknown and post-calculation variables that only include constants and the random variables. User answers are considered incorrect if they are not within ±1 % (at the discretion of the designer) of the designer's answers. Incorrect user answers are highlighted and the user can proceed to edit any item on the screen and recalculate their answers. The user's opportunity to revise a solution based upon feedback is known to achieve deep, lasting learning [19].

5 M-MODEL Pedagogues

M-MODEL is a non-sequential problem solving tool that encourages students to build their own problem mental models with as few restrictions as possible. Although users must complete four of the six models (pre- and post-calculations are optional), they can be completed in any order. They can also be altered before final solution as one model provides further insight into another model. For example, users often add or remove variables as they are writing their equations or refining their graphic model. These cognitions fall under the Analyze (breaking down material or tasks into constituent elements) and Evaluate (making judgments using standards and criteria) classifications of Bloom's taxonomy [20]. Both are at the higher-order cognitions end of Bloom's taxonomy.

M-MODEL allows users to set up and solve a problem in their own terms using solution procedures of their own creation. For example, the problem of Figs. 1, 2 and 3 can be correctly solved by considering T, Cx and Cy as pre-calculation variables or as unknown variables. In the latter case, the user must include additional equations in the equation section. Users may also elect to not use Cx and Cy at all, but rather to replace it with the appropriate moment equation. Other users may elect to use some or all numerical values in lieu of variable labels and values. All of these choices are correct as long as they are consistent and will produce correct answers. Users must then "Create" (Producing alternatives or reorganizing materials in new ways) solutions which is the highest-order cognition in Bloom's taxonomy.

Although M-MODEL promotes procedural and higher-order cognitions, it is not without its penalties. First, students cannot produce correct answers without a through and detailed set of models, and often they need to refine or rebuild their models as their understanding of the problem deepens. This entails additional work on the part of the student which frequently meets with objection. Users need to solve 2–3 problems to become comfortable with the interface. This learning curve can interfere with their learning the content material and may frustrate them. Some of this extra effort is recovered by the equations solver which saves some time. The automatic solving of the equations and lack of computational practice can be problematic during examinations if numerical answers are heavily weighted although the work of Vanlehn et al. [18] suggests that this may not be an issue.

6 Assessment

An experiment was conducted at the US Air Force Academy during the 2010 spring semester to measure changes in student performance attributable to M-MODEL and student attitudes about M-MODEL. This experiment involved 120 students registered in 6 sections taught by 5 instructors of a solid mechanics course. This course is taken during the fourth semester of the Mechanical Engineer/Engineering Mechanics curriculum. Three of the sections (57 students) did one-third of the required homework problems using M-MODEL and the remaining two-thirds using traditional pencil-paper methods. The other three sections only used traditional homework methods. This was done up to the first common departmental examination of the semester; approximately one-third of the semester.

Individual student grades earned on the first examination and individual GPAs were analyzed for changes in student performance. First, a linear-regression analysis of the entire population examination grades as a function of student GPA was done. This regression was then used to predict each student's examination grade given their GPA. Statistical analysis was then done on the difference between the actual examination grade and their GPA predicted grade (DELTA score). When the DELTA score is positive, the student exceeds what one would expect based on their GPA. Averages and standard deviations were then calculated for the treated and untreated students. On a 125 point scale, the treated student DELTA statistics were $N = 57$, mean = 2.02 and SD = 11.01 and for the untreated students they were $N = 73$, mean = -1.38 and SD = 13.54. Treated students scored on the average about 2 points more than one might expect and the untreated students underperformed by 1.38 points. The treated group then scored 3.4 points (2.7 %) better than the untreated students on the average. At the host institution, this equates to about one-third of a letter grade.

Verbal comments were also collected from these and other students. These can be categorized in five groups:

- It takes time to learn the interface

  - M-MODEL version 11 addressed these comments by making the interface more intuitive

- It takes more time to solve homework problems this way

  - Embedded elapsed time data were measured. A typical problem required 10–12 min to complete with a grade ranging from 86 % to 95 %.

- I got lost

  - Hints and tips have been added to address this issue
  - A video tutor was added in version 11

- M-MODEL requires too much detail

  - The author can only presume that students have been missing some of the important details required to understand a problem when solving problems by traditional pencil-paper methods

- I learned more because of feedback and opportunity to immediately correct my mistakes

  - Suggests a better understanding of model building in Engineering Mechanics

Mastering the interface of early versions of M-MODEL was the biggest problem that students reported. Some students elected to go through the optional practice problem for training and some didn't. Most students had no problems with the interface after completing 2–3 problems. This is not atypical of new software interfaces. Current version 11 has addressed the majority of the issues raised by the students.

7 Conclusions

M-MODEL has been demonstrated to be a flexible, computer-based problem solving tool based on the problem statement, graphical, given/find, pre-calculations, post-calculations and equations mental representations of a problem. It is consistent with the model building pedagogue of current mechanics education and problem-solving research findings.

Its design encourages higher-order cognitions required to bridge from linear, simple problems to more ill-defined problems on the engineering intellectual development spectrum as demonstrated by the gains in student performance. The flexibility of its problem solving procedure challenges students to think more deeply about problems and helps them develop the confidence they need to apply their own approach to a problem. This tool also removes the burden of computational procedures so that learners can focus upon model building which is so critical to solving mechanics problems.

A student assessment of M-MODEL was conducted. This experiment demonstrated a gain in student performance on course examinations that is consistent with that measured by Vanlehn et al. [18]. Several interface issues were raised by student users. These issues have been addressed in the current version 11 of M-MODEL. Interested readers may visit a demonstration version of M-MODEL by contacting the author for access privileges.

References

1. Norman, D.A.: Some observations on mental models. In Gentner, D., Stevens, A.L. (eds.) Mental Models. Taylor and Francis Group, New York (1983)
2. Wankat, P.C., Oreovicz, F.S.: Teaching Engineering. McGraw-Hill, New York (1993)
3. Woods, D.R.: An evidence-based strategy for problem solving. J. Eng. Educ. 89(3), 443–459 (2000)
4. Woods, D.R., Hrymak, A.N., Marsall, R.R., Wood, P.E., Crowe, C.M., Hoffman, T.W., Wright, J.D., Taylor, P.A., Woodhouse, K.A., Bouchard, C.G.: Developing problem solving skills: the McMaster problem solving program. J. Eng. Educ. 86(2), 75–91 (1997)
5. Gray, G.L., Costanzo, F.: Interactive dynamics: a collaborative approach to learning undergraduate dynamics. In: Proceedings of the American Society for Engineering Education Annual Conference and Exposition. Charlotte, NC (1999)
6. Mettes, C.T.C., Pilot, A., Roosink, H.J., Kramers-Pals, H.: Teaching and learning problem solving in science. J. Chem. Educ. 58(1), 51–55 (1981)

7. Litzenger, T.P., Vanmeter, M., Wright, M., Kulikowich, J.A.: Cognitive study of modeling during problem solving. In: Proceedings of The American Society for Engineering Education Annual Conference abd Exposition. Chicago, IL (2006)
8. Gramoll, K.: A web-based electronic book (Ebook) for solid mechanics. In: Proceedings of The American Society for Engineering Education Annual Conference and Exposition. Honolulu, Hi (2007)
9. Gramoll, K.: Courses. http://www.ecourses.ou.edu (2009)
10. Dollar, A., Stief, P.S.: Learning modules for the statics classroom. In: Proceedings of The American Society for Engineering Education Annual Conference and Exposition. Nashville, TN (2003)
11. Stief, P., Dollar, A.: An interactive web-based statics course. In: Proceedings of the American Society for Engineering Education Annual Conference and Exposition. Honolulu, Hi (2007)
12. Stief, P., Dollar, A.: Study of usage patterns and learning gains in a web-based interactive static course. J. Eng. Educ. **98**(4), 321–334 (2009)
13. Philpot, T.A.: Mechanics of Materials: An Integrated Learning System. Wiley, New York (2008)
14. Philpot, T.A.: The role of MDSolids in international mechanics of materials education. Int. J. Eng. Educ. **21**(2) (2005)
15. Stanley, R.: A way to increase the engineering student's qualitative understanding of particle kinematics and kinetics by utilizing interactive web-based animation software. In: Proceedings of The American Society for Engineering Education Annual Conference and Exposition. Austin, TX (2009)
16. Gray, G.L., Costanzo, F.: Toward a new approach to teaching problem solving in dynamics. In: Jonassen, D.H. (ed.) Learning to Solve Complex Scientific Problems. Taylor and Francis Group, New York (2007)
17. Schultz, K.G., Shelby, R.N., Treacy, D.J., Wintersgill, M.C., Vanlehn, K., Gertner, A.: Andes: an intelligent tutor for classical physics. J. Electron. Publishing **6**(1) (2000)
18. Vanlehn, K.C., Lynch, K., Schulze, J.A., Shapiro, R., Shelby, L., Taylor, D., Treacy, A., Weinstein, A., Wintersgill, M.: The andes physics tutoring system: lessons learned. Int. J. Artif. Intell. Educ. **15**(3), 147–204 (2005)
19. Suskie, L.: Assessment Student Learning: A Common Sense Guide (2nd ed.). Jossey-Bass, San Francisco (2009)
20. Bloom, B.S., Krathwohl, D.R.: Taxonomy of Educational Objectives: The Classification of Educational Goals. Longmans, Green, And Co., New York (1956)

Development and Evaluation of a Course Model to Prepare Teachers in Mobile-AssisteD-Teaching (MAD-Teaching)

Chun-Yen Chang and Yu-Ta Chien

Abstract This paper introduces the 4-phase cyclic MAGDAIRE model (Modelled Analysis, Guided Development, Articulated Implementation, and Reflected Evaluation) to prepare pre-service teachers for Mobile-AssisteD-teaching (MAD-teaching). In 2012, MAGDAIRE was deployed in a teacher preparation course. Thirty-five pre-service teachers participated in this course. The results suggested that MAGDAIRE efficiently and effectively assisted pre-service teachers' in crafting MAD-teaching modules. A total of 13 MAD-teaching modules applicable to classrooms, including 13 teaching plans, 14 learning sheets, and 6 sets of assessment, were developed. The results of peer assessment indicated that the pre-service teachers' MAD-teaching performances were acceptable, in terms of whether they could fluently use Apps in teaching and whether they could appropriately leverage Apps to support teaching methods and the subject matter to be taught; the mean scores of all the items on the peer assessment were beyond the median level (2.5; Min. = 1 and Max. = 4).

Keywords Mobile learning · Mobile-assisted teaching · Teacher education · Science teaching

1 Introduction

Mobile devices, such as smart phones and tablet PCs, have received substantial attention from the community of educational technology research in the past decade; this rapidly growing interest is due to the great potential mobile devices have in

C.-Y. Chang · Y.-T. Chien
Graduate Institute of Science Education, National Taiwan Normal University, Taipei, Taiwan
e-mail: yutachien@ntnu.edu.tw

C.-Y. Chang · Y.-T. Chien
Science Education Center, National Taiwan Normal University, Taipei, Taiwan

C.-Y. Chang (✉)
Department of Earth Sciences, National Taiwan Normal University, Taipei, Taiwan
e-mail: changcy@ntnu.edu.tw

© Springer International Publishing Switzerland 2015
V.L. Uskov et al. (eds.), *Smart Education and Smart e-Learning*,
Smart Innovation, Systems and Technologies 41,
DOI 10.1007/978-3-319-19875-0_15

163

making learning more flexible than ever before [1–3]. In fact, a large amount Applications (Apps) for mobile devices has been developed to support teaching [4]. With the price drop of tablet PCs in 2012, the USA government has challenged schools and companies to get digital textbooks into students' hands within five years [5]. However, the renovation of teacher education, to foster the teachers who are capable of Mobile-AssisteD-teaching (MAD-teaching), lags far behind the popularization of mobile devices in schools. Thus far, little research has been done to investigate how to build up teachers' capability in the utilization of mobile devices for teaching [1–3, 6]. Therefore, this paper introduces the 4-phase cyclic MAGDAIRE model (abbreviated from Modeled Analysis, Guided Development, Articulated Implementation, and Reflected Evaluation) to prepare pre-service teachers for MAD-teaching. The practical outputs of MAGDAIRE are briefly reported.

2 An Innovative Model to Build up Teachers' ICT Capabilities

MAGDAIRE is based on our practice and research, in the field of teacher education, to build up pre-service teachers' Information and Communication Technology (ICT) capabilities [7–9]. Diverging from the conventional lectures on ICT skills in teacher education programs, MAGDAIRE aims to launch a pre-service-teacher-centred community of ICT-assisted teaching module development. MAGDAIRE engages pre-service teachers in collaboratively designing, developing, and implementing ICT-assisted teaching modules, with the support of a mentoring team which consists of educational researchers, in-service teachers, and educational technology developers. In 2012, MAGDAIRE was revised and deployed in an eighteen-week-long course of the National Taiwan Normal University's (NTNU) teacher preparation program. The course was named "Computers and Teaching," and its main objective was to foster pre-service teachers' capabilities in MAD-teaching. Thirty-five pre-service teachers and a six-man mentoring team participated in this course. Each of the participating pre-service teachers was equipped with an iPad (Apple, Cupertino, CA, USA) to design, develop, and implement MAD-teaching modules. Each phase of MAGDAIRE is briefly described below.

2.1 Modeled Analysis

To begin, the mentoring team collaboratively demonstrates MAD-teaching scenarios of their own design for pre-service teachers to experience. Pre-service teachers are then encouraged to brainstorm with peers and the mentoring team about what features of the iPad are pedagogically powerful in these scenarios. Two of the scenarios are briefly described in the following to sub-sections (in these scenarios, the mentoring team acts as teachers, and pre-service teachers act as students).

MAD-Teaching Scenario I: Teaching Moon Phases with iPad. In this stage, a teacher gives students a lecture about how to operate SkyView Free (Terminal Eleven LLC, Albuquerque, NM, USA), a sky map App for iPad, to trace today's moon in the classroom. The teacher points the camera of iPad to the ceiling of the classroom, and then the real-time 3D graphics of the sky objects are shown on the iPad. As shown in the left panel of Fig. 1, students engage in using SkyView Free to collaboratively predict and record the variation of the moon phases of the coming days. Each student is provided with a learning sheet to write down the data of his/her virtual astronomical observation. Students are encouraged to share their ideas on the cause of moon phase variation with peers and the teacher by using iPad to post messages on Moodle (an online learning management system) and face-to-face discussions. An online quiz is also sent to every student via iPad to stimulate students to monitor their own learning progress.

MAD-Teaching Scenario II: Teaching the Rocks in Life with iPad. In this scenario, a teacher shows students a map of six scenic spots in Taiwan, and asks, "What kinds of rocks might you find at these scenic spots?" Students are asked to use Socrative (Socrative, New York, NY, USA), an instant response system App for iPad, to write down their answers and send their answers to the teacher. The teacher then uses Socrative to list all the students' answers and invite students to poll the best answer. As shown in the right panel of Fig. 1, students engaged in collaboratively using AR Rock (Science Education Center of NTNU, Taipei, Taiwan), an augmented reality display of rocks and minerals for iPad, to verify and categorize their own answers.

Fig. 1 Example of MAD-teaching scenarios: the pre-service teachers experienced

2.2 Guided Development

Pre-service teachers are assigned in groups according to the subject of the teaching license they are pursuing (e.g., the earth science pre-service teachers are arranged into one cohort). Each group chooses a unit of their future teaching subject for developing an MAD-teaching module. Each group then searches appropriate Apps and devises relevant activities as well as assessments to complement the module they plan out. The mentoring team provides pre-service teachers with suggestions and solutions to the problems occurring in module development.

2.3 Articulated Implementation

Each group of pre-service teachers presents and executes their own MAD-teaching module in a classroom setting. The mentoring team induces pre-service teachers to explicate their intentions and reasons for every step they perform. This process forces the pre-service teachers to make their unspoken thoughts explicitly clear in MAD-teaching. This also enables them to experience others' perspectives on MAD-teaching across different contexts.

2.4 Reflected Evaluation

Pre-service teachers are asked to compare their own performances with their peers, and then give comments on others' performances. The mentoring team further encourages each group to reflectively form hypotheses for refining their MAD-teaching modules in the next cycle of MAGDAIRE.

3 Outputs of MAGDAIRE

The effectiveness of MAGDAIRE, to improve pre-service teachers' self-efficacy, knowledge, and skills of ICT-assisted teaching, has been validated in our previous studies conducted in 2010 and 2011 [7, 8]. This section briefly reports the outputs of the deployment of MAGDAIRE in 2012.

3.1 M-Teaching Modules Applicable to Classroom Teaching

A total of 13 MAD-teaching modules applicable to classroom teaching, including 13 specific teaching plans, 14 learning sheets, and 6 sets of assessment, were developed by the pre-service teachers within MAGDAIRE. Three of the MAD-teaching modules are briefly described in Table 1.

Table 1 MAD-teaching modules developed within MAGDAIRE

Module	App used in module	Module characteristic
Torque and barycentre (high school physics)	TrialX2 Winter: Free version, developed by Deemedya (Tel Aviv, Israel)	• Engage students in guiding their riders across ramps, jumps, barrels, and obstacles to get high scores in TrialX2 Winter • Ask students to share their way to get high scores in TrialX2 Winter with peers • Explain the concepts of torque and barycentre to students and encourage students to apply these concepts to get higher scores TrialX2 Winter • Engage students in figuring out what phenomena in TrialX2 Winter contradict Newtons' laws of force
Acid-base reaction (high school chemistry)	Chemist: $4.99 USD, developed by THIX (http://thixlab.com)	• Guide students to be familiar with the instruments displayed in the virtual chemistry lab of Chemist • Show students how to prepare $Ca(OH)_2$ and H_2SO_4 solutions in Chemist • Engage students in mixing $Ca(OH)_2$ and H_2SO_4 solutions and observing the acid–base reaction in Chemist • Let students mix various acids and bases at their will to generate explosions in Chemist! Ask students to record the conditions which will cause explosions for promoting their awareness of lab safety
Atoms, elements, and isotopes (high school chemistry)	Nuclear: Free version, developed by Escape Velocity Limited (Surrey, GU, UK)	• Leverage Nuclear to introduce atomic symbols in the periodic table to students • Show students how to produce the isotopes of hydrogen by manipulating the amount of protons, neutrons, and electrons in Nuclear • Guide students to observe the radiation of tritium • Engage students in manipulating Nuclear to find out the radioactive isotopes of carbon

3.2 M-Teaching Performance Beyond the Median Level

Each group of the pre-service teachers implemented MAD-teaching with their modules in classrooms by the end of the semester. Each of the pre-service teachers was asked to evaluate others' MAD-teaching performances. The focuses of the peer assessments were to determine whether the pre-service teachers could fluently use Apps in teaching and whether the pre-service teachers can appropriately leverage Apps to support teaching methods and the subject matter to be taught. As shown in Table 2, the pre-service teachers' MAD-teaching performances were acceptable; the mean scores of all the items on the peer assessments were beyond the median level (2.5; Min. = 1 and Max. = 4).

Table 2 Mean scores of peer assessment on pre-service teachers' MAD-teaching

Items	Mean[a]	SD
The fluency of the use of Apps in teaching	3.0	0.3
The fitness of Apps for the subject matter to be taught	2.8	0.3
The fitness of Apps for teaching methods	2.7	0.2
The integration of Apps, subject matter, and teaching methods	2.8	0.3

[a]Each item was rated on a 4-point scale (ranged from 1 to 4). The scores of peer assessment highly correlated with the scores judged by the mentoring team (Pearson' $r = 0.7$)

3.3 Awards of Outstanding ICT-Assisted Teaching

Three of the pre-service teachers who participated in MAGDAIRE voluntarily formed a group to enter the 2012 ICT-assisted micro-teaching competition, which was held by the Office of Teacher Education and Careers Service of NTNU. This group leveraged MAD-teaching modules and won second place in the competition. It should be noted that the champion of the competition also participated in MAGDAIRE in 2010.

4 Concluding Remarks

By leveraging the distributed intelligence between pre-service teachers, educational researchers, in-service teachers, and educational technology developers, MAGD-AIRE can efficiently and effectively produce MAD-teaching modules applicable to classrooms. These MAD-teaching modules also can function as paradigms to enrich the scenarios of the Modeled Analysis phase of MAGDAIRE or as examples for pre-service teachers to refine.

The mentoring team is currently building an online platform to make the MAD-teaching modules developed within MAGDAIRE become open-access to in-service teachers. We believe this step could provide in-service teachers with convenient and flexible solutions for classroom teaching. Moreover, it may attract more in-service teachers who are interested in MAD-teaching to contribute their ideas and join our mentoring team. Longitudinal and more in-depth studies, on the impact of MAGDARIE on pre-service teachers' competence in MAD-teaching, are currently being conducted in Taiwan.

References

1. Hung, J.L., Zhang, K.: Examining mobile learning trends 2003–2008: a categorical meta-trend analysis using text mining techniques. J. Comput. Higher Educ. **24**, 1–17 (2012)
2. Hwang, G.J., Tsai, C.C.: Research trends in mobile and ubiquitous learning: a review of publications in selected journals from 2001 to 2010. Br. J. Educ. Technol. **42**, E65–E70 (2011)
3. Wu, W.H., Wu, Y.C., Chen, C.Y., Kao, H.Y., Lin, C.H., Huang, S.H.: Review of trends from mobile learning studies: a meta-analysis. Comput. Educ. **59**, 817–827 (2012)
4. Golinkoff, R.M., Hirsh-Pasek, K.: http://www.huffingtonpost.com/roberta-michnick-golinkoff/educational-apps_b_1632281.html
5. Hefling, K.: http://finance.yahoo.com/news/Challenge-schools-Embracing-apf-3730035119.html
6. Rushby, N.: Editorial: an agenda for mobile learning. Br. J. Educ. Technol. **43**, 355–356 (2012)
7. Chang, C.Y., Chien, Y.T., Chang, Y.H., Lin, C.Y.: MAGDAIRE: a model to foster pre-service teachers' ability in integrating ICT and teaching in Taiwan. Australas. J. Educ. Technol. **28**, 983–999 (2012)
8. Chien, Y.T., Chang, C.Y., Yeh, T.K., Chang, K.E.: Engaging pre-service science teachers to act as active designers of technology integration: a MAGDAIRE framework. Teach. Teach. Educ. **28**, 578–588 (2012)
9. Chien, Y.T., Chang, C.Y.: Developing preservice teachers' sensitivity to the interplay between subject matter, pedagogy, and ICT. In: Hsu, Y.S., Yeh, Y.F. (eds.) Technological Pedagogical Content Knowledge (TPACK)—Science Teachers in Digital Era. Springer, Singapore (In press)

Block Magic: A Prototype Bridging Digital and Physical Educational Materials to Support Children Learning Processes

Andrea di Ferdinando, Raffaele di Fuccio, Michela Ponticorvo and Orazio Miglino

Abstract Block Magic is a prototype for educational materials developed in a successful European research project under the framework of LLP-Comenius programme. It aimed at creating a bridge between physical manipulation and digital technology in education. Block Magic developed a functional prototypal system that enhanced the Logic Blocks Box. The prototype is made up of an active desk/board able to recognise concrete blocks equipped with the RFID passive tag and to communicate with a PC, an augmented reality system. Preliminary trials with Block Magic prototype were run in various schools in Germany, Greece, Italy and Spain, involving children aged 3 to 7. Results confirmed Block Magic educational platform effectiveness in educational context.

Keywords Augmented reality systems · Embodied Cognition · RFID/NFC Technology · Hybrid educational materials · Technology enhanced learning

1 Introduction

A considerable part of learning/teaching processes that involve humans is based on technology. For example the book, the pen, the blackboard that belong to our everyday environment and are used as our mind and body extensions, are indeed technologies designed in the latest periods of human evolutionary history. Technology is a mirror of the times, in education too: evolution (and update) of didactic methods run in parallel with technological evolution. This undeniable fact is even more evident in our historical period. The current acceleration in technological development offers new opportunities to support learning processes in those life periods, as infancy, which are, at present, only partially involved in the systematic application of innovative learning technologies.

A. di Ferdinando · R. di Fuccio · M. Ponticorvo (✉) · O. Miglino
Department of Humanistic Studies, University of Naples "Federico II", Naples, Italy

© Springer International Publishing Switzerland 2015
V.L. Uskov et al. (eds.), *Smart Education and Smart e-Learning*,
Smart Innovation, Systems and Technologies 41,
DOI 10.1007/978-3-319-19875-0_16

1.1 Background

Children know the world around them using their body even before their motor functions (walking) and cognitive functions (logical and linguistic) are fully developed. Every action, touching, moving, pointing, matches and supports their learning processes. Along the growth process, concrete manipulative acts are gradually simulated in human mind and become symbolic and cognitive acts. However the body use, in particular hands use, is a latent and fundamental resource in learning processes that strongly emerges when the environmental conditions allow it. In recent years, the Embodied and Situated Cognition Theory, ESCT, [1] proposed an explanation of how our sensory- motor interactions with the environment structure our neuro-cognitive organization. The ESCT approach, in our opinion, focuses on another fundamental aspect too: these sensory motor interactions happen, especially for humans, in a social and cultural context that proposes objects, artefacts, technologies and concrete or abstract cultural substrates [4].

Well-established psycho-pedagogical approaches (such as the Montessori approach) have proposed well- structured didactic methodologies which focus on children active involvement. These methodologies rely on learning materials (educational toys), which foster both learning and teaching processes by stimulating the manipulation of concrete objects and peer group cooperation. As a result, such approaches stimulate active education, particularly fit for cognitive and social skills acquisition, such as cooperation, collaborative learning, creative and critical thinking, problem solving, digital literacy and so on. These psycho-pedagogical methodologies use structured materials, such as logic blocks, teaching tiles, abacus, physical representations of letters and numbers, etc., aimed at training specific perceptive, cognitive and motor functions. Inside the learning environment, such didactic materials are disposed and used according to specific didactic criteria which foster their spontaneous use by pupils, in order to empower the proactive function in learning processes. In these environments natural cooperative dynamics tend to emerge. However, because the above-mentioned psycho-pedagogical practices are aimed at empowering learning/teaching processes focusing on pupils active participation, as well as on customizing educational interventions, they tend to be extremely expensive. These educational practices imply constant supervision by adults (educators, trainers, teachers, parents, caregivers, etc.), that carry out (discretely) the educational play, and play a role as directors, as well as mediators in the learning processes. Therefore, mass diffusion of this kind of practice is considerably limited.

Some technologies, e.g. augmented reality, RFID/NFC sensors [10] are natural candidates to represent the bridge paradigm [7] to unite the manipulative approach and touchscreen technologies, as widely explored by the Block Magic project [6], from now onwards BM that took place in four European countries. BM is an example of what we can call STELT: Smart Technologies to Enhance Learning and Teaching [8] that links physical and digital applications [9], reported in Fig. 1.

1.2 Research Objectives

In what follows we describe the BM platform and the first trials that were run to test its effectiveness. The research questions that guided our effort were mainly the following: *Is BM tool attractive for children? Does BM contributes to enhance specific cognitive skills? Is BM able to convey social skills learning? Can BM be compared with traditional materials in support children learning processes?*

Fig. 1 STELT platform functional representation

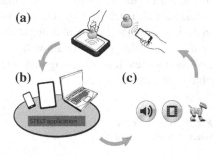

In next sections we are going to illustrate how we tried to aswer these questions and what the reply was. In Fig. 2 the BM kit is shown.

Fig. 2 The BM kit including a set of 48 Magic Blocks, Magic Board, Teachers Manual, CD containing all required software and scripts for all learning activities. See text for explanation

2 Materials and Method

In this section we will describe in detail the BM platform and the trials that were run to test it in four different European countries.

2.1 BM Platform

The BM platform consisted of a set of magic blocks (48 traditional logic blocks), a magic board/tablet device and a specific software (see Fig. 3). It is based on STELT platform introduced above and links together smart technologies and physical material to support children learning processes. It unites the manipulative approach and touch-screen technologies. BM materials derive from structured materials, classically used in education. Structured materials have a fixed numbers of n elements, m categories and rules to connect single parts that represent the structure. Logic blocks, cards, teaching tiles, etc. are structured materials typical examples. These materials promote analytical thought, as they segregate single qualities (e.g. dimension, shape, color, etc.) and allow first to focalize attention on an object single part and then to develop clustering and serialization ability in order to understand the objects features.

Fig. 3 BM platform functional representation

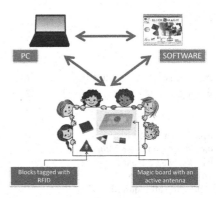

RFID technology The technology used in the BM project is the RFID/NFC, Radio Frequency Identification/Near Field Communication. RFID system consists of an antenna and a transceiver, which is able to read the radio frequency and transfer the information to a device, and a small and low cost tag, which is an integrated circuit containing the RF circuitry and information to be transmitted. These technologies are simple to be used, so they are interesting for applications rather than on technical level.

Magic blocks and the magic board The BM teaching kit consisted of a set of magic blocks (48 traditional logic blocks), a magic board/tablet device. Magin blocks are dervied from Logic Blocks which are didactic materials used worldwide in kindergartens and primary schools [2]. They are made up of a set of blocks (usually 48 pieces) divided in four groups according to different attributes: geometric shape (triangular, squared, rectangular and circular), thickness (thick and thin), color (red, yellow and blue) and dimension (big and small). The BM project proposed an hybrid version that allows an enhancement of traditional logic blocks, equipping them with RFID tags. This configuration permits to a PC or a table, with BM software installed on, to connect with BM Magic Table, another relevant BM material. The Magic Table has an hidden antenna that recognizes each block, sends a signal to the PC/tablets, and produces a feedback coherently with pupils learning path. Each augmented magic block had an integrated/attached passive RFID sensor for wireless identification of each single block. A specially designed wireless RFID reader device, an active board, is used which could read the RFID of a block and transmit the result to the BM software engine.

BM software The BM system aimed to stimulate and teach different skills such as logic, mathematics, languages, etc. therefore the described BM enriched blocks together with the Magic Table are complemented with a software that includes a series of exercise that researchers involved in BM project built on teachers feedback and on their previous experience in pedagogy. The final software was developed through an iterative process: first children and teachers used the exercises and assessed them, then the results were collected together with feedbacks from direct observation and focus groups. This was done by all project partners to define final exercises version. The BM software engine is mainly formed by two parts: the first one is devoted to receiving input from the active board and generating an "action" (aural and visual). These actions implement the direct feedbacks the user can receive interacting with the system. These feedback are regulated by an Adaptive Tutor System embedded that ensures autonomous interaction between the user and the system, receiving active support, corrective indications, feedback and positive reinforcement from the digital assistant on the outcome of the actions performed. Adapting tutoring systems [3, 5] are an Artificial Intelligence application that provides instruction that are tailored on individual learners needs. Traditional applications used in education, indeed, are not individualized to learner needs, but are rather static and rule-based (IF Question X is answered correctly, proceed to question Y, otherwise go to question Z; and so on). The learner abilities are not taken into account. While these kinds of applications may be somewhat effective in helping learners, they do not provide the same kind of individualized attention that a student would receive from a human tutor.

The second software component is devoted to customization too, but it is dedicated to teachers, educators, etc. allowing them to choose the exercises to be proposed to the child, focusing the attention on the skills the child needs to train more. The BM software moreover can collect data about the exercises.

2.2 Participants

Trials involved 4 different schools, 257 students, 2 children with special needs and 10 teachers. Schools were located in 4 different countries (Italy, Germany, Spain and Greece). In detail the trial made in Italy was addressed for children with special needs, with little differences in the protocol and test contexts, described later. Children involved were between 2.5 and 7 years old, attending the early years of primary school and kindergarten.

2.3 Method

To test BM materials in educational context, the BM project included two different scenarios: (1) Individual Game Scenario and (2) Social Game Scenario. In the first one, learners had to solve a task using logical, mathematical, creative, strategic and linguistic skills, whereas in the second one social skills, under group play guise, were required to find game solution. Also teachers were involved with pupils: their role was to create and maintain an adequate environment for BM sessions. The trials were run in a specific setting: a dedicated rooms, different from the classroom where pupils attend traditional lessons. In these rooms large workplaces were prepared with the BM kits, available for free game and manipulation. The teachers, who had already experienced BM platform, set the software choosing the correct level for children in the class. The trials had no pre-defined exercises for children who could skip exercises if they considered it problematic or boring. A trial session typically started with introduction of BM logic blocks and Magic Table device by the teacher, giving pupils the opportunity to play freely with them and use the materials the way they preferred. The trial continuation was different according to the two scenarios introduced above. In the Individual scenario, the teacher acted as an external observer and supported a single pupil when he/she asked for help. This way, the child had to perform exercises autonomously. In the Social scenario groups composed by a minimum of 4 to a maximum of 6 learners were involved and the teacher had a more active role in the session, providing support, observing and/or creating obstacles. Moreover they had to observe children behavior in order to complement these qualitative observations with session results recorded by BM software. This allowed to define a learning curve for each child and to obtain information about intra- group interaction, focusing on team building, leadership, verbal and non-verbal communication. The session had a maximum length of thirty minutes. Data were collected by BM system, including record sheets and videotapes. In the case of children with special needs there was an additional preliminary phase of pre-training that allowed children to understand the task. After each session, the teacher, for all scenarios, had to analyze results using BM software that shows and lists results for each session and for each child. This way she/he could analyze the session and tune the educational goals for every child. For example, if the teacher noted that a pupil lacked in linguistic skills, she/he might modify the proposed exercises to train these skills more intensively. The trial consisted of at least five twenty-minutes long sessions. During sessions, researchers run observations and collected data, without any support role. After sessions, researchers conducted face-to-face interview with teachers based on a semi-structured questionnaire. This questionnaire is formed by 21 items with responses on a Likert scale. Researchers also run collective focus groups with teachers that lasted 60 min. Teachers interviews and focus groups, together with observations were focused on four aspects: BM attractiveness, BM ability to contribute to specific cognitive skills, to strengthen social and working in group skills, BM ergonomics and comparison between BM platform and traditional materials.

3 Results

In this section we report the results we obtained in the four European trials both from teachers (questionnaire and focus groups) and from children.

3.1 Indications from Teachers

BM attractiveness, emerged strongly: the tool is very attractive for children for many reasons. They especially liked to use the tablet and the computer, and were attracted by visual and aural stimulation as well as the mascot Block. It is motivating for the students to use a computer-based system and that the manipulation of real objects makes it even more fun. Teachers think that children found both visual and aural presentation of the tool attractive and the use of text, graphics, sound, and pictures was perceived as balanced so the children didnt get bored. They stayed happily until the end of each session and even wanted to continue playing. Teachers also noticed that children liked to hear their names from the computer and felt like participating in a real game (in the single player scenario). Also receiving an appropriate feedback was crucial to keep an high motivation level. The researchers noticed that in all cases teachers support and encouragement was essential, otherwise when children couldnt solve an exercise and find the solution wanted to abandon their task. Teachers intervened and tried to increase childrens motivation in case they failed in an exercise. This is especially true for 4 years old children and for the social game scenario. Children with special needs find the tool attractive too. In the Italian trial, in fact, teachers stated that the tool is appropriate for children with special needs mainly because they are motivated on activities with multisensory stimulation. Indeed, children were highly motivated. They stayed on their tasks until it finished and none of them appear to be frustrated. In their opinion BM is an interesting tool as it reduces prompting of the teachers and the engagement with BM produces a large increase in stimulation level. Moreover, teachers stated that computer based learning materials enables students with special needs to learn more effectively because they prefer to play independently and they can increase the quality of life through input and output opportunities: as the other children involved in the trials, children with special needs found the verbal reinforcement that encouraged them to play very appealing.

The researchers investigated BM use in relation with specific skills: in teacherss opinion contribute to develop specific cognitive skills: in particular it is fit to improve mathematical and logical skills. All teachers accepted that the tool offers a variety of activities that encourage children to develop mathematical skills. However, in order to find the solution to mathematical problems children, especially the younger ones, needed teachers help. Besides that, teachers believe that children at this age cannot develop and follow a strategy to solve a problem. For this reason, they underlined that children often used a trial-and-error strategy. Also imagination is stimulated by BM. In detail the Creative Drawing, Logic Train and Slice the Shape were identified as the exercises that motivates children to use their imagination. On the other side, this tool doesnt seem to be effective about creativity. The social scenario provide effective stimuli to improve social skills and the ability to work in groups, even if teachers believe that BM platform is less effective in this respect.

Another point that was treated was BM ergonomics: children are able to interact autonomously with the tool, but there is some difference about age. There are some problems with 4 years old children, as they better interact with BM when they are 5 years old. Younger children need a lot of work with a real adult, either teachers or parents. On the other side, teachers suggest that some exercises could be fit even for children above the age of 7, especially those that have problems and/or special needs. Some students cannot follow the pace of the others and such system can be a suitable additional support for their learning.

About BM ergonomics, many teachers suggested to improve game instructions that were often too long or complex to be fully understood, to solve the problem with noise in the classroom and to fix some technical problems the prototype had.

Then the questionnaire and the focus group compared the use of BM with traditional blocks: teachers agreed that the most relevant aspect was the feedback provided by BM system that allowed many children, especially the older ones to interact autonomously with exercises. Block Magic meets the new IT generation and is an interactive tool but there are constraints in terms of creativity and imagination, that, on the contrary, are better stimulated by traditional blocks. Moreover teachers underlined that Block Magic creates more possibilities for the teachers than the traditional logical blocks: there is a wider variety of exercises, some of which are more difficult to play in traditional settings. With Block Magic children can work (almost) unattended and they can spend more time with it. It is useful to underline that these aspects emerged in all trials, thus meaning that children interact with BM platform in a similar way across different countries and cultures.

3.2 Children Improvement Along Sessions

Children interacting with BM get better along the sessions, both in individual and social scenario. The preliminary data we present refer to all trials.

Figure 4 shows childrens performance during the first and the last day sessions. This comparison is complicated by the fact that children didnt play the same exercises the first and the last days, but it is possible to see the improvement tendency. Both the group and a single example child have increased the rate of perfectly solved exercises along sessions over the time. The group increased the rate of perfectly solved exercises from 45.0 % to 80.0 % and the single child increased it from 54.0 % to 71.0 %. In the social game scenario we have compared the group means in the first and the last day sessions: the difference is statistically significant $t(15) = 2,537; P > 0.05$.

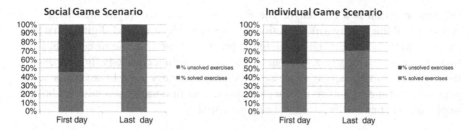

Fig. 4 Percentage of solved and unsolved exercises in first day and last day sessions. On the left results about groups and on the right results about single child are reported

The BM system also allows to record how many exercises have been completed or skipped during the trials by children. Children skipped the exercises either because they failed to solve them or did not listen to the instructions again. As underlined before, some instructions were not fully understood by children. Among the exercises that children played completely some were solved perfectly (without any mistake) and some others were solved faulty (1–9 mistakes). Figure 5 shows the distribution of skipped exercise, faulty and perfectly solved exercises.

Fig. 5 Distribution of skipped exercise, faulty and perfectly solved exercises

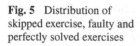

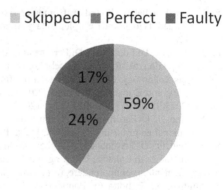

4 Discussion

In this paper we have described the BM platform, its concept and implementation and the results coming from the first trials with children. These data indicate that BM can be an interesting and effective tool to be integrated in school curricula. Even if it is clear that an help from the teacher is still needed, especially for younger children, it is nevertheless true that BM allows to run activities with logic block in a more autonmous and therefore economic way. Moreover BM is very appealing for children and this aspect cannot be neglected in a school context where capturing children attention is a challenge. The BM system exploits Augmented Reality in order to building-up hybrid educational materials (physical and digital) that link together well-known psycho-pedagogical practices based on the direct manipulation of concrete objects (not just touching a screen) with technology thus enhancing the overall learning/teaching processes for children in early ages. In next phases, our goal is to test the prototype on a wider sample and in comparison with traditional materials to better clarify its educational potential.

Acknowledgments The Block Magic project has been funded with support from the European Commission under Life Long Learning programme Comenius.

References

1. Clark, A.: Supersizing the Mind: Embodiment, Action, and Cognitive Extension: Embodiment, Action, and Cognitive Extension. Oxford University Press, Oxford (2008)
2. Dienes, Z.P.: Large Plastic Set, and Learning Logic, Logical Games. Herder and Herder, New York (1972)
3. Freedman, R.: What is an intelligent tutoring system? Intelligence **11**(3), 1516 (2000)
4. Jonassen, D., Land, S. (eds.): Theoretical Foundations of Learning Environments. Routledge (2012)
5. Larkin, J., Chabay, R. (eds.): Computer Assisted Instruction and Intelligent Tutoring Systems: Shared Goals and Complementary Approaches. Lawrence Erlbaum Associates, Hillsdale (1992)
6. Miglino, O., Di Fuccio, R., Di Ferdinando, A., Barajas, M., Trifonova, A., Ceccarani, P., Dimitrakoupoulou, D., Ricci, C.: BlockMagic: enhancing traditional didactic materials with smart objects technology. In: Proceedings of the International Academic Conference on Education, Teaching and E-learning in Prague 2013 (IAC-ETeL 2013), Prague, Czech Republic, 17–18 Oct 2013. ISBN 978-80-905442-1-5 (2013)
7. Miglino, O., Di Ferdinando, A., Di Fuccio, R., Rega, A., Ricci, C.: Bridging digital and physical educational games using RFID/NFC technologies. J. e-Learn. Knowl. Soc. **10**(3) (2014)
8. Miglino, O., Di Ferdinando, A., Schembri, M., Caretti, M., Rega, A., Ricci, C.: STELT (Smart Technologies to Enhance Learning and Teaching): una piattaforma per realizzare ambienti di realt aumentata per apprendere, insegnare e giocare. Sistemi intelligenti **25**(2), 397–404 (2013)
9. Miglino, O., Gigliotta, O., Ponticorvo, M., Nolfi, S.: Breedbot: an evolutionary robotics application in digital content. Electron. Libr. **26**(3), 363–373 (2008)
10. Shepard, S.: RFID: Radio Frequency Identification. McGraw Hill Professional, New York (2005)

Part III
Smart e-Learning

Game Mechanics Used for Achieving Better Results of Massive Online Courses

Liubov S. Lisitsyna, Alexander A. Pershin and Matvey A. Kazakov

Abstract This paper describes an approach to achieving better results of massive online courses by using game mechanics in the process of online education. The results of creating and using an interactive massive online training course for web developers in HTML5 and CSS3 in Russia are presented. The course uses new game mechanics of perfectionism and cohort ratings. The results of experimental research confirming positive effect of such game mechanics on the course performance (time of course completion, achieving better results in training, percentage of successful course completion) are provided.

Keywords MOOC · Cohort ratings · Perfectionism · HTML5 · CSS3 · Training of user web interface developers

1 Introduction

With online education development and worldwide market of the education services using the MOOC technology (Massie Open Online Courses) emerging, the attention of the developers of such courses is drawn to the issues of achieving better results (percentage of successful trainees of such courses is not higher than 5 % [1]). The situation in the IT labor market which shows constant lack of qualified

L.S. Lisitsyna (✉) · A.A. Pershin · M.A. Kazakov
ITMO University, Kronverkskiy Pr. 49, Saint Petersburg 197101, Russia
e-mail: lisizina@mail.ifmo.ru

A.A. Pershin
e-mail: pershin_zan@mail.ru

M.A. Kazakov
e-mail: matvey.kazakov@gmail.com

© Springer International Publishing Switzerland 2015
V.L. Uskov et al. (eds.), *Smart Education and Smart e-Learning*,
Smart Innovation, Systems and Technologies 41,
DOI 10.1007/978-3-319-19875-0_17

personnel requires creating massive online courses that can fulfill the task of personnel training and continuing professional development on tight schedule. The main feature of such courses is focus on practical work with source code, creating stable skills for performing typical professional tasks. Usual video lectures for theoretical knowledge presentation and computer testing for knowledge control purposes fade into insignificance in this type of courses. Certainly, the course content should be actual and interesting. It is a necessary, but not always a sufficient condition for the course's success. The course should contain mechanisms that motivate the trainees to complete the course successfully.

Among the approaches that improve training motivation the one using game mechanics which make the process fascinating and exciting, stimulating to achieve better marks in the training should be distinguished. Unfortunately, modern online courses use only simple game mechanics, though the potential of game mechanics use for the course's performance is much higher. While creating an online course, strategic monitoring of course completion speed and final results should be done in order to compare and timely inform the trainees. The monitoring task is complicated by the fact that online courses are used not only in synchronous, but also in asynchronous format. In this case the training process does not have fixed terms and regular schedule, which increases the risk of abandoning the course by the trainee. This circumstance warrants new monitoring approaches development and correct monitoring results processing in the asynchronous training course.

This paper reflects the results of a research conducted at ITMO University on development and implementation of new game mechanics in interactive online course for web interface developers.

2 Statement

Nowadays the following tree types of MOOC courses can be distinguished.

1. Courses for career-guidance and subject field promotion.
2. Courses close in content to base disciplines of higher education institutions' programs (mathematics, physics, informatics, etc.)
3. Courses for applied technologies studies and practical skills training (such courses are mostly created for IT sphere).

All three types of MOOC courses use the so-called game mechanics for studies motivation improvement ("gamification" of the course), and amount and complexity of game mechanics is much higher in third type courses. The following game mechanics are used in MOOC courses [2, 3]:

1. points;
2. badges and achievements;
3. levels;
4. storytelling;
5. best students ratings;

6. gradual material presentation;
7. gradual complication.

Mechanics 1–4 are quite primitive: they affect external motivation of the trainees [4] and come down to encouragement by rewards for certain actions or best training materials execution, they are designed to create positive emotions in the trainees. Ratings of best students are ineffective due to massive character of the courses (tens of thousands of trainees). Mechanics 6 and 7 can be very effective, but their in-course implementation depends on the course's content.

The existing MOOC courses use many of the mentioned mechanics. But the implementation of these mechanics often does not bring the expected results. For example, Codecademy [5] uses only one type of practical exercises which offers the trainee to apply several stages of the source code alteration. It should be noted that the trainee has to click the check button after every alteration to the source code is made, which significantly reduces the effectiveness of the gradual material presentation. Moreover, manual start of the checking process reduces the speed of training. And absence of more complicated tasks in the course decreases the trainee's involvement. That's why the mechanic of gradual complication proves ineffective in this course. Furthermore, Codecademy does not have trainee ratings that can raise competitiveness in the trainees and accelerate the training process.

In total, it may be concluded that the existing online courses implement game mechanics that can influence only external motivation. That's why for improving internal motivation in online courses we set and accomplished the mission to design new effective game mechanics, built in an interactive online course "Creating web interfaces using HTML and CSS".

3 Course Content and Structure

The course "Creating web interfaces using HTML and CSS" contains 16 chapters, 276 practical exercises and 27 "challenges" (practical exercises close to typical professional tasks). The course is developed in asynchronous format, it is accessible at http://htmlacademy.ru. By January 2015, the course had more than 68 thousand trainees, which completed totally more than 4,6 million exercises and challenges.

This course was developed on the basis of methodology for expected training results planning [6] and ensures development of the practical skills for a web interface developer [7].

The course uses two types of training interfaces. Practical exercises interface (Fig. 1) contains:

1. pop-up window with theoretical information;
2. code editors on the left side of the page;
3. result visualizer ("mini-browser") on the right side;
4. sub-tasks block in the lower part of the visualizer;
5. control and navigation buttons.

The interface of practical tasks is similar to those of other online courses. However, it has two major differences. Firstly, this interface has automatic start of visualization and code checking which considerably accelerates the training process. Secondly, sub-tasks are checked separately, which allows successful implementation of the gradual material presentation mechanic.

Challenge interface (Fig. 2) differs from the practical exercises interface in that it has no subtask block and has two new blocks:

- completion scale which reflects example and result matching;
- block with example picture and differences map.

Challenges are a more complicated type of practical task that offers the trainee to achieve matching the example by any means possible. To make an analogy with video games, challenges are similar to "bosses". By using challenges one can create courses with difficulty level curve very close to that of a game and use the gradual complication mechanic to maximum effect.

Unlike practical tasks, challenges completion is assessed using the scale from 0 to 100 %. This type of assessment in challenges allows using involving game mechanic of perfectionism.

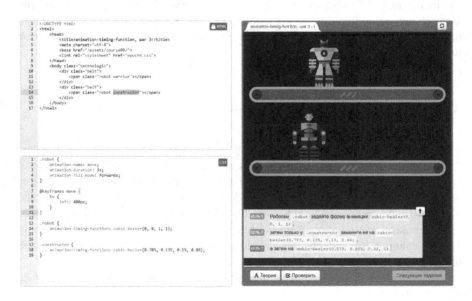

Fig. 1 Example of practical exercises interface

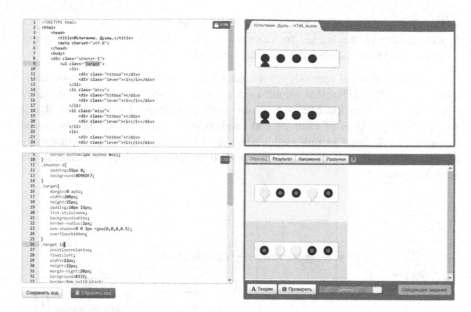

Fig. 2 Example of challenge interface

4 Perfectionism Mechanic in Online Course

Perfectionism mechanic works as follows: in order to pass the challenge, the trainee has to reach 90 % matching level between the result and the example, achieving 100 % match is not necessary. However, despite that, many trainees try to reach 100 % result, which is confirmed by challenge completion statistics. Badge "Complete 20 challenges" was received about 6400 times, and badge "Complete 20 challenges to 100 %" was received 4200 times. Perfectionism mechanic allows achieving deep involvement of the trainee into task completion, because for the trainee working with the task after its non-ideal completion is similar to a game, which is characterized by the following conditions:

1. absence of pressure (the task is completed already, further task performing can be done for pleasure);
2. absence of punishment (nobody will scold you if you do it better) [8].

While designing this course a new authentic methodology for assessing challenge results in the field of web interfaces development using HTML and CSS based upon comparison with the offered example was developed. It contains the following steps schematically shown in Fig. 3:

1. Adjustment of the so-called "browser engine" parameters which forms the pictures from HTML and CSS codes, so that the size of the images is the same.
2. Forming an example image from the code provided by the author of the challenge.

3. Forming an image from the code provided by the trainee.
4. Comparison of the images by calculating their similarity using a root-mean-square deviation method and forming a differences map.

From the traditional methods for images comparison – root-mean-square deviation method (RMSD), peak signal to noise ratio (PSNR), structural similarity index (SSIM) – the first one was chosen after experimental comparison of test images. Minimal level of example and result match which allows the task to be considered as fulfilled in this methodology is 90 %, which is due to commonly accepted limits of web development standard deviation (3-5 pixels against the original layout).

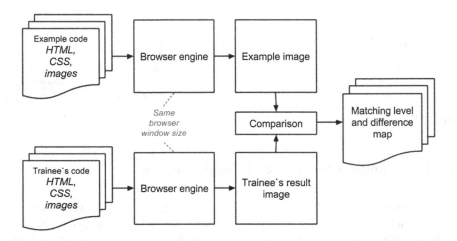

Fig. 3 Scheme of challenge assessment

5 Cohort Ratings in Online Course

Best students' rating is a game mechanic that adds an element of competition [9] to the studies and motivates the trainees to complete the courses up to the end or to fulfill the tasks better. There are some disadvantages of the existing ratings.

1. Excessive competition caused by big quantity of the rating participants in massive online courses. The difficulty of getting to the top of the rating reduces motivation to compete.
2. Big differences in the progress of the rating participants, starting the training at different time (in asynchronous training format). Impossibility to chase the leader reduces motivation to take part in the rating and can negatively affect the motivation for the training.

That's why for avoiding these disadvantages the approach to ratings based on cohort analysis methods was designed (further – cohort ratings).

The new method for building cohort ratings in an asynchronous interactive online course is based on dividing the whole amount of trainees to non-overlapping sets (cohorts) with a defined max power for each set. Trainees with short range of the course starting dates are joined in cohorts.

This algorithm allows to get rid of the following disadvantages of common ratings in asynchronous courses:

1. big differences in progress of the trainees;
2. excessive competition caused by big quantity of the ratings participants.

Best trainees rating in each cohort is generated by the following pattern: the trainees are graded to the total quantity of tasks completed, then to the average challenge completion result, and then to course completion speed. The rating is reflected in the trainee's personal profile.

This rating algorithm affects not only the speed of training, but also motivates to complete the tasks better, because this allows receiving higher rating. Consequently, two mechanics are combined: ratings encourage the perfectionism mechanic, and perfectionism makes the competition process more interesting.

6 Experiment Results

The first experiment studied the influence of perfectionism on the course's modules and the whole course completion percentage. For that purpose the statistics of course completion by the trainees which started the course from 01.04.2014 till 20.07.2014 was studied. From the total of 11319 trainees 417 completed the course, which is 3.7 % of the total amount. After that two sub groups were distinguished in this group:

(a) Trainees that completed first three challenges of the course up to 100 % ("idealists")
(b) Trainees that completed first three challenges of the course, but not all three of them were completed up to 100 % ("non-idealists").

The first sub group counted 2830 trainees, 367 of them completed the full course which is 13 % of the group amount. The second sub group counted 836 people, 50 of them completed the full course, which is 6 % of the group amount. Figure 4 shows the course completions graph for different sub groups. The graph (Fig. 4) reflects that the trainees who used the game mechanic of perfectionism (idealists) showed better marks of the test results.

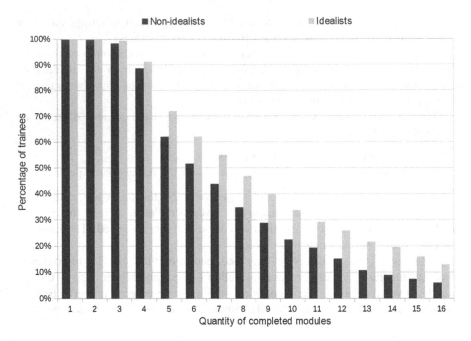

Fig. 4 Course results in different experiment groups

The second experiment studied the influence of perfectionism on time spent additionally on ideal completion of the challenge (Fig. 5a), on searching for additional materials (Fig. 5b), on achieving better challenge results (Fig. 5c). For that purpose a poll was created. There we 199 participants to the poll that passed at least 5 challenges. From the graphs presented on Fig. 5 you can see that the offered method motivates the trainees to spend more time in order to receive better results.

The third experiment studied the influence of cohort ratings on the speed of completion and results of the asynchronous course. This experiment was conducted using A/B testing on two groups of trainees: the first group was able to see the cohort rating on the personal profile page, and the second group wasn't. The group that had access to cohort rating showed better performance in the course – 4.7 % against 3.6 % of the group without cohort rating. The cohort rating presence had a more noticeable effect on the general course completion speed which is shown on the graph in Fig. 6. The trainees with cohort ratings reached the middle of the course on the average 50 h faster than those without the cohort rating.

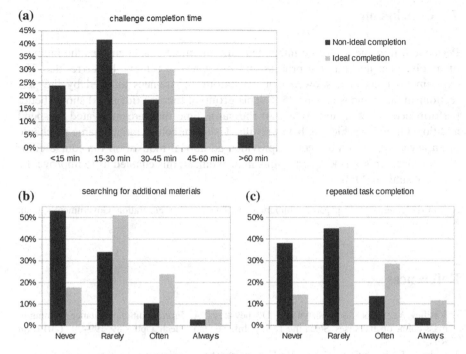

Fig. 5 Polling results in the second experiment

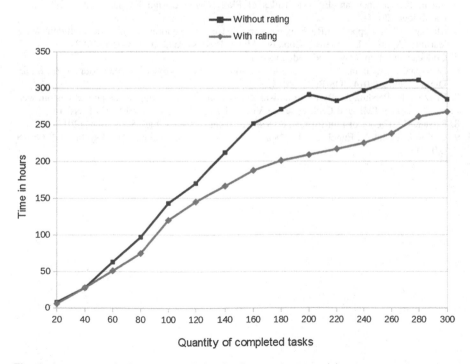

Fig. 6 Average speed of course completion by the experiment participants

7 Conclusion

Perfectionism game mechanic integrated into an online course ensures the increase of involvement into training process and motivates to achieve better results. The experiments conducted showed that the amount of trainees affected by the perfectionism mechanic is at least 25 %. This group of trainees (idealists) shows better performance (13 % against 6 %). At the same time they are motivated to spend additional time for achieving better results. Usage of cohort ratings game mechanic in an asynchronous online course provides additional motivation for course completion speed improvement and results improvement; this enabled us to improve the course results to 29 %.

Acknowledgments This paper is supported by The Russian Federation Government's grant # 074-U01.

References

1. Vasiliev, V., Stafeev, S., Lisitsyna, L., Ol'shevskaya, A.: From traditional distance learning to mass online open courses. Scientific Tech. J. Inf. Technol. Mech. Optics **89**, 199–205 (2014) (in Russian)
2. Lisizina, L., Pershin, A., Uskov, V.: New approaches to efficiency of massive online course. Sci. Tech. J. Inf. Technol. Mech. Optics **93**, 164–171 (2014). (in Russian)
3. Salen, K., Zimmerman, E.: Eric Rules of Play: Game Design Fundamentals. MIT Press, Cambridge (2004)
4. Malone, T.W., Lepper, M.R.: Making learning fun: a taxonomy of intrinsic motivations for learning. Aptitude Learn. Instr. Conative Affect. Process Anal. **3**, 223–253 (1987)
5. Codecademy. http://www.codecademy.com
6. Lisitsyna, L., Lyamin, A.: Approach to development of effective e-learning courses. In: Smart Digital Futures 2014, IOS Press (2014)
7. Lisizina, L., Pershin, A.: By the establishment of Fab lab for training developers of custom web interfaces in HTML and CSS. Distance Virtual Learn. **79**, 32–38 (2014). (in Russian)
8. Koster, R.: Theory of Fun for Game Design. O'Reilly Media, Inc., Sebastopol (2013)
9. Granic I., Lobel A., Engels R.: The benefits of playing video games. Am. Psychol. **69**, 66–78 (2014)

Maintaining Online Engagement in e-Learning Through Games Based Learning and Gamification Techniques

Geraint Lang

Abstract The development of e-Learning over the last ten years has provided extended opportunities for many to study partially or wholly online, and provided opportunities for those unable to attend full-time degree courses to study at a distance from their host university. This paper will discuss how the advent of smart devices has greatly extended the reach of online anytime anywhere learning. However, sustaining engagement online throughout the lifetime of a degree course can be challenging for the isolated student, who does not experience the face to face contact with fellow students and tutors. This paper looks at some of the research underpinning the use of games based learning and more recently gamification along with some of the reasons why the lure of playing games is so appealing. The paper also suggests possible ways that their utilization could be considered within the design of e-Learning courses and their facilitation.

Keywords Smart devices · Ultraversity · Online learning · Game based learning · Gamification

1 Introduction

Portable smart connectivity devices spawn not only new technology and their associate nomenclature such as Android and iOS that identify particular smart devices, but bring with them new lifestyle practices, for example recording the user's personal daily health, fitness and dietary records within specific apps downloaded to their device. Combine these lifestyle adaptations with their growing proliferation in society and they are now replacing such familiar practices as browsing through magazines while awaiting treatment in the dentist's surgery, and instead a more familiar sight is observing people fully engrossed interacting with

G. Lang (✉)
University of Chester, Chester, UK
e-mail: g.lang@chester.ac.uk

© Springer International Publishing Switzerland 2015   193
V.L. Uskov et al. (eds.), *Smart Education and Smart e-Learning*,
Smart Innovation, Systems and Technologies 41,
DOI 10.1007/978-3-319-19875-0_18

their handheld smart device, browsing the Web or connecting with friends and family on Facebook.

This all pervasive technology now permits almost 'always on' connectivity and contactibility. The ubiquitous smart phone (highly portable and reasonably compact) enables at a social level peers, families and couples to stay connected and locate one another through their devices' inbuilt GPS, working seamlessly with a mapping app; in a learning and teaching situation smart devices enable students on field trips to geo-tag locations for further study purposes, in addition to utilizing their devices' built-in cameras [1]; investigative project directors can share updates on their Twitter feeds, and in a related area, university lecturers can update their students about changes to lecture locations via social media, and more individually by sending emails directly to their students' smart phones or tablets, thus freeing them from the constraints of a computer hard-wired to a fixed data point.

Other examples of the growing revolution in smart device use come to us from the field of medicine, where some practitioners are now using this technology to consult with patients who may live many miles away from their doctor's consulting room using Skype and Facetime [2]. Reciprocally patients in turn are able to share vital health data such as blood pressure readings wirelessly captured on their smart phones with their doctors. Commercially, our shopping habits which a decade ago confined most of us to journeying to a shopping mall in order to make purchases are gradually changing as more of our retail transactions are taking place online. These undertakings can often be supported by shopping Apps such as Red Laser [3] which is able to compare the availability and cheapest price of a particular product identified in the first instance by scanning its barcode, then via Internet connectivity revealing the location and best price for that item, be that online or within the smart phone user's local area. This all relies on being able to locate and then utilise good Internet connectivity either through a mobile phone provider's mobile broadband service, or alternatively through a WiFi hotspot. Although major Internet Providers such as BT in the United Kingdom boast one of the largest number of available free WiFi locations throughout the UK, through their BT WiFi-with-FON network, nevertheless users are warned that the service is not secure. Therefore those who use an unsecured, non-password-protected WiFi network, in order to enact financial transactions on a smart device must be confident that their sent data is encrypted, and therefore cannot be read by an unscrupulous third party hacking and thus intercepting their sensitive data.

The growing use of mobile internet on smart devices has for many become commonplace, and part of their everyday life. Periods of escapism from reality are also possible through users accessing games on their smart devices. Increasingly the potential of harnessing the underlying elements of games play is being investigated by the business sector [4], educators and governments in order to better engage and improve the user experience. Companies such as Microsoft have leveraged gamification techniques over their competitors in order to keep their products at the forefront of the market place, specifically in the case of the development of their leading Office suite, where reviewers of the software translated into world languages are relieved of the tedious routine of laboriously searching line after line of code for mistakes by amassing rewards and tokens. This paper will seek to elaborate on these areas. In particular, could the use of gamification within an e-learning environment not only stimulate student engagement, but sustain their online presence, and so improve the overall student experience?

This positional paper will illustrate through examples where the use of games and gamification techniques have been utilised to motivate and sustain learner engagement.

2 Early Online Learning - The Ultraversity Model

The author gained first-hand experience of establishing and developing wholly online degree courses in a British university. From one particular course's inception in 2003, namely the Ultraversity degree programme, valuable insights were gained into the necessary pre-requisite features for such a course to run, particularly in order to first embed online student participation, and more importantly to maintain their engagement throughout the duration of the course.

Ultraversity is Anglia Ruskin University's wholly online undergraduate degree course, and has been running since 2003 [5]. It was originally the brainchild of Professor Stephen Heppell [6], then the director of the university's learning technology research department, Ultralab [7]. In addition to offering a unique three year online degree course, accessible from anywhere at any time via the Internet, Heppell and his team sought to attract older undergraduates who for a variety of reasons had been previously unable to undertake a full time university undergraduate degree. Action research was at the heart of the course, and each student's final dissertation work centered around the production of research focusing on an aspect of their workplace practice. At the outset of each student's course employers were encouraged to support and mentor their employee's progress through their degree, and champion their ongoing research. In many cases at the completion of their course, the findings of graduates' research was shared within their workplaces [8].

One of Ultralab's early discoveries from the Ultraversity course was the vital importance of nurturing critical peer support and the opportunities for shared learning which the tutors (known as learning facilitators) [9] encouraged and fostered within each cohort. Developing a team spirit within each cohort instilled in it a unique sense of community, and brought new meaning to Wenger's concept of 'communities of practice' [10]. The potential for developing a work based learning degree with collaborative peer working at its core evolved from Ultralab's work on behalf of the United Kingdom's Department for Education during the early 2000s. That work involved establishing and developing two distinct but related online education professional communities; the first was Talking Heads, and the second was Virtual Heads [11]. The former community began life early in 2000 as a pilot community of 1200 newly appointed headteachers in England, who were encouraged by a core team of facilitators to become a self-supporting online community of practice, which later expanded to include all headteachers in England. It was later supported and administered by the National College for Teaching and Leadership. The second online community was appropriately called 'Virtual Heads', as its core membership consisted of aspiring school leaders from across England who were enrolled on the Department for Education's National Professional Qualification for Headship (NPQH) programme.

Early into the life of the NPQH community it soon became apparent to Ultralab researchers that shared situational learning was becoming a core aspect of the Virtual Heads members' interactions within their online community of practice. In much the same way that the refined and tuneful output of a concert orchestra depends greatly upon its conductor to draw out a collective musical masterpiece, so too an online learning community relies upon the facilitation skills of its tutor to summarise and prompt dialogue, encourage members' contributions and in overall terms drive the community forward towards the completion of a module of study, this being the central and overriding motivating factor for each student participating in the undergraduate degree programme.

Despite drawing undergraduates from quite diverse backgrounds of employment, typically an IT consultant located in Germany and a car manufacturer's engine plant manager domiciled in Essex for example, both students shared a common bond, namely that they were intentionally devoting considerable amounts of their study time in an online forum with other students along with a tutor/facilitator, undertaking a shared learning journey. Throughout the life of the Ultraversity degree, the technology has continued to evolve, and the sharing of learning has become easier, due in part to the growing availability of more affordable hardware such as smart devices, alongside the evolution of mobile internet [12]. Increasingly institutions of higher education are engaging with their students through social media, and this practice is assimilated by today's students in much the same way as they download a new app for their smart device. Drawing upon personal observations the author believes that current students are more technologically able than most members of undergraduate cohorts ten years ago, in terms of their ability to use the Internet not only for social, gaming and retail purposes, but also as a tool for learning. A study undertaken by Selwyn, Gorran and Furlong in 2006 cited a lack of access to the Internet as a barrier to adult learning [13].

2.1 Harnessing the Rewarding Experience from Indulging in Game Play

Given society's increasing opportunities to access the World Wide Web compared to a decade ago perhaps through the increased availability of smart devices, Higher education institutions representative of one sector have accordingly increased their web footprint. However, in terms of the reality of the overall student experience, with its main goal the pursuit and achievement of an academic qualification, not all students share the same level of motivation, or the determination to achieve the highest possible grades. Retention of student numbers is also a priority for academic programme leaders, and in particular for those offering online courses. As mentioned earlier, many distance learners who are in full time employment and undertake their course work online in their spare time require particularly high levels of self discipline in order to satisfactorily complete a course of study. Often recreational activities are sacrificed and replaced by study time, giving new meaning to the age old saying, 'all work and no play makes Jack/Jill a dull person'. However, there may be a way to combine play with learning that might lighten the learning load and simultaneously increase student motivation and sustained engagement with their course work.

In 2008 research was published that had investigated the extent to which new media such as online gaming had become a regular feature in the everyday lives of young people. In addition that research indicated how this inclusion of technology had affected youth mindsets and attitudes with regards to learning. The same research revealed that when those young people surveyed weren't consciously subjecting themselves to the adrenaline highs generated from playing computer games that they became disengaged particularly within a formal learning situation [14].

Others were also beginning to see the potential of games for providing a stimulus for learning.

In his book, Teaching Digital Natives, Mark Prensky introduced the concept of 'total creative engagement', and related it in particular to the then youth attitudes to learning in schools [15]. Prensky also partly echoed one of the findings of the 2008 research, attributing to some extent the perceived disengagement experienced by some young people within formal education due in part to them regularly experiencing adventurous, challenging, creative, engaging, and motivational experiences through their immersion in video games.

In Scotland since 2008, the government there has been supporting the use of games based learning in schools, through the appointment of Derek Robertson as the country's National Advisor for Emerging Technologies and Learning. He initially ran a test across a selection of Scottish schools in utilizing a brain–training game on the Nintendo DS. The purpose of this test was to explore the potential of games based learning. Thirty two schools took part, and over 600 pupils were involved. Prior to half the children using the game for twenty minutes at the start of the school day for nine weeks, all those involved in the future survey undertook a maths test. At the end of the nine week trial period, all the pupils were again tested, and all showed improvement in their maths scores. Significantly those pupils who had engaged in the brain training game had made a 50 percent improvement in their maths scores compared to the control group who had not used the game. Robertson has also utilised other popular computer-based games as the foundation for engaging pupils in cross-curricular project–based learning [16].

McGonigal believes that including games in the education process alone is not enough. In her book 'Reality is Broken' [17] she argues that the inclusion of a number of games within the overall educational experience of a young person's time spent in compulsory schooling is not sufficient to make a significant difference in more fully engaging them in the overall learning process. She supports the views of educational innovators such as Prensky that our current concept of what schools are about should change, so that the whole school experience should become a game in its entirety. Rather than this being a radical and theoretical concept, this is apparently a reality in one New York City school. That institution is called Quest to Learn. Following two years of curriculum development planning, the school opened in 2009. It caters for grades 6–12 (in United Kingdom terms this equates to Key Stages 3 and 4, namely the secondary phase of compulsory schooling.) Although the school is led by a long serving teacher and administrator, its overall curriculum development is the responsibility of an experienced person from the games industry, who has a specific research focus on how learning can be positively influenced by games play [17, 18]. McGonigal cites an example of a sixth grade student's day, who is able to earn points for achieving particular reading levels, and 'expert' recognition for pursuing a learning specialism of their choice, which in turn fosters collaboration with older students elsewhere in the school. The author of this paper has also personally witnessed the positive engagement in formal school work displayed by school students when similarly motivated by the possibility of earning points gained for work achieved. Although not directly comparable with the New York City school example, nonetheless some school teachers elsewhere are aware of the motivational influence that rewards can bring their students. McGonigal asserts that much of the school's unique approach is founded upon gamification.

2.2 Gamification as Opposed to Serious Educational Games and Games Based Learning

According to Werbach and Hunter, gamification may be defined as 'the use of game elements and game-design techniques in non-game contexts'. (Werbach and Hunter [19], p. 26). They attribute the first use of the term 'gamification' to a British game developer who in 2003 established a business designing game interfaces for electronic devices. They further assert that although the term only came into more mainstream use in 2010, it is generally not widely understood, and should not be confused with serious games or games based learning [19].

In the context of this paper it is important to understand the difference between serious gaming as utilised within the educational context, and in particular games based learning, and more recently the emergence of gamification. McGonigal [17] maintains that all games contain four components or traits. These are respectively the goal, rules, feedback system and voluntary participation.

The **goal** is the end result, namely what the player or gamer is setting out to achieve, be it as an individual or as a member of a team, competing against another team of players.

A game is bound by a set of **rules.** The person playing the game must abide by certain set of rules. These are clear guidelines that govern the play of the game.

The rules clearly limit the obvious ways for the gamer to achieve the goal, therefore requiring the game player to be more creative in their thinking.

The player needs to receive **feedback** about how they are playing, in order to make possible changes to the way they are playing in order to improve their game and ultimately achieve the goal.

Finally **participation** in the game is usually **voluntary**, unlike the real game of life! This voluntary participation automatically assumes that the gamer will abide by the rules of the game, and accept the feedback.

McGonigal then applies these four traits to the game of golf, and compares it to the online game 'Portal', maintaining that the latter also comprises of the four traits. She argues that good games promote positive emotion, and are thus motivational. The Playful Learning website [20] states that games based learning is a form of learning where students may learn by trial and error, by role-playing and by treating certain topics not as content but as a set of rules or system of choices and consequences. Playful Learning is a United States Schools' initiative, which has brought University researchers and school teachers together to share good practice in the use of games to enhance and enrich learning. To support this aim, their website lists several subject specific games. It is important therefore to emphasise the distinction between serious games primarily for learning (often associated with the concept of games based learning), which can be resource – heavy in terms of content, software and design (and therefore potentially costly to produce), as opposed to gamification, where only elements of gaming are utilised in a non-gaming context. Granic, Lobel and Engels [21] add a note of caution in relation to gamification, stating that there is a dearth of associated research currently available as to its efficacy.

In November 2014 the University of Chester's Riverside Innovation Centre, along with Coventry University's Disruptive Media Learning Lab, Cookie Box (Spain), Helsinki University, and colleagues from Romania, Rome, Catalan and Israel embarked upon a collaborative 'Leonardo' European Union (EU)–funded research project exploring the potential for utilising gamification in order to engage with hard to reach adults from each of the participating countries [22]. The project builds upon the earlier 'Directing Life Change' (DLC) EU-funded project [23]. Gamification for Hard to Reach Adults is a 2 – year pan European Project building on the knowledge, skills, and competences of the above partner organisations from a variety of research disciplines and from the business and municipality sectors. The expense incurred from embarking upon an undergraduate degree course in the United Kingdom has become a costly undertaking, and now must rank as one of the most expensive events in life! The author has tutored many mature undergraduates who have benefitted from undertaking a fully online degree, principally because of its reduced cost compared to that of a full time residentially based degree, and students' ability to study online from virtually anywhere. Some of the author's students were located a considerable distance from the host university, in some cases many thousands of miles away on another continent, and with their course being based online there was no requirement to travel to the university campus, thus further reducing costs to the individual student. Thanks to smart devices it is possible to conduct a one-to-one video tutorial between tutor and student in such situations, with fewer restrictions for both parties in terms of being tied to a fixed computer as in the past. However for some students who by virtue of personal circumstances may only be able to study online, maintaining their interest and keeping them sufficiently motivated in order to complete a three-year distance learning degree can at best be challenging. This paper has provided some of the current thinking and associated examples relating to how games based learning and in particular the thinking behind gamification is beginning to influence and enhance student engagement in their learning in both the USA and UK.

2.3 Motivation to Learn Online

Maintaining student motivation levels so as to hopefully encourage them to continue to study online, without the benefits of face to face contact that full-time students located within the physical space of a university campus often take for granted, is a particular challenge facing university course designers. Clearly online students need greater engagement in order to stay the course, particularly as a number have commented to the author that they find the online experience isolating. The greater connectivity afforded by smart devices has provided more opportunities for both student and tutor to interact in terms of increased contact between both parties, and from the author's own experience this has enabled quicker response times to students' queries. However some of the research findings [17, 24] point to the potential benefits of harnessing the intrinsic motivational elements present when engaging in game play, particularly if applied within the formal educational setting.

Ryan and Deci [25] point to the need to understand the difference between intrinsic and extrinsic motivators, where for instance students engaged in tertiary education may well be extrinsically motivated by the thought of successfully completing a degree course in order to obtain a well-paid job. On the other hand research undertaken at the University of Paisley in Scotland in 2007 confirmed that nearly 70 % of students played computer games mainly for enjoyment and relaxation purposes [26]. The motivation here is intrinsic. Another finding derived from the Paisley research revealed that out of 250 respondents to the question as to whether computer games could be used within higher education teaching, 169 (69 %) responded positively. The same research also revealed particular skills desired by both the workplace and higher education were developed and reinforced through the students' use of immersive games that emphasised strategy, adventure and role-playing activities.

Rymer [27] succinctly summarises the intrinsic motivation that lies behind our voluntary participation in games, either as solitary or team players, thus:

$$Wanting + Liking = Rewarding$$

According to Werbach and Hunter [19], people 'work best when they are deeply engaged and even passionate about what they do' (Werbach and Hunter [19],

Goal Structuring

(After Rymer 2011)

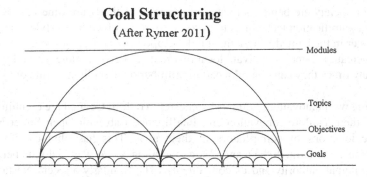

Fig. 1 The overarching module

p. 45). This accords with the earlier research conducted by Mihaly Csikszentmih-alyi in 1975, who studied the concept of a certain type of happiness he named 'flow'. This he described as 'the satisfying, exhilarating feeling of creative accomplishment and heightened functioning' [28]. He particularly associated it with the feelings that game players experience, particularly where the game in question has clear goals, self imposed rules and provides regular feedback to the player (namely three of the four main elements of a game identified earlier in this paper).

Applied to an e-learning course, typically at university undergraduate degree level, both Rymer [27] and Muntean [29] suggest that in establishing initial goals, these should be viewed in a series of linear steps, as in Fig. 1. It illustrates the overarching module, which in order to be completed it is suggested that at the commencement of a module, it must first be split into manageable goals, which in turn go to make up a smaller number of associated objectives, which then must be completed before a higher order of topics can be finished.

Upon completion of each objective, students should be encouraged to undertake a test, such as the quiz feature present in Moodle. This complemented by regular and timely tutor feedback would not only reinforce learning, but hopefully motivate the students and maintain their engagement. Muntean further postulates that posi-tive student engagement within an e-learning community forum should be rewar-ded, at the very least by extrinsic acknowledgement by the tutor, and in the author's experience by clearly stating in the course guidelines that contributing and sharing knowledge within the forum should be directly referenced in assignments, thus contributing to a student's overall grade mark.

2.4 Societal Engagement

If there are now greater opportunities to utilise the Web through better access and more affordable smart devices, there are also emerging examples of how certain

sectors of society are being motivated to combine their leisure time interests coupled with gamification techniques to become more involved in lifestyle changes. In
Manchester in the United Kingdom, app developers are accessing street light data in
order to create a game to motivate local residents to become more active by seeing
how many times they can 'claim' a certain number of street lights while out jogging
[30].

Staying with innovations within Manchester, Trafford Council are combining the
work of their Trafford Innovation and Intelligence Lab with Open Street Map to
map the locations of defibrillators in their area. This has obvious life-saving
implications, and through the team's blog, encourages two-way dialogue between a
local municipal authority and its constituents [31]. Similarly a recent Nesta article
reported that there was potential for citizens to utilise their smart devices in order to
report on environmental problems that could adversely affect their daily lives. The
article concludes with a cautionary note that participants in such a future scheme
shouldn't be confined merely to providing data alone, but should be encouraged to
more actively engaged in bringing about positive change within their localities [32].

3 Conclusion

The inclusion of games for learning, and incorporating the thinking associated with
gamification in designing online learning course may not be universal panaceas for
educators, nor a sure-fire means of attracting a loyal customer base for retailers, and
in political spheres a means for reaching out to communities and voters. Neither can
games nor gamification claim to be the solution to motivating and re-engaging
bored students, undergraduates or reaching disaffected adults across the European
Union. There is a possibility that the results of research derived from studies
undertaken with school children in Scotland discussed earlier in this paper, along
with the increasing application of games based learning in the United States point to
a possible powerful alternative approach to increasing personal motivation particularly when applied to learning, specifically by utilizing some of the elements
associated with gaming in non game contexts.

One of the main obstacles identified by Selwyn, Gorrard and Furlong's research
in 2006 mentioned previously in this paper was the lack of access to the Internet for
potential adult learners. Given the extent to which smart devices are increasing in
usage, follow up research should be conducted to ascertain whether current day
potential adult learners may be more inclined to consider undertaking online
learning courses, particularly through short-duration programmes of study, thus
opening up a potential new income stream for higher education institutions. A
possible future research question might be 'has the increasing availability of mobile
internet access and the growing range of smart devices increased the number of
mature learners accessing further education?'

This is a salient point for partners in the EU-Leonardo project mentioned earlier
to bear in mind. Its future research findings must demonstrate to the European

Union and its member states that the inclusion of gamification within government discussion forums, easily accessed through smart devices will not only establish and maintain two-way communication with hard to reach adults, but also the wider population across the European Union. Taking this future source of research into consideration, smart device users generally could be encouraged to become more actively engaged in local and national politics, where elected representatives could capitalize on increased opportunities to engage with their voters, drawing upon a basic feature of gamification that 'rewards' online engagement with increasing status (through the inclusion of a participation counter each time a voter makes an online contribution) as an individual's participation levels increase. Informal feedback from this author's students indicates that students' online forum contributions motivate them to continue to participate in those discussions as they derive intrinsic satisfaction from regular engagement with their peers and tutors. However such online forms would need to be moderated.

The examples cited earlier in this paper clearly indicate that that the inclusion of gamification techniques can increase the motivation levels of those engaged in learning. If the 'feel-good' factor benefits associated with engaging in games can be more widely communicated to online undergraduate course designers in such a way that they incorporate such intrinsic rewards within course designs, then one of the main reasons for students disengaging with learning could be addressed. The establishment of the aforementioned European – wide research project to study the use of gamification techniques in engaging not only hard to reach adults, but also the wider population in the daily affairs that govern their lives must provide a springboard for further research in order to provide greater opportunities for members of society to assume more responsibility for their lives, thus devolving greater decision-making away from central governments to more localized control.

Elements from this paper will be used to introduce the concept of gamification to partner institutions, local government organisations and voluntary groups associated with the European 'Gamification for Hard to Reach Adults' project [33]. Emerging research from that project will inform and generate future research papers, particularly for dissemination at international conferences such as KES 15.

References

1. Welsh, K., France, D.: Smart phones and Fieldwork. Geography **97**(Part 1), 47–51 (2012)
2. The Doctor will Skype you now. http://tinyurl.com/nbv77uv
3. Red Laser https://itunes.apple.com/gb/app/redlaser-barcode-scanner-deal/id474902001?%20mt=8
4. Microsoft Language Quality Game. http://social.technet.microsoft.com/wiki/contents/articles/9299.language-quality-game.aspx
5. Anglia Ruskin University: The University where everyone's a stranger http://ww2.anglia.ac.uk/ruskin/en/home/news/press_office.html
6. Heppell www.heppell.net
7. CEMP Stephen Heppell, short biography, including reference to Ultralab http://www.cemp.ac.uk/people/stephenheppell.php

8. Lang, G.: Ultraversity-the use of technology in adult education, in 'Integrating Adult Learning and Technology for Effective Education: Strategic Approaches'. IGI Global, Hershey, PA, USA http://www.igi-global.com/chapter/ultraversity-integrating-technology-adult-education/41849 (2010)
9. Lang, G.: Facilitation of online teaching and learning, pp. 626–642. In: Wang, V.X. (ed.) Encyclopedia of Information Communication Technologies and Adult Education Integration. http://www.igi-global.com/chapter/ultraversity-integrating-technology-adult-education/41849 (2011)
10. Wenger, E.: Communities of Practice-Learning, Meaning, and Identity. Cambridge University Press, Cambridge (1998)
11. Lang, G.: Pen-i-Ben: an online community of newly appointed head teachers and their mentors in Wales, UK. In: International Conference for Information Technology in Education ICICTE 2012 Rhodes. http://www.icicte.org/ICICTE12index.htm (2012)
12. Lang, G.: From home computer to smart device, vol. 262. In: Smart Digital Futures 2014 (Ebook) pp. 611–623. doi:10.3233/978-1-61499-405-3-611 (2014)
13. Selwyn, N., Gorrard, S., Furlong, J.: Adult Learning in the Digital Age. Routledge, Oxford (2006)
14. Ito, M., Horst, H., Bittanti, M., Boyd, D., Herr-Stephenson, B., Lange, P.G., Pascoe, C.J., Robinson, L.: Living and Learning with New Media: Summary of Findings from the Digital Youth Project. http://tinyurl.com/p4rrc34 (2008)
15. Prensky, M.: Teaching Digital Natives. Corwin, Sage, California. ISBN 978-1-4129-7541-4 (2010)
16. Miller, D.J., Robertson, D.P.: Using a games console in the primary classroom: effects of 'Brain Training' programme on computation and self-esteem. British J. Educ. Technol. **41**(2), 242–255 (2010)
17. McGonigal, J.: Reality is Broken. Why Games Make Us Better and How They Can Change the World. Vintage Books, London. ISBN 9780099540281 (2011)
18. Quest to Learn (2014) http://q2l.org/
19. Werbach, K., Hunter, D.; For The Win. How Game Thinking can Revolutionize Your Business. Wharton Digital Press. Philadelphia. ISBN 978-1-61363-023-5. (2012)
20. Playful Learning. http://playfullearning.com/portfolio/what-is-gbl/
21. Granic, I., Lobel, A., Engels, R.C.M.E.: The benefits of playing video games. Am. Psychol. **69**(1), 66–78 (2014)
22. H2R Project. Riverside Innovation Centre. http://riversideinnovationcentre.co.uk/h2r-project/
23. Directing Life Change. www.directlifechange.eu
24. Kopfler, E., Osterwell, S., Salen, K.: Moving Learning Games Forward. The Education Arcade. Massachusetts Institute of Technology. http://education.mit.edu/papers/MovingLearningGamesForward_EdArcade.pdf (2009)
25. Ryan, R.M., Deci, E.L.: Self-determination theory and the facilitation of intrinsic motivation, social development, and well-being. Am. Psychol. **55**(1), 68–78 (2000). http://dx.doi.org/10.1037/0003-066X.55.1.68
26. Connelly, T., Boyle, L., Hainey, T.: A Survey of Students' Motivations for Playing Computer Games: A Comparative Analysis. http://tinyurl.com/ljvzn4z (2007)
27. Rymer, R.: Gamification: Using Game Mechanics to Enhance eLearning. http://elearnmag.acm.org/archive.cfm?aid=2031772 (2011)
28. Csikszentmihalyi, M.: Play and intrinsic rewards. J. Humanist. Psychol. **15**(3), 41–63 (1975). http://dx.doi.org/10.1177/002216787501500306
29. Muntean, C.I.: Raising engagement in e-learning through gamification. http://tinyurl.com/psauwk7 (2011)
30. Gallagher, P.: Friends Bright idea to fight obesity. Manchester Evening News Online (2014). http://tinyurl.com/lv6vjdd (2014)

31. Whyte, J.: Putting our Data onto Open Street Map (OSM). https://traffordinnovationlab. wordpress.com/ (2014)
32. Nesta: Citizens as sensors. Monday 27th October. http://www.nesta.org.uk/blog/citizens-sensors (2014)
33. ASSE, Gamification for hard-to-reach adults – New horizons for re-engaging and re-mobilizing hard-to-reach adults in long-term unemployment situations http://tinyurl.com/pj4dym2 (2014)

Learning by Doing History with Smart Historical Gaming

Brian M. Slator, Otto Borchert and Guy Hokanson

Abstract History teaching reform asks that students learn to identify, analyze, respond morally, and display their knowledge about the past. However, most historical games focus on a single individual viewpoint or quiz students on disjointed facts. Blackwood offers an immersive multi-user simulation which provides the means and motivation for players to learn world history while playing a retailing game set in the historical "Old West" of the USA. The system is being developed as a learning laboratory to test the hypothesis that having a role-based experience where history is presented as current events is an effective means for learning history. Future studies will focus on the effects of group learning and peer-mentoring on understanding history.

Keywords Smart historical gaming · Learning laboratory · Learning-by-doing

1 Introduction

The Blackwood simulation [1] (http://cs.ndsu.edu/~slator/html/blackwood/) was developed as a learning laboratory to study questions like "Is gaming a good way to teach history?" To accomplish this, a multi-player online simulation of an Old West town was constructed, giving students an authentic context within which to assume a role. In Blackwood, each player assumes the role of a specific type of shop-keeper: blacksmith, leather-maker, wheelwright, etc. The aim of each player is

B.M. Slator (✉) · O. Borchert · G. Hokanson
Department of Computer Science, North Dakota State University, Fargo, ND, USA
e-mail: brian.slator@ndsu.edu

O. Borchert
e-mail: otto.borchert@ndsu.edu

G. Hokanson
e-mail: guy.hokanson@ndsu.edu

© Springer International Publishing Switzerland 2015   207
V.L. Uskov et al. (eds.), *Smart Education and Smart e-Learning*,
Smart Innovation, Systems and Technologies 41,
DOI 10.1007/978-3-319-19875-0_19

to manage a successful retail operation; an agent-based simulation provides demand [2, 3].

Online games have been shown to motivate student learning, and economic simulations provide a built-in premise for competition. The interesting challenge of this project was to find a way to teach American History, within the context of the game, that would itself motivate student learning. Simply providing the opportunity to read about historical events would not provide sufficient impetus, and re-enacting historical events in the simulation would be too confining.

The problem was solved by implementing a set of intelligent dialog-based software agents that engage players on historical topics. Successful interaction with these agents serves to enhance a player's status in the game, leading to greater success at their primary goal: earning profit in the economic simulation.

2 Background: Research Parameters

The World Wide Web Instructional Committee (WWWIC) at North Dakota State University (NDSU) is an ad hoc committee of faculty, staff, and students working to advance education through the use of IVEs [4]. WWWIC has implemented educational IVEs for a wide range of scientific disciplines [2, 5–7]. These IVEs are designed using a common set of guidelines [8]:

- Collaborative and multi-user
- Constructed with the help of content experts
- Strong cognitive and pedagogical features to assist the players in learning content
- Consistent evaluation of learning outcomes

2.1 Context: The Blackwood Design

In 1998, WWWIC began work on a virtual educational environment [9] combining microeconomics with Western history. The aim was to provide an engaging context for role-active immersive distance education and a platform to teach business-oriented problem-solving in a learn-by-doing pedagogical style [10–12]. Design discussions focused on the implementation of a virtual environment to simulate a 19th Century Western town. This would be populated with intelligent software agents to simulate an economic environment representative of the times. The idea was to implement this as a spatially oriented virtual environment, borrowing freely from historical records and employing digital images from online archives and other sources.

The educational game would be one where players join the simulation and accept a role in the virtual environment. Rather than everyone vying for a portion of the same economic market, roles were to be variable and specific. For example the blacksmiths in Blackwood must compete against each other with a goal of bringing shoppers into their business. In a simple sense, this competition reduces to making themselves more 'attractive' to the shoppers, where attraction is computed by a formula that includes variable influences, depending on shopper profile. However, every shopper, to some degree, is attracted by level of service, quality of goods, visibility as a function of advertising, travel time, and other intangibles, discussed below.

2.2 Background: Historical Games

A number of games to teach history have been developed, from computer to board games. Some of these games include: Professor Noggin's History of the US, American Trivia Junior, Wyatt Earp's Old West, and Settlers of America Trails to Rails. These games focus on either teaching independent trivia facts or on a resource strategy gameplay mechanic, not on establishing historical thinking in students [13, 14].

Barton and Levstik have identified four primary activities that students perform when they "do history". Students should be able to identify, analyze, respond morally, and display [15]. That is, students should be able to relate and identify with individuals in the past. They should be able to analyze multiple viewpoints and understand individual intention, motive, plans, and purposes. They should be able to make moral judgments about events in a reasoned, multifaceted way. Finally, they should be able to exhibit knowledge they have gained.

While instructors learn inquiry-based techniques for teaching these higher levels of historical cognition in their pre-service methods courses, implementing it in the classroom remains elusive [16]. Developing curricular resources for teachers to easily implement in their classrooms would be beneficial. Blackwood represents a number of specific history content standards [17], including specific Era 6 standards:

- The student understands how the "second industrial revolution" changed the nature and conditions of work. (Standard 3A)
- The student understands the rise of national labor unions and the role of state and federal governments in labor conflicts. (Standard 3B)
- Analyze the causes and effects of escalating labor conflict. (Standard 3B3)
- Explain how commercial farming differed in the Northeast, South, Great Plains, and West in terms of crop production, farm labor, financing, and transportation. (Standard 1C4)

3 Synthesis

By combining an economic simulation, where each player is expected to compete for a slice of the retail pie, with an authentic historical simulation, we hope to engage learners in two aspects of role-based learning: microeconomic strategizing combined with historical enculturation. To support this melding, we pursued a line of agent-based research [1, 5]. This research explores knowledge engineering and representation issues involved with developing intelligent agents of the following types:

- "atmosphere" agents - an agent that simply lends to the local color. For example the Dollar Bay Retailing Game features a street magician, a beat poet, a lost child, a political activist, a street vendor, a beat cop, and a door-to-door salesman;
- "infrastructure" agents - an agent who contributes in some way to the gameplay: For example the Dollar Bay Retailing Game features a banker, a market analyst, wholesalers, employees, a financial planner, and an advertising consultant;
- "tutoring" agents - an agent that monitors player moves, visits players to give them advice in the form of expert stories and cases, and ultimately assists players in learning to play. These cases and stories represent expertise of other players.

4 Geography

The mythical town of Blackwood is situated somewhere in the American West. Its location is not precisely known, but it lies somewhere in the Great Plains region west of the Mississippi River, south of the Canadian border, east of the Rocky Mountains, and north of the Mason-Dixon line; i.e. somewhere in the Montana, Colorado, Dakota, Wyoming vicinity.

Blackwood is uniquely situated in a convergence of geological and historical influences. The Black River flows northward and meets Wood Creek just to the north of town, and glacial movement in the Ice Ages has left rich soil for farming to the north and west. The same glacial activity created a hilly moraine region to the northeast, and there is some evidence of mineral riches, particularly native silver, in the hills in that direction. To the south of town, and east of the Black River, there is a dense mixed forest that provides much of the lumber used for building in the area. The land is flat and rocky to the south and west making it better for cattle ranching (see Fig. 1).

By 1880, Blackwood had a population of 2,500, but the Northwestern Railroad was in the process of building a line across the region and planned to cross the Black River at the Blackwood ford, very near town. By the Spring of that year, a lumber town had sprung up in advance of the railroads farthest progress, and the

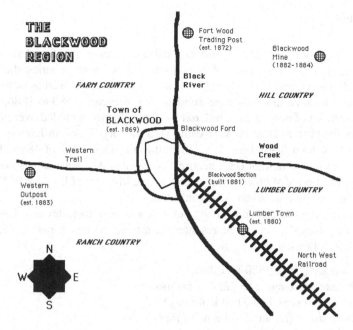

Fig. 1 Map of the region. When the game starts in 1880, Blackwood has been established and the municipality known as Lumber Town is just coming into existence

arrival of the railroad promised to bring an even greater influx of people to the greater Blackwood area.

When the railroad arrived in 1881, the population of Blackwood, which had been steadily growing, exploded with railroad workers and passengers from the east. Then, when silver is discovered in the area in 1882, another large increase in population occurred. However, the North West Railroad encountered financing difficulties and westward construction was halted for five years. Nonetheless, the population of Blackwood continued to grow by 50 % a year as more and more families settled in the area and business continued to expand with the influx of westward travelers, small land owners, and miners. The population peaked at 25,000 in 1884, but then the silver mines were exhausted and the crest of the boom ended.

The final chapter of Blackwood development occurred in 1886, the year of the "Great White Ruin", a terrible winter that left 300 people dead on the Great Plains (along with thousands of cattle). The combination of a heavy winter snow and an extremely wet spring caused the Black River to slowly overflow its banks and inundate the town. Virtually every home and business was damaged by the flood, and few of the townsfolk remained to rebuild. The call of the west was strong, and most people pack up and venture into the new frontier. By the fall of 1886 Blackwood was once again a small and remote town; a mere whistle stop on the rail line heading west.

5 Assets

The first necessity is a reliable historical database to serve as a foundation for the software agents. This was achieved by a number of students scouring the internet and assembling resources from the US Library of Congress as well as other places. This initially resulted in a web page divided into the years, 1880 to 1886, with an uncategorized set of news events, ordered by date. This list included over 500 news items from the birth of Douglas MacArthur (US military leader) in January of 1880 to the death of Sam Starr (legendary US outlaw) in December of 1886. The collection tends to be somewhat USA-centric, but worldwide events are also recorded, such as the election of William Gladstone as Prime Minister of England in 1880 and the discovery of gold in South Africa in 1886.

A knowledge representation for the news stories was then devised. Each news object has a 'stories' property, represented as a table, and each row in the stories table has the following fields:

- Year: integer (range 1880 to 1886)
- Month: integer (range 1 to 12, 0 if no month)
- Day: integer (range 1 to 31, 0 if no day)
- Headline: string (required to be non-empty)
- Story: string (possibly empty)
- Topic: list of strings (Primary, Secondary, Tertiary, Person, Place)

The second necessity is an ontology to index story topics. This was devised empirically, and from the bottom up, by reviewing all the stories and determining the key (tertiary) features of each story, then abstracting a smaller set of set of 'Secondary' and finally 'Primary' categories. For the sake of this project, it was decided not to import a generic set of categories but to induce them ourselves, keeping the search space as small as possible while covering all concepts. The Primary keys and thus the highest level concepts are Birth, Culture, Death, Disaster, Economics, Entertainment, Household, Government, Law, Marriage, Military, Religion, and Science

The main value of ontology is for inference and association strength. Thus, with this representation, an event like the death of outlaw Jesse James on the 3rd of April in 1882 would be coded as:

```
{1882, 4, 3, "Outlaw Jesse James killed",
"Outlaw Jesse James was shot in the back and killed in St.
Joseph, Mo., by Robert Ford, a member of his own gang for a
$5,000 reward. Jesse and Frank, the bank robbing James
brothers, were born as Woodson and Alexander.", "Law",
"Crime", "Outlaw", "Jesse James", "St. Joseph, MO"}
```

With an algorithm to calculate "semantic distance" in the topic hierarchy the representation allows us to search for semantically similar stories such as

```
{1881, 4, 28, "Billy the Kid escapes",
"Billy the Kid was held in Lincoln County Courthouse jail,
near Carizozo N.M., but escaped and killed two guards. He used
an 1876 single-action army revolver made by Samuel Colt. The
gun sold for $46,000 in 1998.", "Law", "Crime", "Outlaw",
"Billy the Kid", "Carizozo, NM"}
```

6 Discussing History

Armed with the news objects, and a knowledgebase of stories indexed and organized by a custom ontology, it became time to build news into the game. First, a
newspaper office was created with a morgue for stories, classified by the Primary
keys, and then creating a 'news ticker' object for capturing stories by category.

```
Morgue
You are in the Morgue of the Newspaper where all the old sto-
ries are saved. You can read stories by clicking on the news
ticker machines in each room.
Obvious exits are: Newsoffice, Birth, Culture, Death, Disas-
ter, Economics, Entertainment, Government, Law, Marriage,
Military, Religion, and Science
```

For example, going into the Law wing of the Newspaper Office Morgue will
bring you into proximity with objects named 'ticker-1880-law', 'ticker-1881-law'
and so forth. This allows players to visit a well-organized collection of news stories
for the purpose of individual study. The problem is motivating students to study.

7 Motivating Students to Study

Learning the facts of American History is not intrinsically motivating for many,
perhaps a majority, of students. However, when there is an extrinsic motivation it
can lend appeal to an otherwise onerous task.

For example, fans of a sports team will follow their team and study their players,
learning facts and statistics because of their interest in the sport, and the desire to
discuss related topics with others. People are motivated to study biography, statistics, physiology, psychology, even physics, because they are related to the sport,
its players and their performance.

In Blackwood, it is retailing. To implement extrinsic motivation we draw on the
'sports bar' metaphor, where different stores and restaurants will 'brand' themselves
in order to target a niche in the market. The metaphor is further employed in the
strategy for implementation.

Every blacksmith in Blackwood is in competition with all the others, attempting
to attract shoppers to their place of business. One of the intangibles mentioned

above is news knowledge. As described below, if a player can demonstrate knowledge of, say, stories about the Law in 1882, they are awarded a badge of honor for that. Shoppers are attracted to stores with badges of honor, the more the better.

8   Story Agents in Blackwood

As sports fans enjoy discussing their team, we attempt to mimic a dialog between software agents specializing in news stories and the players who wish to attract more shoppers. This takes the form of a dialog that begins with (1) an invitation to play, (2) an acceptance of the offer, (3) presentation of a story in the form of a question, (4) acceptance of an answer to the question, (5) reporting whether the answer is correct, and concludes with (6) following up with a related story.

Turning a story into a question takes the following steps. First, a story is chosen:

{1882, 4, 3, "Outlaw Jesse James killed", "Outlaw Jesse James was shot in the back and killed in St. Joseph, Mo., by Robert Ford, a member of his own gang for a $5,000 reward ... ", "Law", "Crime", "Outlaw", "Jesse James", "St. Joseph, MO"}

Either the 9th or 10th field of the story (Person "Jesse James" or Place "St. Joseph, MO) is randomly chosen, and a question is generated using string substitution. As in:

"Outlaw ---???--- was shot in the back and killed in St. Joseph, Mo., by Robert Ford, a member of his own gang for a $5,000 reward ... "

Then the knowledgebase is searched heuristically for similar stories using the Primary, Secondary, and Tertiary keys. The best matching stories provide 'wrong answers' in their Person field, and these are randomly combined with the correct answer, which is presented to the player in the form of a multiple choice question,

A. Billy the Kid
B. Belle Starr
C. Jesse James
D. John Lee

To do this, a sequence has been programmed, as follows.

```
1. When a player visits a location with a story agent,
     they are told to "click on me to take a quiz"
2. If the player clicks on the agent
   2.1 The agent inspects the player history for story
         questions they have answered
```

```
2.2 Choose a news story the player has not answered
  2.2.1 The story is converted into a question
  2.2.2 Search for 'wrong answers'
  2.2.2 The question is presented to the player
2.3 The player chooses an answer
    2.3.1 With a correct answer, the player is credited
      2.3.1.1 If the story is the last one on the ticker,
                the player gets a badge for that ticker
    2.3.2 With incorrect answer, the player is informed
2.4 Find a similar story, on a different ticker
```

9 Conclusion and Future Work

Blackwood was designed to address various issues by enforcing role-based constraints, which also supports the necessity of player interaction for goods and services. The blacksmith needs to buy supplies from the Dry Goods store.

Blackwood is also intended to stretch the boundaries of 'Learning by Doing'. It has long been recognized that many field-based and laboratory sciences lend themselves to a learn-by-doing approach; this is why geologists in training go to field camp, and biologists in training operate laboratory instruments. Mastering the tools of the trade is a way to demonstrate competence.

But, what about History? What does it mean to "do" history? Clearly, Blackwood does not offer the stereotypical experience. Historians do their work in libraries and on historic sites. They sift through various and sometimes incompatible versions of events, seeking simple truth and interpretation. Blackwood offers an alternative view, where students are able to immerse themselves in individual roles in a historical simulation. After multiple play-throughs, students should be able to offer compare and contrast viewpoints of individual roles. Rather than focusing on the development of disjointed facts, this will allow students to identify with characters in the game, analyze diverse viewpoints of a point in history, and exhibit knowledge learned through the story agents. Future modules may explore activities that would inspire a moral response (e.g. the tradeoffs of organized labor).

The effectiveness of this approach remains to be examined. We plan to conduct a series of experiments, where high school students are asked to either play Blackwood, or read a series of narrative texts or textbook descriptions of relevant historical events. The Blackwood group would be asked to play through the game in different roles, and create historical narratives. Students would read other narratives to gain insight into individual intention, motive, plan, and purpose, with the ability to ask the other players who wrote that narrative about their experiences. These experiences should help students learn how to "do history" in immersive, inquiry driven ways. Analysis of these narratives would provide insight into what students learned from story agents and other aspects of the gameplay.

What we are attempting, and still remain to prove, is the hypothesis that embedding historical knowledge within a competitive gaming context, and then making that historical knowledge an asset in the competition, is a good way to present history, and that players will learn it as well or better than otherwise. The experiments we plan will confirm or deny our predictions.

Acknowledgments The original 'story to question' idea and the first version of the matching code were due to Adam Helsene.

References

1. Slator, B.M. with members of CSci345.: Rushing headlong into the past: the Blackwood simulation. In: Proceedings of the Fifth IASTED International Conference on Internet and Multimedia Systems and Applications (IMSA 2001), pp. 318–323. International Association of Science and Technology for Development, Honolulu, HI (2001)
2. Farooque, G., Slator, B.M.: The Agents in an Agent-based Economic Simulation Model. In: Proceedings of the 11th International Conference on Computer Applications in Industry And Engineering (CAINE-98). International Society for Computers and Their Applications, Las Vegas, NV (1998)
3. Hooker, R., Slator, B.M.: A Model of Consumer Decision Making for a Mud Based Game. In: Proceedings of the Simulation-Based Learning Technology Workshop at the Third International Conference on Intelligent Tutoring Systems (ITS'96), pp. 49-58. Montreal (1996)
4. Slator, B.M., Beckwith, R., Brandt, L., Chaput, H., Clark, J.T., Daniels, L.M., Hill, C., McClean, P., Opgrande, J., Saini-Eidukat, B., Schwert, D.P., Vender, B., White, A.R.: Electric Worlds in the Classroom: Teaching and Learning with Role-Based Computer Games. Teachers College Press, New York (2006)
5. Saini-Eidukat, B., Schwert, D.P., Slator, B.M.: Geology explorer: virtual geologic mapping and interpretation. Comput. Geosci. **28**, 1167–1176 (2001)
6. Hokanson, G., Borchert, O., Slator, B.M., Terpstra, J., Clark, J.T., Daniels, L.M., Anderson, H.R., Bergstrom, A., Hanson, T.A., Reber, J., Reetz, D., Weis, K.L., White, R., Williams, L.: Studying native American culture in an immersive virtual environment. In: Proceedings of the IEEE International Conference on Advanced Learning Technologies (ICALT-2008), pp. 788–792. IEEE Computer Society Press, Santander, Spain (2008)
7. White, A.R., McClean, P., Slator, B.M.: The virtual cell: a virtual environment for learning cell biology. In: Tenth International Conference on College Teaching and Learning. Center for the Advancement of Teaching and Learning, Jacksonville, FL (1999)
8. Borchert, O., Brandt, L., Hokanson, G., Slator, B.M., Vender, B., Gutierrez, E.J.: Principles and Signatures in Serious Games for Science Education. In: Van Eck, R. (ed.) Gaming and Cognition: Theories and Practice from the Learning Sciences, pp. 312–338. IGI Global, Hershey, PA (2010)
9. Curtis, P.: Mudding: Social phenomena in text-based virtual realities. In: Ludlow, P. (ed.) High Noon on the Electronic Frontier: Conceptual Issues in Cyberspace, pp. 347–373. MIT Press, Cambridge, MA (1996)
10. Duffy, T.M., Lowyck, J., Jonassen, D.H.: Designing Environments for Constructive Learning. Springer, New York (1983)
11. Norman, D.: The Psychology of Everyday Things. Basic Books, New York (1988)

12. Hill, C., Slator, B.M.: Virtual lecture, virtual laboratory, or virtual lesson. In: Proceedings of the Small College Computing Symposium (SCCS98), pp. 159–173. Small College Computing Symposium, Fargo-Moorhead (1998)
13. Smart Kid Educational Games: American History Games. http://www.smart-kid-educational-games.com/american-history-games.html (2015)
14. Moby Games: Wyatt Earp's Old West, http://www.mobygames.com/game/wyatt-earps-old-west (2015)
15. Barton, K.C., Levstik, L.S.: Teaching History for the Common Good. Lawrence Erlbaum Associates, Mahwah (2004)
16. Van Hover, S.D., Yeager, E.A.: Challenges facing beginning history teachers: an exploratory study. Int. J. Social Educ. **19**, 8–21 (2004)
17. National Center for History in the Schools: United States History Content Standards for Grades 5-12, http://www.nchs.ucla.edu/history-standards/us-history-content-standards (2015)

Smart Educational Environment as a Platform for Individualized Learning Adjusted to Student's Cultural-Cognitive Profile

Zinaida K. Avdeeva, Naida O. Omarova and Yulia V. Taratuhina

Abstract In this paper, the authors attempt to describe a possibility of an individual approach to learning within multicultural electronic educational space. To solve this problem the authors proposed criteria for determining learner's cultural and cognitive profile and ways of adaptation of educational content and interface as a pre-requisite for creating "smart educational environment". An important factor in the design of educational media environment is to focus on the personal and socio-cultural approaches to learning information typical for a selected cultural group, resulting in the possibility of constructing an individual educational trajectory. The authors propose to design a prototype of "smart educational environment" whose interface and content could be adjusted to a student's cultural-cognitive profile.

Keywords International educational communication · Personal cultural-cognitive profile · Adaptation of interface

1 Introduction

Global education combines different educational systems and models, which are based on differentiated cultural, ideological, religious, philosophical, and axiological worldviews. Undoubtedly, integration processes are an integral part of globalization - the world becomes "integrated" one way or another. Formation of a unified educational space is one of the priority tasks of the near future. However, this is not an

Z.K. Avdeeva (✉) · Y.V. Taratuhina
Higher School of Economics National Research University, Moscow, Russia
e-mail: avdeeva@hse.ru

Y.V. Taratuhina
e-mail: jtaratuhina@hse.ru

N.O. Omarova
Institute of National Problem of Education of Russian Academy of Education, Mahachkala, Russia

© Springer International Publishing Switzerland 2015
V.L. Uskov et al. (eds.), *Smart Education and Smart e-Learning*,
Smart Innovation, Systems and Technologies 41,
DOI 10.1007/978-3-319-19875-0_20

easy task, which is based on the dichotomy of, on the one hand, preserving national identity, and processes of cultural and educational integration, on the other. Now we can observe the transition of a number of educational processes into the online context and, as a consequence of openness, the acquisition of partially or completely multicultural principles. Of course, in a multicultural context, cultural enrichment, expansion of general and professional outlook, teachers and students face a number of problems, mainly with pragmatic reasons. Despite the processes of globalization and integration, culture of each country, anyway, is reflected in the educational process and, in many respects, its causes, which, in turn, entails the specificity of training content, values and objectives of education, teaching methods, pedagogical discourse, the specifics of building educational trajectory, etc.

Today's concept of education means lifelong learning. These processes cause the knowledge-based approach in education to be gradually replaced by competence-based one. Nowadays we can witness a number of educational processes migrating into the Internet and, consequently, their becoming more transparent and more or less multi-cultural.

Currently, a significant part of educational process transforms into online format. Obviously, in the modern Educational Environment (EEE), tutor's didactic functions will be modified, and the problem of an individual approach, motivation and commitment of student in the process of learning will rise most sharply. The learning process in the MEE is mostly auto-didactic. So it is necessary to determine the manner, in which an individual approach may be realized in EEE, where the nature of communication is predominantly mediated, before designing an individual educational trajectory. In our opinion, an individual approach in a multicultural e-Educational space will look like an adaptation of content and interface to a student's personal cultural and cognitive profile. The purpose of moderne education is to create conditions in which educational path could be formed and adjusted with the help of a trainee. We believe it is appropriate to establish the criteria for determining cultural cognitive profile of a person, in order to make possible further technical and methodical adaptation of educational content and interface. Control of knowledge and qualification boundary will be realized by "competence-based profile of students". It is also possible to create advisory service facilitating "build-up" of professional competence according to cultural-cognitive profile of an individual. To improve "cultural intelligence" of tutors, we suggest to use online cultural assimilator.

2   On Communicative and Didactic Aspects of Mapping an Individual Study Pathway in Electronic Educational Environment

It is obvious that with the application of Electronic Educational Environment (EEE) the didactic functions of a tutor will be changed, and the whole educational process will become autodidactic. And, before mapping an individual study pathway, one must decide how this individualistic approach will be applied in EEE where there is no immediate communication between students and tutors. When studying via the information technologies, a student is supposed to develop individual learning skills and to get well acquainted with the up-to-date on-line education technologies, which means that the student's self-motivation becomes a more important factor.

A tutor's role here will be taken by the EEE itself. However, for this purpose it is necessary to select and set the criteria for running individual study process (for in-stance, a student's cultural and cognitive profile, consisting of the emotional and activities components).

Individual Study Pathway (ISP) should be mapped and adjusted through rec-ommendation services which will suggest best suitable courses for a student and, if a student decides to take courses outside the recommended range, identify possible pitfalls. Evaluation of task and study materials in the major courses of the education path can be done similarly. Thus, the mastering of a course is supposed to follow the path best suitable for a student's information processing skills. We must also point out that this evaluation method can be applied when assessing both major and out-side-the-range courses, thus formulating recommendations for the student on how to develop their competence profile on the basis of the courses provided in the system, as well as doing those in free access. Summing it up, implementing the individual study pathway will pro-duce a specialist's competence profile.

Developing and implementing ISP is a complex process that includes the fol-lowing components: (1) Forming an individual information space. (2) Personalizing educational resources. (3) Personalizing educational objectives and finding meats for their achievement. (4) Adapting educational content and interface in EEE. (5) Achieving synergy effect through combining individual reflection and self-organization capacities.

In order to develop custom approach in education for an electronic education platform it is necessary to set several modules (stages) for composition of most optimal model of education for each student. In the first stage of interaction with digital educational platform, it would be most appropriate to conduct criterion-oriented test, which would not only determine cultural-cognitive profile, but also collect information about other basic characteristics, such as: motivation, educational back-ground, informational and communication technology skills, professional interests. However, it is necessary to take into consideration the pos-sible dynamic nature of before mentioned variables. It would be rational to retest an individual periodically, and make corrections to the selected education course. Collected information should be used to compose custom courses with most appropriate educational method selected for each student.

As for didactic functions of EEE we would like to point out a possibility for a student to "teach" the system. One variant of such system is when intellectual component of system analyses both collected data from conducted tests and stu-dent's reflections about courses he/she has completed. If necessary, it amends selected educational strategy based on analysis results. Further aggregation of statistical parameters in database will allow to perceive various trends. Moreover, accumulated data would be used in the process of tailoring education experience to new students in the earliest phases of customization. In other words, it would be used in the stage of adaption of learning method and content with a consideration of peculiarities of cultural cognitive profile of a student. We assume that educational content in EEE would be presented according to personal output rules, formed by cultural cognitive profile of students.

Principles of Personal Cultural-Cognitive Profile Design. In our opinion educational activities consist of operational and cognitive components. In order to describe cross-cultural differences we have to consider cultural models by G. Hofstede [6], R. Nisbett [9], E. Hall [5], M. Kholodnaya [7] etc. [1, 3, 4, 8, 18, 19].

We can define a number of the following parameters, underlying analysis of the culture-related aspects of behavior, mentality, activity and determining specificity of cultural-cognitive personality profile (Fig. 1): specific nature of activity; specific features of information representation; specific features of mentality and attention; specific features of social communications; dominant values.

COGNITIVE STYLE		Reflexive	Blended	Impulsive
Cognitive parameters	Specifics of working with information	*Attention to context* :Hi *Information structure*: trees *Type of thinking*: holistic	*Information structure*: blocks with a surface bond	*Attention to context*: Low *Information structure*: systematically organized by atomic units *Type of thinking*: analytical
	Attention specifics	Attention to «a frame»	Attention to objects into a frame. Frame plays linkage function	Attention to objects
	Decision-making specifics	Orientation to the authoritative opinion, the inclusion of others in the decision-making process, uncertainty avoidance	Orientation to the free choice of a permitted framework of society	Orientation to their own opinion, loyalty to the uncertainty
	Creativity specifics	Interpretation within the existing tradition	Creating a new, more advanced in framework of tradition	Innovativeness
Contextual parameters	Discourse specifics	Unity with collective, maintaining harmony	Variability	Expression of individuality
	Relation to the rules	Universalism	Variability	Particularism
	Relation to code of conduct	Closeness	Variability	Openness
Activity-related parameters	Specifics of activity	Reactive	Polyactive	Monoactive
	Relation to time	Time – nonlinear value (Cyclical)	Understanding the limitations of time as a resource. Cost of time: Low	Time – linear value. Cost of time: High
	Attitude to society	*Type of culture*: collectivist *Power distance*: high	*Type of culture*: hierarchical structure Power distance: average	*Type of culture*: individualism Power distance: low
	Status specifics	Significance of the origin	Depends on the situation	Significance of personal achievement
	Specifics of communication	*Attention to context*: high . *Style of communication*: branched argument. *Reasoning*: deduction *Genre*: narrative	*Attention to context*: average . *Style of communication*: mixed. *Reasoning*: intuitive *Genre*: mixed	*Attention to context*: low . *Style of communication*: cognitive, linear reasoning based on facts. *Reasoning*: induction *Genre*: discussions and debates

Fig. 1 Basic parameters of cultural-cognitive personality profile design

In fact, advanced "cultural intelligence" is an important component of adaptive educational process [14, 15]. Teacher with mature "cultural intelligence" will be able to identify cultural-cognitive profile of person and find appropriate communication strategy, and in case of strategic planning - an individual approach to education with suitable methods and training materials. Since we consider mainly the educational processes of e-learning environment, we are also interested particularly in those possessing "cultural intelligence." In our opinion, e-learning environment with "cultural intelligence" will enable adaptation of interface and educational content to a cultural-cognitive profile of the individual.

3 Models and Techniques of Interactive Didactic Support of Learners in Virtual Learning Environment

Support of modeling individual educational trajectory is mainly aimed at forming an environment for mapping a personal success as professional fulfillment through the following [16]:

- Forming virtual educational space which interlinks off-line courses, off-line events and on-line materials and events, generated by active experts and practition-ers.
- Tutor's support in forming a specialty's competence profile and modeling the labour market.
- Forming a student's competence profile and modeling this profile on the ba-sis of targeted preferences.
- Personalizing the educational environment and content.

The major service function for each user or participant group (Fig. 2) in the active learning process is to form a virtual educational environment as a system of interre-lated study courses through interlinked competence profiles of the courses and im-portant events both for a university campus, and for online experts in a course's field of application [15].

3.1 Basic Models for Realizing The «Smart Tutor» Service

Model of educational program, $MK_p(U_i, K_o)$, includes a system of study courses U_i, interlinked with each other and with the specialty's competence profile. [17]

Model of competence profile of curriculum $MK_p(U_i, K_o)$ includes: (1) Model of relations between courses U_i from the perspective of mastering certain competence. In particular, for such a multidisciplinary profession as "business informatics", oriented to prepare a range of specialists - this would be presented as a network model that can be constructed using the methods of data-mining, handling text courses.

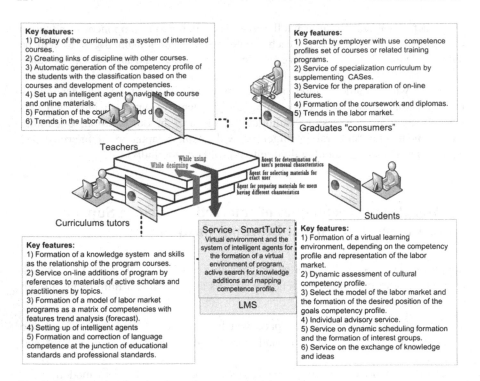

Fig. 2 Major service function for each users or participants group

(2) Model of accumulation of amount and quality of competences in relation to each of the possible specialization of labor market of a particular course; (3) Model of competence profile of curriculum U_i is a model of the relationship between knowledge, skills obtained and competencies. (4) Model of competence-profile of labor market of the curriculum graduates is a "summary" of competencies that determine the possibility of taking every competence on the labor market.

Accordingly, if the selection criteria of various academic disciplines are inputted, it becomes possible to receive a slice of the full set of vacancies. Specific professions on the market can be considered as a qualitative scale, which measures the level of professional success. Moreover, apart from such representation of the future professional activity of the student, we would like to implement a system that would take into account some personal parameters associated with the psychological, cognitive, communicative personality characteristics, which influence the choice of life goals.

At the same time, every study course of the program, U_i, must be represented within the three-element model of knowledge and skills, connected with the competences, K_i: knowledge and skills of the major course; additional theoretical knowledge and skills related to the course's objectives and based on competences, generated by active groups of experts currently developing theoretical and practical knowledge; practical knowledge of the level of application of skills, formulated by typical organizations representing the market (Fig. 2). In particular, the major university departments, being representatives of the labor market for the graduates, provide synchronization of competences acquired during the education, with those demanded in the market.

Developing a set of models of a study course's competence profile must be based, on the one hand, on the ontological approach to designing a model of the courses' links with each other, and on the other, on the methods of the decision taking theory for building a student's motivation function with flexible selection criteria, and on the association network theory for describing the principles of forming a student's knowledge model.

The following major methods of artificial intelligence have been analyzed: knowledge representation methods, argumentation modeling methods, education modeling and methods of knowledge acquisition by intellectual systems. The most suitable approach to solving creative tasks is the logics-semantics approach, showing a task as a structured model with links between elements [2, 10–14].

Application of up-to-date scientific and didactic approaches allows a regulating of the workload of a student. At the same time, the course presentation form must also change: the traditional linear structure must be substituted by hypertext and hypermedia form, allowing increase in the volume of the course materials, more forms of its' presentation, as well as better search for the necessary information. These approaches help develop self-study and creativity skills.

The basic knowledge representation models have different ideas underlying them. Empiric models are based on studying into the principles of human memory and task-solving processes. The second group of models is theoretical ones. They are based on formal logics and combinative models. A semantic net is a directed graph. Its nodes are concepts and objects, its edges are links between objects. Semantic networks fall into extension and intension ones. Intension semantic network describes names of object classes. Extension one describes the relations in a given situation.

The proposed approach to structuring and classifying educational tasks contributes to a student's forming knowledge system in the field of physics and mathematics. In the course of the research, we have developed semantic models for nature studies disciplines, providing rational sequence of developing study path models, fulfillment of requirements of the curriculum of certain courses.

Here is an example of semantic network on the topic "Matrices and their operations", "Linear Algebra" course (Fig. 3). Semantic model, representing the logical structure of the study materials and links between the concepts, allows better understanding of the topic. The model represents major concepts and causality relation between them.

In open education systems it is necessary to design an environment capable of integrating resources of various automated educational systems. This environment is called an information educational system for open learning. Such systems can be represented by a system of university departments and sub-faculties, educational institutions and training courses (majors).

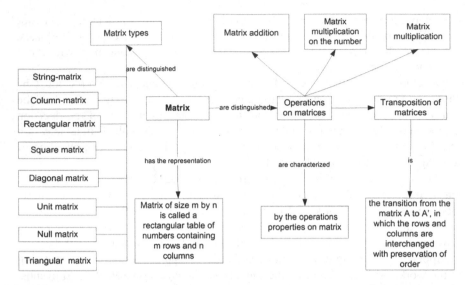

Fig. 3 Semantic model of structured knowledge for "Linear Algebra", topic "Matrices and their operations"

Interactive e-courses are an important means for a student's self-study process, however better results can be acquired through multimedia training systems, both separately and within the traditional education form. The innovative course is aimed at acquiring both generally cultural and professional competences, allowing self-learning when necessary.

3.2 A Possibility of Building a Learning Trajectory Based on Culture Specific Features

Based on the processes above, the process of cultural adaptation of multimedia content for each student according to his cultural characteristics appears. It means that, the initially identified student's cultural and cognitive profile will determine the specifics of the learning process, preferred learning tasks, working with educational information, and methods of getting feedback. We suppose that the difficulties caused by the distant form of educational process can be overcome, if subjects of pedagogical process have cross-cultural competency and there is synchronization of activities. In fact, consideration of the student's cultural and cognitive profile specific and, as a consequence, the nature of its educational activities will meet the expectations of students and make the learning process more effective. We also suppose that two main areas must be studied for creation of smart EEE: the selection and design of educational content according to SCCP and selection of interface options, text and illustrations (including information resource pages). [15, 16].

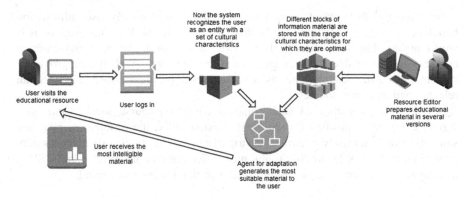

Fig. 4 Scheme of the formation of individual pages for users based on their cultural specificity

Everything, from evaluation of user's cultural parameters to generating personalized pages for him, can be automated. At the initial stage, we can query the user, and then select the most appropriate information for him as follows (Fig. 4).

It is important to say that not only internal but also external courses and educational programs may be assessed this way to provide student with recommendations for development his competence profile by both internal and public external courses. External courses should be selected on the basis of the possibility of certification for already selected learning directions and to improve knowledge in areas, which student is apt for. Regular analysis of SCCP development trends should be used as the basis for such an advisory service.

4 Conclusion

This paper formulates the major principles of constructive learning in an educational environment that contains 3 levels:

- human-to-human level
- human-to-EEE level
- adaptive educational content level

Designing a smart environment is, first of all, based on adapting and filtering the educational environment to fit a student's cultural-cognitive and competence profile. Thus, we can see, that on the 1st level it is essential to develop cultural intellect of the participants of educational communication. The 2nd level involves designing the EEE, with its own cultural intellect and capable of presenting knowledge according to a competence profile. The 3rd level means composing an adaptive - in some cases invariable - educational content (courses' semantic maps, minimal thesauri for disciplines).

We assume, that with this approach a student will mostly use information handling techniques best fitted for his/her own style in the learning process. It is worth noting that both internal (from own faculty or university) and external courses could be graded and presented to a student this way, with system providing student with recommendation on his/her competence profile improvement using all relevant and available sources.

In the end, completion of such personal education course would form up a certain competence profile of a specialist. Undoubtedly, the problem of mapping students' individual study pathways is currently one of the topical issues in modern education, and, in order to choose the most effective models of designing ISP, a thorough research into the global experience in this sphere is essential.

References

1. Beamer, L.: Learning intercultural communication competence. J. Bus. Commun. **29**, 285–303 (1992)
2. Berners Lee, T., Lassila, O.: The semantic web. Sci. Am. (2001). http://www.sciam.com/article.cfm?articleID
3. Blanchard, E., Frasson, C.: Making intelligent tutoring systems culturally aware: the use of Hofstede's cultural dimensions. In: International Conference on Artificial Intelligence, Las Vegas, pp. 644–649 (2005)
4. Gonçalves, V.: Facebook in the learning process: a case study. In: International Conference of Education, Research and Innovation: Abstracts. Madrid https://bibliotecadigital.ipb.pt/handle/10198/7439 (2010)
5. Hall, E.T.: The silent language in overseas business. J. Harvard Bus. Rev., 87–95 (1960)
6. Hofstede, G.: Culture's Consequences, International Differences in Work Related Values. Sage Publications, London (1980)
7. Holodnaya, M.A.: Cognitive Styles. About Nature of the Individual Mind, 2nd edn. Peter Publishing house, St. Petersburg (2004)
8. Lewis, R.: When Cultures Collide: Managing Successfully Across Cultures. Nicholas Brealey Publishing, London (1999)
9. Nisbett, R.E.: The Geography of Thought. Free Press, New York (2003)
10. Osipov, G.S.: Artificial Intelligence Techniques. Fizmatlit, Moscow (2011)
11. Osuga, S., Saeki, S., Sudzuki, H. et al.: Knowledge acquisition: trans. from Jap, Mir, Moscow (1990)
12. Robert, I.V.: Theory and Methods of Education Informatization (Psycho-Pedagogical and Technological Aspects). IRO RAE, Moscow (2007)
13. Shihnabieva, T.: On the representation of knowledge and control in automated training systems. J. Inf. Educ. **10**, 55–59 (2008)
14. Shihnabieva, T., Omarova, N.: Using adaptive semantic models in physics and mathematics education. J. Sci. Note **35**, 25–32 (2011)
15. Taratuhina, Y.V.: Choice of appropriate multimedia technology and teaching methods for different culture groups. Univ. J. Educ. Res. **2**(2), 200–205 (2014)
16. Taratuhina, Y.V., Aldunin, D.: Specificity of web user interface (WUI) organization in different cultures. World J. Comput. Appl. Technol. **1**(3), 59–66 (2014)

17. Taratuhina, Y.V., Avdeeva, Z.K., Mirishli, D.F.: The principles and approach support the mapping of the personal study pathway in electronic educational environments. J. Procedia Comput. Sci. **35**, 560–569 (2014)
18. Triandis, H.: Culture and Social Behavior. McGraw-Hill, New York (1994)
19. Trompenaars, F., Hampden-Turner, C.: Managing People Across Cultures (Culture for Business Series). Capstone Wiley Publishing Ltd, London (2004)

E-Learning Tools for a Software Platform

Dumitru Dan Burdescu

Abstract This paper presents the design and development of new tools for an innovative e-learning platform, with implications and analysis from human computer interaction (HCI) point of view. The goal of tools is to improve the educational process by helping professors to form a correct mental model of each leaner's performance. The main functions offered by the developed tools are evaluation of level of understanding of course material by a student and analysis of difficulty level of questions proposed by a professor for an exam. An important part of pre-education process is text analysis and learning concepts' extraction from uploaded learning content. The developed tools will provide an instructor with the ability to control difficulty level of test and exam questions and perform corresponding functions – to add, delete, or modify test/exam questions.

Keywords E-learning process · Distance learning · HCI · Assessment · Tools

1 Introduction and Related Works

E-learning has become a solution for people who want an acceptable and rationale way to learn from the point of view of time and space. The contemporary era has enhanced the accomplishment of new and complementary competencies eliminating borders between people and knowledge through the proliferation of e-Learning worldwide. During the last years, the interaction between professors and learners in online educational environments has been considerably improved, especially by developing new tools and implementing different functionalities that integrate intelligent data analysis techniques. More and more software companies specialized in e-learning and research centers use computer-supported cooperative tools to

D.D. Burdescu (✉)
Computers and Information Technology Department,
University of Craiova, Craiova, Dolj, Romania
e-mail: dburdescu@yahoo.com

© Springer International Publishing Switzerland 2015
V.L. Uskov et al. (eds.), *Smart Education and Smart e-Learning*,
Smart Innovation, Systems and Technologies 41,
DOI 10.1007/978-3-319-19875-0_21

231

overcome the geographical distance and benefit from access to a qualified resource pool and a reduction in development costs. Companies tend to adopt new and improved ways of using technology in the training process of employees as one of the most innovative methods currently used is mobile learning [1]. The e-learning performance is also visible in the rising number of research materials and papers on this topic as "the number of papers is increasing peaking in 25 % of the total in 2010" [2]. Development of an e-learning platform is based on the evaluation of e-learning requests and their match to process outcomes.

Recent standardization efforts in e-learning are focused on the reuse of learning material and functions. The TESYS software system [3, 4] is built on the idea that e-learning system is a collection of activities that put learners in contact with resources (e-learning courses, quizzes, etc.). The functionality of the TESYS e-learning software system is seen as a pool of implemented activities that may be performed by students/learners/trainees. These activities are grouped in accordance with a role of a particular user: administrator, secretary, professor/instructor or student/learner. Those activities may be seen as Web services offered by TESYS software system [5, 6].

The most important characteristics of e-learning environment should be usefulness, effectiveness, learnability, flexibility and success [7]. The context for which the tools are developed is related to online education. From this point of view, the developed tools are actually Web applications that integrate various technologies in order to achieve its business goal. The ultimate goal of those tools is to extend system's usability with respect to the particularities of e-learning environment. Additionally, one of the main goals of the educational research is an identification of learner's current level of understanding [8].

Cognitonics (cognitive psychology for information society and advanced educational methods) [9, 10] is another scientific discipline which contribution is needed for effective e-learning. Cognitive aspects are playing an important role in the information society and also in the particular case of e-Learning applications. From cognitonics point of view, the developed applications for sustaining online courses (or other related activities) should develop creativity, support cognitive-emotional sphere and appreciate the roots of the national cultures. One of the main goals of this quite new research discipline it is the development of a new generation of tools for online learning that compensate the broadly observed negative distortions [11].

Among many application domains where cognitonics finds a suitable place is e-learning. From this point of view, e-learning tends to have inputs from various research domains in order to improve its effectiveness and control the main negative side effects regarding linguistic ability, phonological ability, social relationships, etc.

Another research domain that is highly connected with the discussed issues is HCI. Currently, there are numerous research efforts that deal with user interface adaptation in e-learning software systems, adaptable interfaces featuring multiple views and finally integration of usability assessment frameworks that are designed and refined for the context in which they are applied. HCI issues related to e-learning are user-centered design [12] and user sensitive design [13]. From this perspective, adaptation of knowledge presentation, of interaction style regard specific issues like domain knowledge base generation, user/system interaction modeling, interface evaluation. Poor interaction in various online educational activities (e.g. evaluation of exercises after class, quiz games, intelligence analysis, etc.) may find proper solution by employing specific HCI research methodologies related to usefulness, effectiveness, learnability, flexibility and success.

2 Architecture of the TESYS Platform

These days the active use of e-learning software platform is a requirement for every university that wants to align to the existing trends and to have a high quality of educational processes and activities. TESYS e-learning software platform [3] provides flexible e-learning solutions and addresses needs of our university. Custom knowledge representation will enable designing context-aware environments, and, therefore, create the premises for shifting towards smart e-learning environments. The benefits of this approach for professors include a help to prepare courses and high quality e-learning content. Learners will benefit from this approach by being able to access needed learning material. The architecture of the TESYS software system [4] is presented on Fig. 1.

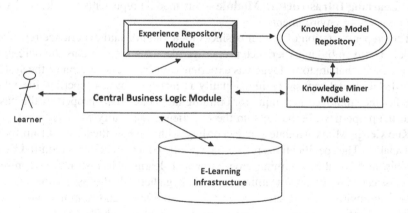

Fig. 1 TESYS system's architecture

The main TESYS system's components are the following ones.

Central Business Logic Module. This module contains the logic for accessing the e-learning infrastructure and for sending queries and receiving responses from the Knowledge Miner module. It is the main entry point into the system and HCI research; it also contains all general operation logic and tools.

Experience Repository Module. This component collects all data regarding the actions performed by learners. The research of HCI and cognitive psychology issues are cornerstone in shifting gears from "technology that solves problems" towards "design that emphasizes the user's needs". These general research areas have a great impact on the field of e-learning due to the wide range media that can produce cognitive affection at various industries. Text, voice, picture, video, or virtual reality are among the most common options.

Knowledge Model Repository. This module manages the current knowledge model representation; it is used whenever the intelligent character of an action is needed. Its usability significantly depends on particularities of e-learning environments. That is why a general fundamental usability evaluation formulated by Nielsen [14] needs a proper specific adjustment for e-learning environments.

Knowledge Miner Module. This unit collects business logic for querying the knowledge model repository. Many research hours have been allocated to the purpose of extracting key concepts from course materials, messages, questions and finding ways of using them for enhancing the teaching and learning processes. Also, a considerable amount of work has been put into discovering the similarity between concepts. The Natural Language Processing is another major research area, with a strong focus on documents (text and diagrams). In [15] the author presents the construction of an English-Romanian tree-bank, a bilingual parallel corpora with syntactic tree-based annotation on both sides, also called a parallel tree-bank. Tree-banks can be used to train or test parsers, syntax-based machine translation systems, and other statistically based natural language applications.

E-Learning Infrastructure Module – this module represents the classical view of an e-learning environment.

The proper operation of central business logic module and experience repository module is driven by an experience properties file. This file contains the definitions of the actions that are to be logged as experience during the operation of the system. Knowledge model repository functionality is managed by a properties file which specifies the employed technique for building the model. This properties file has as input the properties file that sets up the experience repository module.

Knowledge Miner Module runs according with the specifications set up by the data analyst. The specifications regard the specific educational goals required by the administrators of the e-learning environment. E-learning infrastructure represents the classical view of an e-learning system. It gathers all the assets managed by e-learning environment: users (e.g. learners, professors, and administrators), disciplines, chapters, course documents, quizzes. It also embeds the needed functionalities for proper running, like security, course downloading, communication, testing or examination. Central business logic module along with e-learning infrastructure represents the classical structure of an e-learning environment. Experience repository module, knowledge model repository and knowledge miner module may be regarded as an intelligent component that runs along the e-learning environment in order to enhance it.

Starting from the course documents that were previously uploaded by the professor on the platform, the software system extracts the concepts, using a custom concept extraction module, which incorporates a stemming algorithm and formulas. The obtained data is then transferred into the XML files. The five most relevant concepts are also implemented in the TESYS database. As soon as the professor uploads the test questions and specifies each concept's weight for every question, the student's activity monitoring process can begin. Afterwards, using the concept-weight association, student's responses to the test questions and taking into consideration the performances of learner's colleagues, the software system will be able to show relevant statistics to the professor, so he can understand each learner's learning difficulties as well as the general level of the class.

The tools are designed to review the difficulty of the proposed exam questions and advise the professor on lowering or increasing the exam difficulty. The entire process is supervised by a professor, who takes the final decision. Many issues appear when applications contain a mixture of data access code, business logic code, and presentation code. Such applications are difficult to maintain, because interdependencies between all of the components cause strong ripple effects whenever a change is made anywhere. The Model-View-Controller (MVC) design pattern solves these problems by decoupling data access, business logic, and data presentation and user interaction. This architecture of the software platform allows development of the e-learning application using MVC architecture. This three-tier model makes the software development process a little more complicated but the advantages of having a web application that produces web pages in a dynamic manner is a worthy accomplishment. The model is represented by DBMS (Data Base Management System) that in our case is represented by MySQL. The controller, which represents the business logic of the TESYS platform is Java based, being build around Java Servlet Technology. In this case, Apache Tomcat 5.0 is used as servlet container.

3   Tools and Technologies

The main purposes of the tools are to help the professor understand and analyze learning's activity without having any face-to-face activity. The first step for a learner is to retrieve a text from documents. A key feature of the module is an extraction of the most important concepts from every chapter that belongs to a course. This part of the module is divided into two steps: stemming and computing TF-IDF values. Each concept extracted from the course chapter's document is represented as an element in the *.xml file. The stemmed form of the word is stored as the element name, and its original form, TF value and IDF value are stored in the element attributes. The first five concepts that have the highest TF-IDF value are inserted in the database, for further use on the software platform. Figure 2 illustrates a part of the interface available to a professor to manage the concepts. A professor will be provided with a list of extracted concepts and several options such as (a) add new concepts, (b) modify the existing ones in case concepts are extracted incorrectly, delete the irrelevant concepts.

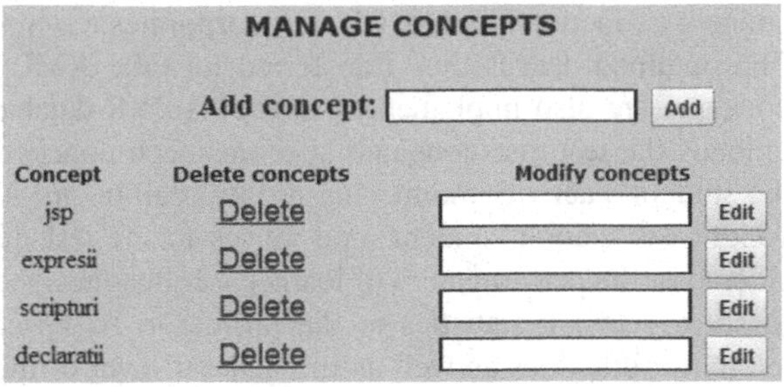

Fig. 2 Management of concepts

Several tools and algorithms have been developed for English word stemming [16], but for the Romanian language this research area is still at the beginning, therefore we developed our own tool and set of rules to accomplish this task. After the stemming process, the system will use the TF-IDF formulas for every word in the document; the obtained data are stored in *.xml file.

Within the e-learning platform a discipline has the following structure: Chapters, Test Questions, Exam Questions and Concepts. The chapters are documents uploaded by the professors, which can have one of the following extensions: .pdf, .doc, .docx. These documents are parsed and stemmed, and, as a result, are able to generate a list of concepts (Fig. 3).

Test questions are the questions used by learners throughout a semester for assessment of their current knowledge. As presented in Fig. 3, a list of concepts is available for each question – those concepts were extracted from each chapter relevant to that question.

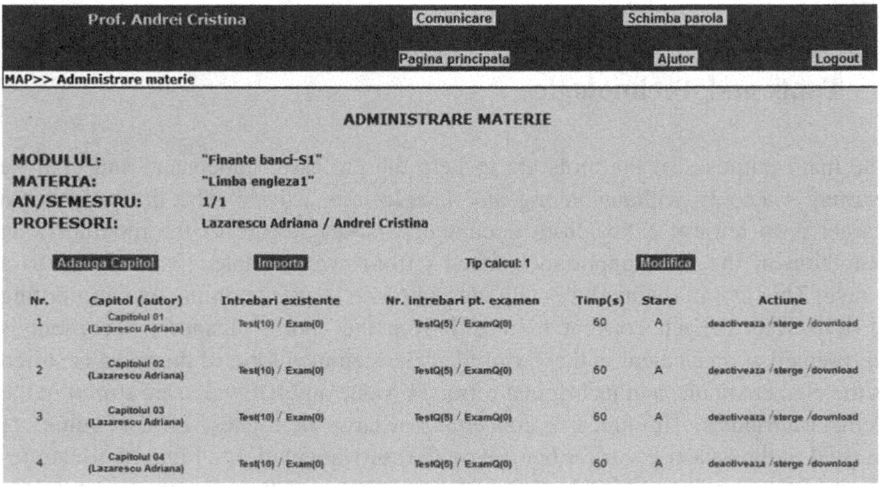

Fig. 3 Administration of discipline

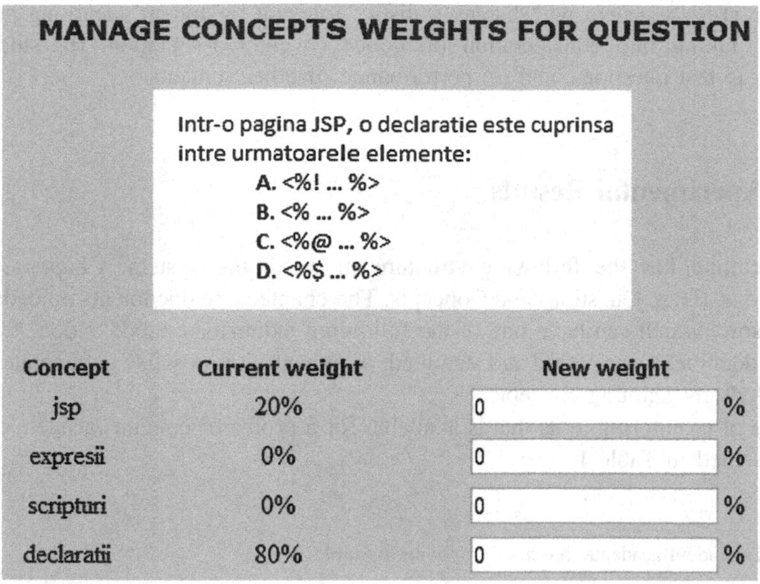

Fig. 4 Management of weights

A special feature was added to the e-learning platform – this feature allows a professor to select a concept and assign corresponding weights (Fig. 4).

This feature is important because it helps professor (a) to determine student's ability of understanding course material and (b) to identify student's progress. The most relevant information about a student can be obtained by evaluating the correctness of student answers to test questions, taking into consideration a concept associated with those questions and their given weights. Additionally, this approach will help a professor to figure out concepts that are difficult to learners and ways to help a learner.

Usability evaluation is the final and critical step within the lifecycle of any software system. An application of general heuristics (i.e. with no special tuning to educational context) may be a reasonable option; however, the use of approaches that are adapted to e-learning may offer even greater progress [18]. We can discus only constant data found on the server and additionally core business logic that includes concept's extraction, activity monitoring and recommender modules [17]. The TESYS system automatically extracts the underlying concepts and then transfers data into XML files. As soon as the professor uploads test questions and specifies each concept's weight for every question, a monitoring of learner's activity will begin.

The TESYS system will display statistics relevant to student's learning performance, taking into consideration (a) concept-weight association, (b) student's answer to test questions, and (c) performance of other students.

4 Experimental Results

A discipline has the following structure in e-learning system: Chapters, Test Questions, Exam Questions and Concepts. The chapters are documents uploaded by professors, which can have one of the following extensions: *.pdf, *.doc, *.docx. These documents are parsed and stemmed; as a result, it is possible to generate a list of underlying learning concepts.

One of monitoring tools that is available for a professor contains academic data as presented in Table 1.

Table 1 Student academic progress: a monitoring table

#	Student identificator in a class	Overall level of under-standing by a student	Number of questions answered by a student	Learning concept 1	Learning concept 2	Learning concept 3	Learning concept 4
1	21	78.33	25	100	70	100	100
2	18	76.42	26	100	80	100	100
3	16	65.37	32	100	83	100	100
4	14	56.45	18	0	73	0	0
5	12	53.23	33	0	66	0	0
6	7	51.53	28	100	60	100	80

Professor can watch academic progress of each learner, be informed about student's level of interest in online course content and identify concepts that may be somewhat difficult for students. From cognitive perspective, the visual stimuli refer to the following processes: visual search, find, identify, recognize and memory search [19]. From this perspective, specific issues may be obtained within the TESYS e-learning system. The presented educational assets (e.g. quizzes, concepts, etc.) can be characterized by someone as confusing, not findable, etc. and, thus, reduce the unwanted side effects regarding the distortions in perception of learning content.

In order to better explain how the system works, we will consider a sample scenario; it is based on example in [4] also. The main steps of the scenario are:

Professor: concept setup. Let's assume that professor P is a professor on the e-learning platform, and has a course with two chapters. Professor should upload the documents of the chapters first. Immediately after that the TESYS system parses the documents, applies the stemming algorithm and the TF-IDF formulas, and ultimately extracts the most important five key concepts from each file: C11-C15, C21-C25, etc. The list of concepts can be accessed from the course page; a professor can edit, modify, add, delete concepts.

Professor: test questions setup. This step is performed when professor P loads test questions for students to answer. Each question is assigned with weights of the extracted concepts, denoting the relevance level of every concept for each particular question. The weights have values in the range of [0.0,1.0]. The cells corresponding to the concepts that have absolutely no relevance to a question have a weight of 0.0 and are left blank. A professor is able to assign the weights as percentages, as previously presented in Fig. 4. These weights can be updated at any time, and the progress of the learners will be modified accordingly.

Student: take a test. Let us consider student S1. For example, the first test a student takes contains questions 1, 2, 3, 5, 7. In [4] we presented answers given by a learner to designated questions. S1 is a learner under analysis, S2 to Sn are the other learners that answered test questions. It is assumed that the tests contain only single choice questions, so the answer can be only evaluated as CORRECT or INCORRECT. After computing the weights and results, the TESYS system will provide instructor with the following statistics.

1. Learner's level of understanding of each concept:

$$LU_c = \sum_{i=1}^{n-p} W(c, qi) \Big/ \sum_{j=1}^{n} (c, qj)$$

   In this formula, LUc represents the level of understanding of concept 'c' and w (c, qi) is the weight associated with this concept for question qi. The numerator of the fraction is therefore the sum of the weights of the concept for the correctly resolved questions, and the denominator is the maximum amount that could be obtained if the responses to all questions were correct.

2. Learner's performance relative to his/her classmates:

$$Performance = \sum_{i=1}^{k} (LU\, ci) / Avr(Lu\, ci).$$

3. The difficulty level of every test question

   $$Difficulty = (number\ of\ correct\ answers) / (total\ number\ of\ answers)$$

4. The difficulty level of each concept.

After analyzing these obtained data for each learner, professor will be able to start creating a mental model regarding learner's current level of understanding of course material and his/her place among other students. Also, if necessary, instructor might decide to modify the course material, for example, add some extra information on a particular concept that the students have trouble to understand. Another possible action by a professor is to increase or decrease a general difficulty level for test questions.

5   Conclusions and Future Steps

With the growth of the Internet, online education has removed space and time constraints from learning and expanded opportunities for learners. Furthermore, with the advent of massive open online courses online education serves as an effective counter measure for expensive colleges. While there are many significant beneficial aspects of tools for e-learning, our experience clearly shows that those benefits are not easily reachable.

This paper describes a use case of building and integration of various tools for existing TESYS e-learning software platform. The architectural model of the TESYS is presented as well as descriptions of TESYS main structural components. We also presented descriptions of functionality and graphic users interfaces of several developed tools for TESYS.

Another important issue discussed in this paper is the analysis of cognitonics and HCI related issues as an attempt to offer a higher quality interaction design that minimizes the cognitive side effects.

We believe a continuous use of the developed tools within TESYS e-learning platform will provide positive evidence of proposed ideas, approaches, and design and development software solutions.

The next steps in this research, design and development project deal with

(1) detailed analysis and identification of structure (i.e. components and links) of underlying knowledge and data (e.g. concepts, weights, formulas);
(2) detailed analysis of TESYS features in terms of high quality HCI and cognitonics perspectives;
(3) design and development of Romanian language focused stemmer;
(4) study of additional techniques and algorithms to determine learning concepts, weights for test questions and overall student's knowledge level;
(5) analysis of existing e-learning tools in terms of HCI and cognitonics perspectives in order to arrive with a design and development of a comparative framework for e-learning systems.

References

1. Miller, L.: State of industry report: organizations continue to invest in workplace learning. Am. Soc. Train. Dev. 42–48 (2012)
2. Wu, B., Xu, W.X., Ge, J.: Innovation research in e-learning. Phys. Procedia **24**, 2059–2066 (2012)
3. Burdescu, D.D.: TESYS—An e-learning system. In: The 9th International Scientific Conference "Quality and Efficiency in e-Learning", vol. 3. ISSN 2066-026X (2013)
4. Burdescu, D.D. A Platform for e-Learning, IOS Press - Frontiers in Artificial Intelligence and Applications, vol. 262 ISBN: 978-1-61499-404, pp. 633–640 (2014)
5. Vossen, G., Westerkamp, P.: E-learning as a Web service (extended abstract). In: Proceedings of the 7th International Conference on Database Engineering and Applications (IDEAS), Hong

Kong, China, IEEE Computer Society Press, pp. 242–249; full version available as Technical Report No. 92, IS Dept., University of Muenster (2003)

6. Burdescu, D.D., Mihaescu, C.R.: TESYS: E-learning application built on a web platform. In: International Conference on E-Business, Setubal, Portugal, pp. 315–319. ISBN 978-972-8865-62-7 (2006)

7. Randy Garrison, D.: E-Learning in the 21st Century: A Framework for Research and Practice. Routledge Press, London (2011)

8. Mihaescu, C., Tacu, M.G., Burdescu, D.D.: Use case of cognitive and HCI analysis for an E-learning tool. Informatica—Int. J.Comput. Inf. Spec. Issue: Front. Netw. Syst. Appl. **38**(3), 273–280 (2014). ISSN 0350-5596

9. Bohanec, M., Gams, M., Mladenic, D. et al (eds.): Proceedings of the 14th International Multi-conference Information Society—IS 2011, vol. A, Slovenia, Ljubljana, 10–14 Oct 2011. The Conference Kognitonika/Cognitonics. Jozef Stefan Institute, http://is.ijs.si/is/is2011/zborniki.asp?lang=eng, pp. 347–430 (2011)

10. Gams, M., Piltaver, R., Mladenic, D. et al. (eds.): Proceedings of the 16th International Multi-conference Information Society—IS 2013, Slovenia, Ljubljana, 7–11 October 2013. The Conference Kognitonika/Cognitonics. Jozef Stefan Institute, http://is.ijs.si/is/is2013/zborniki.asp?lang=eng, pp. 403–482 (2013)

11. Fomichov, V.A., Fomichova, O.S.: Cognitonics as a new science and its significance for informatics and information society. Informatica. Int. J. Comput. Inf. (Slovenia) **30**(4), 387–398 (2006). http://www.informatica.si/vol30.htm#No4

12. Norman, D., Draper, S.W. (eds.): User Centered System Design. Earlbaum, Hillsdale (1986)

13. Granić, A., Glavinić, V.: Automatic adaptation of user interfaces for computerized educational systems. In: Zabalawi, I. (ed.) Proceedings of the 10th IEEE International Conference on Electronics, Circuits and Systems (ICECS 2003), Sharjah, Dubai, pp. 1232–1235 (2003)

14. Nielsen, J., Molich, R.: Heuristic evaluation of user interfaces. In: Jane Carrasco Chew and John Whiteside (eds.) Proceedings of the SIGCHI Conference on Human Factors, Computing Systems (CHI '90). ACM, New York, NY, USA, pp. 249–256 (1990)

15. Colhon, M.: Language engineering for syntactic knowledge transfer. Comput. Sci. Inf. Syst. **9**(3), 1231–1248 (2012)

16. Sharma, D.: Stemming algorithms: a comparative study and their analysis. Int. J. Appl. Inf. Syst. **4**, 7–12 (2012)

17. Burlea Schiopoiu, A., Badica, A., Radu, C.: The evolution of e-learning platform TESYS user preferences during the training processes, (ECEL2011), 11–12 Nov 2011, Brighton, UK (2011)

18. Squires, D., Preece, J.: Predicting quality in educational software: evaluating for learning, usability and the synergy between them. Interact. Comput. **11**, 467–483 (1999)

19. Zhang, S., Zhan, Q., Du, H.: Research on the human computer interaction of e-learning. In: Proceedings of the International Conference on Artificial Intelligence and Education (ICAIE) (2010)

Transferring Smart E-Learning Strategies into Online Graduate Courses

Laurie F. Ruberg

Abstract This study is designed to increase understanding of how learner-centered instruction and emerging technologies can be applied to improve online courses and increase student satisfaction and performance at the higher education level. The research questions examine predictors of student achievement, persistence, and satisfaction in online learning and investigate what strategies are most effective at building learner success and optimal online learning contexts. Three models for systematic design of online learning courses are considered. Two of these models are used in a reflective data analysis process that uses data from three recent online courses as test cases. Analysis of results shows that an integrated use of the two models featured provides a useful design approach to support on-going innovation and systemic analysis of online course implementation. Recommendations for expanding upon this research are provided.

Keywords Online learning · Reflective teaching · Instructional design

1 Introduction

As today's colleges and universities are experiencing many changes, there is great need to study ways to improve online courses. Research [1, 2] demonstrates that rising economic pressures heavily weigh on student choices to pursue graduate studies online. Unlike traditional pathways of returning to the university full-time, more and more students today are working towards advanced degrees while working fulltime. In addition, research shows that online teaching and learning have pedagogical benefits, such as learner control in terms of learning pace as well as path, and interactions between learners and instructors, or learners and learners [3–5].

L.F. Ruberg (✉)
College of Education and Human Services, Learning Sciences and Human Development Department, West Virginia University, Morgantown, WV, USA
e-mail: lfruberg@mail.wvu.edu

© Springer International Publishing Switzerland 2015   243
V.L. Uskov et al. (eds.), *Smart Education and Smart e-Learning*,
Smart Innovation, Systems and Technologies 41,
DOI 10.1007/978-3-319-19875-0_22

Looking more closely at these changes in higher education demographics we see the following trends. Between 2001 and 2011, enrollment in degree-granting institutions increased 32 %, from 15.9 million to 21.0 million [6]. In recent years, the percentage increase in the number of students age 25 and over has been larger than the percentage increase in the number of younger students. Between 2000 and 2011, the enrollment of students under age 25 increased by 35 %. Enrollment of students 25 and over rose 41 % during the same period [6]. The increasing number of part-time students attend school while carrying full-time jobs, many students are married, have children, and juggle other responsibilities. To address the needs of increasing numbers of part-time and non-traditional students, gives the online classroom more attention and importance [1, 2].

Previous research shows that students are attracted to the convenience, flexibility to customize their coursework schedule to fit their life style, access to a greater variety of course offerings, and semi-anonymous social interaction that is offered through online learning. This study examines student perceptions and preferences to identify ways to incorporate smart learning tools and strategies to better serve learners in higher education online degree courses. Specifically, this research applies an iterative, systematic instructional design process to present an integrated approach based on evidence to guide improvement of online courses.

A variety of traditional classroom teaching techniques are easily adapted to online learning, but effective online teaching online requires integration of new learner-centered pedagogies. As online education is an increasingly viable option for many students, the field of online instruction is changing so rapidly that the techniques for teaching online must be revisited regularly [7]. This research addresses the problem of how to integrate pedagogical guidelines and smart technologies in a manageable, systematic way by applying knowledge transfer and ranking of research evidence elements that are then prioritized in a systemized cue for curriculum integration [8].

The purpose of this paper is to increase understanding of how learner-centered instruction and emerging technologies can be applied in a systematic design process to improve online courses and increase student satisfaction and performance at the higher education level. The research questions listed below were used to guide the literature review, research, methodology, data collection, analysis, and discussion of results.

1. What factors are the strongest predictors of student achievement, persistence, and satisfaction in online learning?
2. What design strategies are most effective at building on these learner success factors to create optimal online learning contexts?

2 Literature Review

2.1 Predictors Suggesting Success in Online Learning

Recent research [3] of learner satisfaction, achievement, and persistence in graduate-level online learning found that: (a) locus of control, self-efficacy, and learner perception of task value were significant predictors of learner satisfaction with online learning; (b) self-efficacy and learner perception of task value were reliable predictors of student achievement; (c) learner perception of task value, satisfaction, and achievement were significant predictors of student persistence; and (d) learner satisfaction with the online learning was significantly mediated by locus of control, self-efficacy, learner perception of task value, and persistence.

Individual behavior is viewed as a function of his/her expectancies and the value of the goal toward which the individual is working. Thus, task value is related to the belief that the task one is performing is valuable to oneself. From this context and based on the expectancy-value theory [9], task value tends to predict the online learner's decision regarding whether to pursue learning further.

The most basic measure of student satisfaction is reflected in students self-assessment that they are successful in the learning experience and are pleased with their experience" (p. 74) [10]. In their empirical investigation of critical factors influencing learner satisfaction, Sun [4] describes survey results that point to these seven factors as having the greatest influence learner perceived satisfaction: learner computer anxiety, instructor attitude toward e-Learning, e-Learning course flexibility, e-Learning course quality, perceived usefulness, perceived ease of use, and diversity in assessments.

The persistence rate refers to the percentage of students who return to college at any institution for their second year. The retention rate is the percentage of students who return to the same institution for their second year [6]. Ashwin [11] suggest that persistence in online learning is related to student perceptions of their enjoyment and personal value of their learning.

The factor labeled institutional support refers to strategies that go beyond specific coursework designed to encourage learners and help them be successful. Some of these institutional support factors include offering facilities and/or coordination services for students to form (face-to-face or virtual) study groups or tutoring services. Other support services include course-targeted library assistance, and software resources.

Student achievement is defined by is the status of subject-matter knowledge, understandings, and skills at one point in time. In most online courses, student achievement is defined by their mid-term, project, final exam, and overall project grade. Ashwin [11] addresses achievement within the context of learning outcomes and suggests that educational experiences should lead to development in knowledge, concepts, skills and attitudes.

2.2 Design Strategies to Create Optimal Online Learning Contexts

Butler et al. [12] describe a classroom study in which they identified three simple powerful principles from cognitive science that could be combined to improve learning. The three principles are: repeated retrieval practice (students solved three sets of problems on each topic presented), spacing (distributing practice over time produces better long-term retention), and feedback (immediate feedback after assignment deadline). This research found that these principles from cognitive science increased retention and promoted transfer of learning regardless of the content area. This design strategy should be implemented across all of the courses discussed in this paper.

Koehler and Mishra [13] introduced the Technological Pedagogical Content Knowledge (TPCK later renamed TPACK) as a conceptual framework to describe the knowledge base teachers need to effectively teach with technology. The TPACK framework represents the interdependence of the interdependence of content knowledge, pedagogical knowledge and technology knowledge. Niess [14] suggests using the TPACK (also known as TPCK) framework as a means for integrating the development of knowledge of subject matter with the development of technology and knowledge of teaching and learning to support educators teaching in their subject matter with technology.

For this study, the TPACK model as represented by Koehler & Mishra [13] provides a helpful reminder of the relationships and interplay between the three domains that must be considered in online course design and development: Technological Knowledge (TK), Pedagogical Knowledge (PK), and Content Knowledge (CK). However, the TPACK framework is complex and problematic as conceptualized by Koehler and Mishra [13], and has been the subject of much scholarly debate [15]. While the TPACK model is quite useful as a conceptual visualization of the triage of knowledge components that must be applied for effective technology-rich learning contexts, the reflective learning framework developed by Ashwin [11] is a less problematic and more straight forward conceptual tool to guide the design on online course curriculum, pedagogy, and assessment.

The ideas that form the reflective teaching conceptual framework are that enduring educational issues concerning educational aims, learning contexts, classroom processes, and learning outcomes are expressed in practice through curriculum, pedagogy, and assessment. The reflective teaching conceptual framework presents an generative, narrative tool for designing online learning course components and fits the criteria posed by Butler [12] that designers much identify and use models that encourage easy and meaningful integration of evidence-based research. This model will be used to guide the systematic design process described in the methodology section.

Using the reflective teacher conceptual framework discussed by Ashwin [11] in combination with the media wheel presented in Fig. 1, demonstrates two components of a systematic instructional design processes that emerge from online learning literature. A truly systematic design process is multi-dimensional and requires distinct conceptual tools that can support considerations based on curricular, pedagogical, and assessment concepts, which are reinforced by using the reflective conceptual framework. This level of analysis, however, does not address the course designer's need to consider learner perceptions of the online experience as described by Sun [4] and to select media-based learning experiences as depicted in the media wheel (Fig. 1).

Teaching practices continually examine student perception of online course experiences and factors that influence e-Learner satisfaction [4]. Learner centered teaching means subjecting every teaching activity (method, assignment or assessment) to the test of a single question: "Given the context of my students, course and classroom, will this teaching action optimize my students' opportunity to learn?" The Media Wheel presented in Fig. 1 provides a visual map of flexible online learning experiences. The four major divisions of the wheel, communicative media, productive media, interactive media, and narrative media illustrate the modalities that online course designers can use to categorize and then assess the relevance and usefulness of their online course presentation in supporting the predictive/positive factors for student satisfaction, achievement, and persistence outlined above.

To create an online learning community that nurtures students' sense of social presence and involvement with the ongoing communications events, researchers have identified strategies for facilitating and guiding online discussion to support community-building and nurture student learning and knowledge construction [16]. Online learning communities shift the ways in which learning is delivered and alter the roles and responsibilities for students and teachers in regard to social, cognitive, and teaching presence. Role adjustments should be addressed at the beginning of each course so that students know what to expect and how to perform socially in the unique context of the course-specific learning forum [17].

Researchers have analyzed student use of online synchronous and asynchronous discussion tools and have found that these different tools can be strategically used to serve practical, coordination, and social functions or can foster in-depth learning, knowledge construction, and high-level synthesis of ideas—depending on the instructor guidelines and presentation of material [16, 18–20].

Online surveys comparing student success in online courses with instructional and environmental factors found that student success in online classes was primarily based on their perception of the organization, clarity of activities, and grading policies of the course [21]. Effective online learning environments focus on constructive and cumulative learning. The learning context should strive to achieve authentic and enduring understanding, include cooperative approaches, offer lessons that students can complete at their pace, and clearly align with overall course goals and objectives [21].

The learner satisfaction and social constructionist approaches to learning are recognized in the communication media section of Fig. 1. Learner social needs are addressed in the classroom processes categories of the reflective teaching conceptual framework (Table 1).

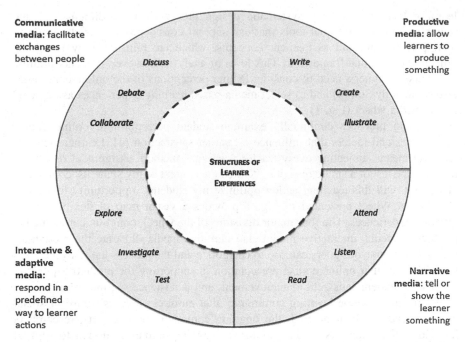

Fig. 1 The media wheel was used to visually map online learning experiences that are selected to enhance learner experiences and learning outcomes (Source: hedbergskan.wordpress.com)

3 Methodology

An interpretive qualitative approach was used to study and enable understanding of the course design and the reflective teaching conceptual framework [11]. The media wheel (Fig. 1) can be used to improve online course design and implementation at the higher education level. The multimedia data sources, content management system used, course curriculum goals and objectives, survey instruments, and observation techniques are detailed in Table 2.

The following steps were followed within the interpretive design processes: (1) Review student comments from all courses and aligning them with conceptual framework presented in Fig. 1 to identify areas where changes are in progress, difficulties encountered, and sources of support. Table 1 provides an abbreviated summary of these data sources; (2) Classify selected new activities from across the courses and show how these activities align within the different Media Map domains. (This data is presented in Table 2.); and (3) Characterize and partition each course's knowledge elements within the Media Map structure. The list of changes in progress and areas of improvement are compared with the predictors of student achievement, persistence, and satisfaction in online learning to identify alignment, which suggests areas most likely to increase student satisfaction.

Table 1 Reflective conceptual framework applied to online course systematic design

Factor	Changes needed	Difficulties encountered	Sources of support
Instructor	Make course structure, clear for students	Design options on LMS limited	Incorporate social media external to LMS
Student	Class includes different generations, degree programs, experience levels	Student technology issues prevent consistent course experience	Offer live Collaborate (videoconference) sessions with pedagogical content & open Q/A
Educational Aims	Be more specific about knowledge, concepts & skills, to be learned in each course & across degree program	More specific knowledge, concepts & skills are less broadly applicable to diverse students	Apply research-based instructional strategies to make key elements of course meaningful for diverse skills & abilities
Learning Contexts	Students want more critical feedback from peer reviews & value this communications when it's effective	Students vary in their confidence & experience in giving and receiving peer feedback	Show students research rationale for peer interactions as well as case-based examples of positive peer exchanges
Classroom Processes	Apply principles from cognitive science research [12] to promote optimum long-term retention and transfer of course content	Apply principles of repeated retrieval practice, spacing & feedback to improve curriculum design	Share Butler's [12] research study with students so they can incorporate these evidence-based principles into their learning practices
Learning Outcomes	Curriculum needs to offer learner sequence with remedial and advanced activities	It is difficult to measure student variability until after module assessment	Begin new content modules with a course-level or sequence-based pre-assessment

Learning Context. The learning context is an online graduate degree program in instructional design and technology. Giving students authentic exposure and hands-on meaningful experience with smart technology tools is part of the course goals and objectives. The student population participating in the online degree program represents widely varied demographic backgrounds and careers, and includes a wide range of content and technical skills and abilities. The diversity of participant backgrounds and interests make the class discussions and project work very engaging, but also create challenges in how to offer an organized and objectives-based curriculum and grading policy that is equitable, doable, and challenging for all students.

4 Analysis

Below is a narrative review of the online course that is used as the focus of this study. The course included self-paced learning modules, web-based surveys, self-tests, assignment timelines, sequenced activities, synchronous and asynchronous communication, social presence, discussion boards, scaffolding tools and prompts, videos, screencasts, videoconferencing, interactive simulations, and web links.

Table 2 Aligning course activities to media wheel categories

Learning activities	Communicative	Productive	Narrative	Interactive/adaptive
Discuss: *technocentricism and technology integration*	X		X	
Join: twitter group	X			X
Design and share: lesson that requires social media	X	X	X	
Discuss: *digital literacies*; *bloom's taxonomy*	X		X	
Discuss: design models	X		X	X
Test tools: *kodu, scratch, excel*	X	X	X	X
Design/share: ID project	X	X	X	
Review: presentations	X	X	X	X

The Technology Integration course represents the most recently adapted syllabus and includes the most complete application of the Media Map approach. Students rated their peer interactions high as well as their peer evaluation of student-selected new technologies. The instructor sustained the context and disseminated weekly tasks, but students found their peer interactions most meaningful to them and these peer activities gave students the greatest satisfaction.

The over-arching objectives for this course were twofold. First, it was important for students to **explore** current research literature and new technologies to gain understandings of central issues surrounding integrating instructional technologies successfully and effectively. Second, students would have opportunities to **develop** learning spaces tailored toward the addressing 21st Century skills. The course was designed to also address eight additional objectives related to learning technology integration theories, concepts, processes, and practices.

This is a graduate-level course and was structured to support student learning and achievement of all course objectives through (1) weekly readings and discussions about central issues related to integration of instructional technologies; (2) technology evaluations which involved analysis of the potential for a new tool to solve a real or hypothetical instructional problem; (3) peer feedback which required comparison of technology reviews within small groups where classmates have different perspectives, experiences, and expertise; and (4) culminating in a final project that involved the creation of a technology-rich and innovative learning space that utilizes technologies to solve an identified instructional problem.

Students completed an end of course survey of key components of the class. The results of this survey were and shared with the students before the student evaluations were due. Data from both of these sources help me to assess course readings, course-required technology tools, peer selected technology evaluations, class activities, achievement of course objectives, student perceptions of their improvements in skills and abilities, and recommendations for improving future course offerings.

It was also important for students to see the results of the course review survey so that they could compare their responses with the rest of the class. The survey results were a good way to reinforce for students that one technology solution does not work equally well for everyone. Realizing that a successful plan for a program-wide implementation of the smart e-learning integration requires a plan to transfer knowledge from graduating students to novice ones, student interactions were structured to mix experienced online students with those brand new to online courses.

Delivery mode/Class format –This syllabus required weekly design, development, and resource collection for each week's activities, assignments, and the final project. The text, Invent to Learn: Making, Tinkering, and Engineering in the Classroom [22] was added because this book introduces current issues involved in creating 21st Century learning environments and reinforces each of the specific course learning objectives with authentic case references.

Technologies used – Since this was an online course, eCampus [a campus adaptation of Blackboard] was used as the course management system—consistent with other online courses. By the end of this course, students reviewed and/or used **24 different technologies** through hands-on application with guidance from the instructor, course assignments, and the peer technology evaluation presentations.

Pedagogical innovations introduced into course. An underlying theme for this course was an attempt to bring examples of effective technology integration projects into the class activities and discussions. The textbook that was chosen for this class provided a critical link to active innovations underway in "cities of learning" throughout the United States. Pittsburgh is one of the cities of learning hubs, which is inspired by the ideas presented in the *Invent to Learn* book.

The pedagogical approach was guided by wanting to make sure that each weekly activity was also tied to a theory-based rationale. The course began with a discussion of the on-going dialectical struggle between the attraction of new technologies that offer exciting and engaging affordances, but that risk being used for superficial purposes and which may even distract from the contextual learning objectives.

4.1 Reflective Teacher Conceptual Framework

The reflective teacher conceptual framework was shown to be a helpful model to guide the design and implementation of online courses. This tool includes definitions embedded within the model as well as guiding questions that help prompt the designer to think both within and outside the course context for strategies and tools to extend the learning experience. The tool also encourages the designer to build curriculum, pedagogy, and assessments based on multiple learning theories based on the targeted questions asked across the enduring issues through more pragmatic curricular, pedagogy, assessment levels of design. This tool guides the designer through a logically structured, multidimensional approach to planning their online course aims, learning context, classroom processes, and learning outcomes while thinking through curriculum, pedagogy, and assessments. The reflective teacher tool addresses instructional design considerations that foster extended learning, metacognitive learning, and efficacy through questions, which are typically disregarded in syllabus development such as "...Is the pedagogic repertoire successful in enhancing learning disposition, capabilities, and...a confident sense of personal identify?

Table 1 data provides a case-based representation of how the Reflective Teacher tool can be applied to support on-going design improvements. Column 1 displays the factor being examined. Column two identifies changes needed that are gathered from student exit surveys, observations of online class-wide or peer group discussions, or analysis of student performance. The needs described are then assessed to identify any difficulties or barriers to addressing the changes needed. The difficulties encountered are described in column 3. Column 4 identifies possible solutions to the difficulties or sources of support to overcome barriers prohibiting change. The reflective teaching conceptual framework is used as a tool used in the data analysis process to guide systematic online course design across multiple dimensions. Referring to related categories on the reflective teaching conceptual framework expands the design space to consider specific course issues within the larger social, research, and methods of practice contexts.

4.2 Media Wheel Model

Application of the Media Wheel as a tool to guide online program updates and enhancements across courses proved to be very useful. As Table 2 illustrates, use of the media wheel (Fig. 1) as a tool to categorize instructional activities by media types gives a useful quick overview of the types of activity engagement offered in the course over time. A quick comparison of the media types used with predictors of student achievement, persistence, and satisfaction with online learning will suggest areas where instructional media should vary and where pedagogical and assessment strategies should be updated to include emerging technologies that more effectively engage learners and address learning goals.

5 Discussion and Recommendations

The interpretive analysis of the reflective teacher conceptual framework and media wheel shows that these models are effective tools to promote learner-centered instruction, integrate new technologies and evidence-based research practices.

Use of the reflective teacher framework helped position the instructional design work in a rich environment where multiple dimensions of educational aims, learning contexts, classroom processes, and learner outcomes could be considered while curriculum, pedagogy, and assessment methods are prepared. The media wheel provided an efficient and easy tool to visualize how to incorporate a variety of media into the weekly course activities. Use of both of these models helps keep the designer engaged in developing valuable, effective and innovative courses.

Regarding what factors are the strongest predictors of student achievement, persistence, and satisfaction in online learning, the data and analysis shows that each of these factors is considered in the systematic design process that results from using each of the design models. Examples provided in Table 1 addressing learning contexts demonstrate how locus of control and self-efficacy factors can be supported in design choices.

Regarding what design strategies are most effective at building on learner success factors to create optimal online learning contexts, analysis shows that the two models used for this case-based design data collection and analysis (Tables 1 and 2) helped the researcher generate design ideas and integrate learner feedback and incorporate new research evidence within efficient and simple guidelines.

Finally, an online course instructor needs to collect survey data about their participants, the course content, social context of the course activities, and overarching goals and objectives of the learner. Systematic protocols for observing online interactions between students and between students and instructor should be established. Within course and across courses comparison of student performance can help provide document artifacts of successful learning outcomes. All of these factors can help course designers understand what successful online learning looks like and areas for continued improvement and enhancements.

References

1. Palloff, R.M., Pratt, K.: Building Online Learning Communities: Effective Strategies for the Virtual Classroom, p. 18. Jossey-Bass, San Francisco (2007)
2. Elam, E.: A quantitative inquiry into the factors that influence success in online classes. A dissertation presented to the faculty of the Curry School of Education, University of Virginia. UMI 3528581. ProQuest LLC, Microform edn. (May 2012)
3. Joo, Y.J., Lim, Y.L., Kim, J.: Locus of control, self-efficacy, and task value as predictors of learning outcome in an online university context. Comput. Educ. 62, 149–158 (2013)
4. Sun, P.-C., Tsai, R.J., Finger, G., Chen, Y., Yeh, D.: What drives a successful e-learning? An empirical investigation of the critical factors influencing learner satisfaction. Comput. Educ. 50 (4), 1183–1202 (2008)
5. Tang, M., Byrne, R.: Regular versus online versus blended: a qualitative description of the advantages of the electronic modes and a quantitative evaluation. Int. J. E-learn. 6(2), 257–266 (2007)
6. U.S. Department of Education (NCES), National Center for Education Statistics. Digest of Education Statistics, 2012 (NCES 2014-015), Chapter 3. (2013) http://nces.ed.gov/fastfacts/display.asp?id=98

7. Motte, K.: Strategies for online educators. Turk. Online J. Distance Educ. **14**(2), 258–267 (2013)
8. Reik, A., Maurer, M.: Consideration of dynamics in knowledge prioritization preparing an efficient company-internal knowledge transfer. Impact: J. Innov. Impact. ISSN: 2051-6002, http://www.inimpact.org **7**(1), 257–265: inkt14-026 (2014)
9. Eccles, J., Adler, T.F., Futterman, R., Goff, S.B., Kaczala, C.M., Meece, J.L., Midgley, C.: Expectancies, values and academic behaviors. In: Spence, J.E. (ed.), Achievement and Achievement Motivation, pp. 75–146. Freeman, San Francisco (1983)
10. Moore, J.C.: A synthesis of Sloan-C effective practices: December 2009. J. Asynchronous Learn. Netw. **13**(4), 73–97 (2009)
11. Ashwin, P., Boud, D., Coate, K., Hallett, F., Keane, E., Krause, K.-L., Leibowitz, B., MacLaren, I., McArthur, J., McCune, V., Tooher, M.: Reflective Teaching in Higher Education. Bloomsbury Academic, London (2015)
12. Butler, A.C., Marsh, E.J., Slavinsky, J.P., Baraniuk, R.G.: Integrating cognitive science and technology improves learning in a STEM classroom. Educ. Psychol. Rev. **26**, 331–340 (2014)
13. Koehler, M.J., Mishra, P.: What happens when teachers design educational technology? The development of technological pedagogical content knowledge. J. Educ. Comput. Res. **32**, 131–152 (2005)
14. Niess, M.L.: Preparing teachers to teach science and mathematics with technology: developing a technology pedagogical content knowledge. Teach. Teach. Educ. **21**, 509–523 (2005)
15. Voogt, J., Fisser, P., Roblin, N.P., Tondeur, J., van Braak, J.: Technological pedagogical content knowledge—a review of the literature. J. Comput. Assist. Learn. (2012). doi:10.111/j. 1365-2729.2012.00487.x
16. Wang, C.H.: Questioning skills facilitate online synchronous discussions. J. Comput. Assist. Living **21**, 303–313 (2005)
17. Garrison, D.R., Cleveland-Innes, M.: Critical factors in students' satisfaction and success: facilitating student role adjustment in online communities of inquiry. In: Bourne, J., Moore, J. C. (eds.) Elements of Quality Online Education, pp. 29–38. Needham, SCOLE (2004)
18. Ajavi, L.: An exploration of pre-service teachers' perceptions of learning to teach while using asynchronous discussion board. Educ. Technol. Soc. **12**(2), 86–100 (2009)
19. Curtis, R.: Analyzing students' conversations in chat room discussion groups. Coll. Teach. **52** (4), 143–149 (2004)
20. So, H.J.: When groups decided to use asynchronous online discussions: collaborative learning and social presence under a voluntary participation structure. J. Comput. Assist. Living **25**, 143–160 (2009)
21. Geijets, P.H., Hesse, F.W.: When are powerful learning environments effective? The role of learner activities and of students' conceptions of educational technology. Int. J. Educ. Res. **41**, 445–465 (2004)
22. Juarez, A.M.: Student success in the online classroom. Doctoral dissertation, Capella University, Minneapolis, MN (2003)
23. Martinez, S.L., Stager, G.: Invent to Learn: Making, Tinkering, and Engineering in the Classroom. Constructing Modern Knowledge Press, Torrance (2013)

The Information System of Distance Learning for People with Impaired Vision on the Basis of Artificial Intelligence Approaches

Galina Samigulina and Assem Shayakhmetova

Abstract In the educational field the intellectual technologies for distance learning are rapidly developing. Particularly these technologies are required by visually impaired people. A lot of research is devoted to the creation of effective intellectual educational technology. Methods of artificial intelligence such as neural networks, genetic algorithms, artificial immune systems, and others are applied for processing multi-dimensional information in real-time, forecasting of learning results. They contribute to increasing the quality of learning and to developing logical thinking, allowing to enhance the process of learning and to carrying out an individual approach to visually impaired people.

Keywords System of distance learning · Information technologies · People with impaired vision · Artificial intelligence approaches

1 Introduction

One of the most actual tasks of the rapidly developing modern information society is the efficient use of the latest IT-technologies, in particular on the basis of intelligent systems (IS).

Currently, there is an increasing trend of the development of artificial intelligence (AI) systems based on neural networks and other biological approaches

G. Samigulina (✉)
Doctor of Technical Sciences, Chief of Laboratory of the Institute of Information
and Computational Technologies, Almaty, Kazakhstan
e-mail: galinasamigulina@mail.ru

A. Shayakhmetova
Ph.D. – doctoral, Kazakh National Technical University named after K. I. Satpayev,
Almaty, Kazakhstan
e-mail: asemshayakhmetova@mail.ru

© Springer International Publishing Switzerland 2015
V.L. Uskov et al. (eds.), *Smart Education and Smart e-Learning*,
Smart Innovation, Systems and Technologies 41,
DOI 10.1007/978-3-319-19875-0_23

(neural networks (NN), genetic algorithms (GA), artificial immune systems (AIS), neuro- fuzzy logic (NFL) etc.).

Currently, there are many publications on this topic. Intelligent e-learning systems (IELS) are systems based on knowledge [1, 2], which developed on the basis of engineering knowledge and the theory of AI. These articles are considered the technical challenges of the design process of intelligent e-learning systems. The combination of these methods provides the design of reliable IELS.

The article [3] presented IS and their implementation in the areas of e-learning and distance education. The main issues relating to intellectual e-learning and its advantages over traditional learning are discussed. Several approaches of possible implementation are presented and the use of intelligent agents for e-learning is demonstrated. Several evolutionary tracks of e-learning technologies are considered: an intelligent tutoring system, intelligent interface and intelligent agent [4].

The article [5] proposes the architectural design, and explains the personality-oriented modules of the development of intellectual, adapted web e-learning system. In order to effectively achieve the objectives, the system addresses a diverse contingent of learners with different intellectual levels. The course content is designed and adapted to the characteristics of each learner. Each learner must answer two well - chosen set of tests to determine the basic knowledge. The system saves the last three sessions of the training process for each learner, even if the latter has not been completed, and the learner can complete the learning process. This helps to improve the level of knowledge and the performance of learners.

The article [6] explains how traditional teaching methods are transformed into E-Learning Systems and rebuilt in intelligent e-learning systems [7]. Methods of AI [8] used in the implementation of intelligent tutoring systems are discussed. In [9] the adaptive and intelligent technologies are reviewed in the development of e-learning course using a modular object-oriented dynamic learning environment (Moodle) based on Petri nets as a modeling formalism, as the classical networks and fuzzy Petri nets are not adapted to reflect changes in the new incoming data, parameters such as Moodle. The authors introduce an adaptive fuzzy Petri net of a higher order, which is dynamically suitable approach to the parameters.

Currently, information technologies are increasingly being implemented in the life of visually impaired people [10, 11]. For many blind people, computer technology has become an indispensable means of work and communication. Learning of people with impaired vision (PWIV) and their adaptation to the environment is one of the important issues of modern education. Building Intelligent system of distance learning for PWIV is relevant today. Distance learning (DL) allows to choose individual trajectory of learning and to work at one's own pace, it solves psychological problems PWIV, removes the temporal and spatial constraints.

The article [12] presented the intelligent e-learning system based on speech, with dual interface: the voice user interface and the web user interface. This system is designed for visually impaired learners. The designed application complements the existing e-learning systems, such as web - learning, m - learning and others. The article [13] presented an interactive learning tool for visually impaired people. The~character recognition algorithms are used applied to convert images to text and text to voice. A visually impaired learner has the ability to understand the information he had received. The information identification system, which consists of a digital camera voice synthesizer functioning and via the voice was developed, interface application.

When creating information technologies of distance learning PWIV, the problem to ensure the complete work on personal computers is actual. In a special boarding school for the blind and visually impaired children named after Ostrovsky in Almaty (Kazakhstan) the screen reader JAWS is used [14]. The program was developed by a group of blind and visually impaired people from Freedom Scientific, Florida, USA. It gives the opportunity to access PWIV, to make free use of a personal computer running Microsoft Windows. With the help of a speech synthesizer, via the audio card of the computer, information from the screen's with the information announced by the voice aloud, enabling a voice access to a variety of programs, applications and with Braille, it allows the unlimited use of the keyboard.

The next structure of the article is proposed: the second chapter discusses the educational environment MOODLE (Modular Object-Oriented Dynamic Learning Environment), which allows to effectively carry out online learning, create and implement modern distance learning courses, is freely available and can be supplemented with a component for working with PWIV. In the third chapter the formulation of the problem of the study is contained. The fourth chapter the architecture of an intelligent system of distance learning is presented. In the fifth chapter the developed intelligent information system of distance learning people with visually impaired on the basis of approaches AI is described.

2 Moodle Educational Environment

One of the open source systems is the Moodle distance learning environment [15], which allows to effectively carry out network learning courses, to create and to implement modern distance education technologies. Using the Internet as a channel for the exchange of information enables the organization of the network educational process through integrated advanced services for interactive dialogue between a student and a teacher, controlling knowledge at different stages of the distance process. The learners have access to many learning courses, and the teachers can organize the learning process using the following options of Moodle: conducting classes, working out of individual assignments, tests, and etc. Opportunities for communication is one of the greatest strengths of Moodle.

Moodle system has its own features of functioning. The system is designed taking into account the achievements of modern pedagogy with an emphasis on the interaction between learners in the form of discussion. It can be used both for distance and for full-time study. It has a simple and efficient web-based interface. The design is modular and can be easily modified. The plugged language packs can achieve complete localization. Learners can edit their personal accounts, add photos. Each user can specify his local time into which all the dates (e.g., timing of tasks) will be transferred. Different course structures are supported. Each course may be further protected by a codeword. A rich set of modules - components for courses: chat, poll, forum, glossary, workbook, lesson, test, profile, Scorn, Survey, Wiki, seminar resource. Almost all the typed texts can be edited by a built-in Wysiwyg - editor. All assessment of tasks can be assembled on the same page. A full report on the work of the user in the system, with timetables and details of communication with different modules is available. It is possible to configure e-mail (newsletters, forums, ratings and comments of teachers) [16].

One of the advantages Moodle is the ability to connect new components. In this paper an intelligent information system of distance learning (IISDL) for PWIV will be integrated into the system MOODLE- on the basis of an educational portal of the distance learning at the educational portal of the Kazakh National Technical University named after K. I. Satbayev (KazNTU, Almaty, Kazakhstan).

3 Statement of the Problem

It is necessary to develop the intelligent tutoring technology and the distance learning system for people with impaired vision (as component of Moodle), based on modern artificial intelligence methods to study the latest technologies on expensive equipment in the laboratories of joint use (LJU).

The proposed system is implemented using the National Open Research Laboratory of Information and Space Technologies of KazNTU named after K. I. Satpaev.

The solution to these problems will create the favorable conditions for PWIV, that contribute to social adaptation in society and ensure competitiveness in education. The feature of the system is the need to deal with the enormous flow of multidimensional data in real time.

4 Architecture of an Intelligent Information System of Distance Learning

Distance learning is one of the most effective and advanced training systems. The architecture of the Intelligent information systems of distance learning is shown in Fig. 1. This system is interacted with PWIV's learners by the audio interface. Learners PWIV are connected through a communications network and registered in the system of distance learning. Upon receipt of the PWIV's command, the system starts the corresponding processes for data processing.

The internet server receives requests from a learner PWIV and transmits the data to the interpreter of server scripts which implements the basic logic of the application, processes the data received from the user. The access to the laboratory of joint use depending on the subject of education and training courses is organized.

The process of studying innovative technologies on the modern equipment in the LJU is performed.

Recently to improve the efficiency of DL intelligent systems, the AI approaches are actively used. Due to its ability to self-organization and learning, these approaches are considered as promising tools when creating the latest technology of DL.

The main feature in the creation of learning systems based on approaches of AI is the need to work with multi-dimensional imprecise data and revealing the hidden knowledge. It should be adaptive, distributed and ambiguous.

Effective results in the development of intelligent systems of DL can be achieved by a combination of different methods of AI such as neural networks, genetic algorithms, artificial immune systems, neuro-fuzzy logic and theory of fuzzy sets. Neuro-fuzzy logic underlies the methods of working with inaccuracy, granulating information, approximate reasoning, computing with words. Neurocomputations reflect the ability to learn, adapt and identify. Genetic algorithms are used to organize random search and to achieve optimal value characteristics. Artificial immune systems predict [17, 18] of PWIV learning outcomes. Probability calculations provide the basis for managing uncertainty.

Fig. 1 Architecture of intellectual system of distance learning

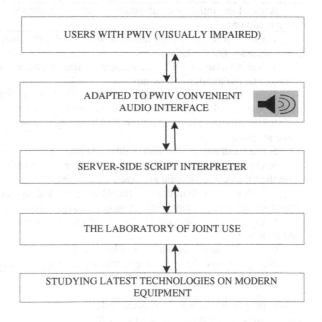

5 Development of Intelligent Information System of Distance Learning for People with Impaired Vision

Development of IISDL based on approaches of AI has modular character. The main modules of IISDL are (Fig. 2):

1. Information block, which contains the methods and means of information storage, includes the development of databases, knowledge bases, online tutorials, digital libraries, catalogs, reference system as a search engine and educational consultations (chat, forum).

2. Intelligent block which processes multi-dimensional data in real-time based on the approaches of AI and predicts to learning results.

3. Learning block, which helps to implement methods, means and forms of transfer of learning information aimed at a particular learner taking into account his individual characteristics. The feature of this unit is to organize the implementation of laboratory and practical works in on-line mode. Accessing the LJU is one of the most difficult tasks for practical implementation.

4. Controlling block is designed to assess learners' knowledge, to track learning progress. The following algorithm to construct an enlarged IISDL for people with impaired vision is proposed.

Algorithm. Intelligent information system of distance learning for people with impaired vision.

1. Creation of the special interface for PWIV with sounded user actions based on screen reader JAWS (USA).

2. Registration of PWIV in distance learning system. The choice of the subject matter and the duration of learning.

3. Building the model of PWIV in view of his special features and development of a database of informative features based on testing.

4. It is adopted of requests from PWIV and transmits the data interpreter of server-side scripting.

5. Pre-treatment of data and intellectual learning.

6. Organization of access to the LJU depending on the chosen model of learning (on the subject of education and learning course).

7. PWIV are studied the theoretical material and implementation of practical, laboratory and independent work on modern equipment in the LJU.

8. Control of knowledge of PWIV. Prediction of learning outcomes based on the AI approaches.

9. Comprehensive assessment of knowledge of PWIV.

10. Operational control of the distance learning process of PWIV.

The proposed system works as follows. After registration, the learner selects a subject of study. Then the system offers to get tested for the model is built of PWIV's learner and to determine their characteristics.

This procedure determines signs of PWIV's learner the following components: the intellectual potential of the personality to determine the level of intelligence and the ability to learning in one form or another; the motivational potential of the personality, which consists of a set of goals and needs of the learner; psychological, physiological and strong-willed potential of the personality, which provides the resulting of difficulties to achieve the objective, helps to quickly and efficiently respond to certain situations.

The input data of the intelligent system are individual signs of PWIV built in the form of time series characterizing each learner.

Using special software provides the reliable and rapid exchange of information and improves the efficiency of the process of acquiring knowledge by PWIV. Implementation of the developed IISDL can improve the quality of learning by PWIV: they will be able to independently use the computer, to find and edit any documents, to work and to communicate on the Internet, to have access to the LJU, to study lectures, to perform laboratory work, to programmer and to test.

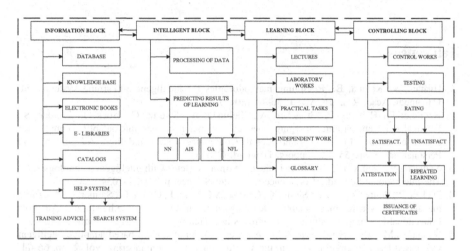

Fig. 2 The main modules of intelligent information system of distance learning

The use of artificial intelligence approaches [19] allows to predict learning outcomes and to efficiently manage the process of learning by PWIV.

6 Conclusion

This intelligent system solves the problem of effective distance learning by PWIV on high-performance computing clusters in the National Open Research Laboratory of Information and Space Technologies (Almaty, Kazakhstan).

The proposed intelligent information system has a number of advantages. The convenient interface for PWIV using sound is developed, which allows to work effectively with IISD. An individual characteristics and peculiarities of perception of learning material by PWIV's learners are taken into account. It makes possible to adapt the learning process taking into account their individual characteristics. An effective processing of multidimensional data based on the AI approaches and forecasting of learning outcomes are implemented. The optimal learning model which adapts to the learner's model is created. Intelligent system enables to carry out laboratory and practical works in real time by the LJU compute clusters for PWIV. The intelligent system makes it possible to perform laboratory and practical works by PWIV in real time on computer clusters in the LJU. The system is able to expand at the expense of the modular structure of IISDL. The possibility of grid computing for the processing of multi-dimensional data is based on AIS.

The developed system allows to effectively introduce the new information technologies by people with impaired vision on expensive equipment and provides the operational control of the learning process based on the use of methods of artificial intelligence.

References

1. Hisham, S., Katoua, B.: Reasoning methodologies for intelligent e-Learning systems. Int. J. Comput. Acad. Res. (IJCAR) 1, 36–44 (2012)
2. Henning, P., Heberle, F., Streicher, A., Zielinski, A., Swertz, C., Bock, J., Zander, S.: Personalized web learning: merging open educational resources into adaptive courses for higher education. In: 22nd International Conference User Modeling, Adaptation, and Personalization, pp. 55–62. Aalborg, Denmark (2014)
3. Nedeval, V., Nedev, D.: Evolution in the e-Learning systems with intelligent technologies. In: The International Scientific Conference Computer Science, pp. 1028–1034 (2008)
4. De Bra, P., Smits, D., van der Sluijs, K., Cristea, A.I., Foss, J., Glahn, C., Steiner, C.: Learning management systems meet adaptive learning environments. In: Intelligent and Adaptive Educational-Learning Systems: Achievements and Trends, pp. 1230–1238. Springer (2013)
5. Nedhal, A.M., Al, Saiyd, Intisar, A.M.: A generic model of student-based adaptive intelligent web-based learning environment. In: the World Congress on Engineering, vol. 2, pp. 66–71. London, U.K. (2013)
6. Jain, P., Sumit, S., Singh, K., Kumar, A.: An innovation towards intelligent e-Learning systems: a study. In: International Journal of Electronics and Computer Science Engineering, pp. 520–523 (2012)
7. Jaryani, F., Ridzuan, M., Ahmad, B., Sahibuddin, S., Jamshidi, J.: How artificial intelligent scheduling techniques support intelligent reflective E-Portfolio. In: The International Conference on Information and Computer Application, vol. 24, pp. 1578–1581 (2012)
8. Jabar, H., Saini, D.K., Hassan, S.: Artificial intelligence in e-learning-Pedagogical and cognitive aspects. In: The World Congress on Engineering, vol. 2, pp. 452–457. London, U.K. (2011)
9. El-Ramly, N., Shebl, D., Amin, M.: Modeling intelligent e-Learning systems based on adaptive fuzzy higher order petri nets. Int. J. Comput. Appl. 25(10), 7–14 (2011)
10. Schelhowe, H., Zare, S.: Intelligent mobile interaction: a learning system for mentally disabled people (IMLIS). Universal Access in Human-Computer Interaction. Addressing Diversity Lecture Notes in Computer Science, vol. 14, pp. 412–421 (2009)
11. David, J.M., Balakrishnan, K.: Performance improvement of fuzzy and neuro fuzzy systems: prediction of learning disabilities in school-age children. Int. J. Intell. Syst. Appl. 12, 34–52 (2013)

12. Azeta, A.A., Ayo, C.K., Atayero, A.A., Ikhu-Omoregbe, N.A.: A case-based reasoning approach for speech-enabled e-learning system. In: 2nd International Conference on Adaptive Science & Technology, pp. 211–217 (2009)
13. Senthamarai, R., Khana, V.: An efficient method for intelligence in e-Learning for visually impaired persons. Int. J. Adv. Res. **1**, 757–767 (2013)
14. The World's Most Popular Windows Screen Reader. http://www.freedomscientific.com/
15. Jalobeanu, M., Naaji, A.: Using Moodle platform in distance education. In: 7th International Scientific Conference e-learning and Software for Education, pp. 402–409. Bucharest (2011)
16. Bierne, J.: Actualizing moodle interactive tools usage within distance learning: need for multilevel approach . Int. J. Inf. Educ. Technol. **3**(1), 44–47 (2013)
17. Samigulina, G.: Development of the decision support systems on the basis of the intellectual technology of the artificial immune systems. In: Automatic and Remote Control, vol. 74, pp. 397–403. Springer (2012)
18. Samigulina, G.A., Samigulina, Z.I.: Intellectual Systems of Forecasting and Control of Complex Objects Based on Artificial Immune Systems, p. 172. Science Book Publishing House, Yelm, Washington (2014)
19. Samigulina, G.A, Shayakhmetova, A.S.: Construction of the intellectual system of distance learning for people with disabilities. In: Problems of Informatics, pp. 87–95. Novosibirsk (2014)

Development of University's Web-Services

Vladimir Tikhomirov, Natalia Dneprovskaya
and Ekaterina Yankovskaya

Abstract Currently many universities face multiple new challenges caused by massive online open courses (MOOCs), financial crisis, and social transformation. However, these challenges could be considered as new prospects of university development. There is an attempt to look at e-learning as connected part of e-business in this paper. The research is dedicated to a new information and communication technologies (ICT) environment, prospective development of educational services, and learning innovations. A design of Web-based educational services is one of the approaches to university sustainability. The outcomes of innovative Web-service – a smart-book - development by MESI University and gained experience are described and analyzed in this paper.

Keywords Web-service · Smart e-learning · Smart-society · Smart-book

1 Introduction

The changing landscape of the business communication, forced by the ICT influence, drives university towards e-business. The rapid penetration of ICT almost in every business activity improves that activity to reach a new level of efficiency. Smart-money, smart-school, smart-technology and other smart-terms appeared

V. Tikhomirov · N. Dneprovskaya (✉) · E. Yankovskaya
Moscow State University of Economics, Statistics and Informatics (MESI),
Moscow, Russia
e-mail: NDneprovskaya@mesi.ru

V. Tikhomirov
e-mail: VPT@mesi.ru

E. Yankovskaya
e-mail: EAJankovskaja@mesi.ru

© Springer International Publishing Switzerland 2015   265
V.L. Uskov et al. (eds.), *Smart Education and Smart e-Learning*,
Smart Innovation, Systems and Technologies 41,
DOI 10.1007/978-3-319-19875-0_24

recently to demonstrate significant changes in the information and communication technology based environment of the world. The ICT is integrating implicitly with social and business activities, and, thus, is transforming the new generation's awareness of the business environment. The e-learning is the essential part of new high-tech environment.

The e-learning is spreading the network by involving consumers, creating the Living Lab and fixing a new type of business communications through new media, LMS, MOOCs. There are some prospects for the e-learning development like the inclusive of consumers, the public-private partnership, and the crowd funding in the context of smart society. The core of smart society is the reasonable choice and implementation of ICT to reach innovative outcomes in education development and new levels of efficiency [1].

The governments of some countries understood the potential of ICT influence on economy and society. Therefore, they launched the governmental policies for the development of information society, for example, Japan in 1971, United States in 1980, Singapore in 1985, United Kingdom in 1988, Finland in 1995, etc. These developed countries got the huge competitive advantage in the global economy due the information society support. However, the intensive implementation of ICT in public environment pushes people out of their usual "comfort zone" to make them to look for new ways sustainable welfare. These days an implementation of ICT is not enough but obligatory condition for sustainability in the modern world. The development of ICT has reached a critical point to change the information environment [2].

The advancements in ICT technologies determine the progress of society as information society in 20th century. These days, new type of technology – smart technology – is critical for a society. Smart technology differs from ICT in way that it adapts to the changing environment instantly.

The new stage of social development focuses on the social and humanitarian issues. The smart technologies based on the humanitarian values of society and education have to enhance the environment. The new effects of the quantity and quality of ICT penetration characterize a smart society [3]. The ICT are used reasonably and appropriately to improve comfort and safety of the society. The ICT in education have to facilitate the flourishing of the personal creative abilities and enable to go well beyond the professional stereotypes and finding new solutions, which are in demand in a smart economy.

With the deep penetration of ICT in the society, the business communication has been increasing by involving the consumers into innovative and marketing activities. For the Internet users, who are interested in interconnection, the role of observer is not enough these days (Fig. 1). The business network has expanded significantly. For example, several years ago, the main communication in business was between entrepreneurs, in education - between educators and students. The current status of advanced communication development is the public-private-people partnership [4].

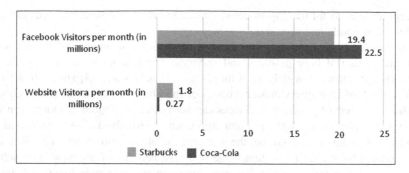

Fig. 1 Online visitors of starbucks and coca-cola facebook pages and websites [5]

The similar effect was achieved due to the wide spread of Web 2.0 technologies that transforms the consumer online service or content in its products. With the advent of Web 2.0, a vector of business communications added an opportunity of instant and effective communication between (a) producers and consumers, and (b) producers (professional association) and between consumers (social networks and communities of consumers) [6].

This change is well illustrated by the Coursera MOOCs platform, where hundreds of thousands of online students study online courses from leading professors around the world. Such a platform provides almost unlimited number of Internet users; online students provide and support the learning process of each other through the horizontal evaluation of individual works (peer-to-peer). The research shows a great demand of social media as knowledge source for students [7].

The key resource of the development of the Web 2.0 services is in the interconnection between the variety sources and communities via social networks, communities, wikis, and virtual laboratories. However, it should be noted that not only the rapid development of technology, but also the rapid transformation of social demand should be taken into consideration.

In accordance with [8], the most influential websites in the world include Wikipedia (#1), YouTube (#2), Flickr (#3), Twitter (#4), etc. meanwhile, The New York Times appeared only in 9th position of rankings, and, thus, being replaced by services where content is predominantly created by Internet users [8].

2 Living Lab Methodology

The main advantage of the information society is the providing a new type of communication between citizens, producers and public administration to ensure the quick information exchange. Initially, the idea of the smooth information flow for sharing of new developments and proposals to support the development of innovation was embodied in the concept of open innovation. The implementation of the concept allowed to fill the gap of communication between researchers, consumers, producers of goods and services.

Manufacturers are certainly interested in attracting consumers for the development and testing of their products and services. The practice of improvements of goods by consumers is widely used these days. OECD study [9] shows that from 10 % to 40 % of consumers make revisions or modifications of the acquired goods.

The consumers of goods and services can be a valuable source of innovation for a company. The increasingly popular approach – crowdsourcing – is aimed at finding new solutions based on the concept of open innovation. The Russian organizations have already evaluated the potential benefits of the new approach.

Based on the "open innovation" concept, we implemented an innovative "Living Lab" methodology. It was designed and developed to combine the individual initiatives of citizens, producers and governments to improve their overall living space - a formation of Smart Society. The main objective of the establishment and development of Living Lab innovations is to ensure faster and more efficient way to the market in comparison with traditional approaches to the development and innovation. The Living Lab methodology is defined by the European Commission as a partnership of public, producers and people for the purpose of co-building innovations designed to meet the needs of citizens.

The basis of the Living Lab methodology is the information environment; it integrates all actors involved in a particular range of applications. In this case the environment acts as a major infrastructure element and the catalyst of communication between consumers, producers, and government officials. It provides real experimental space where users and manufacturers work together to create an open innovation (Fig. 2).

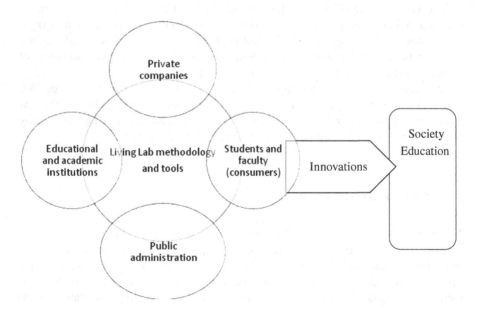

Fig. 2 Smart University Living Lab

The Living Lab tasks include but are not limited to

- facilitation of the effective communication between all involved parties;
- creation of conditions for co-working and collaboration;
- identification and dissemination of best practices;
- formation of communities of users and manufacturers.

The development and implementation of Living Lab methodology is supported by the European Commission. Universities are active participants of the Living Lab initiative. Due to university's authority among producers and consumers, it enables an organization of effective partnership with governments to create an open innovation process. At the same time, university's Living Lab is an essential element of the learning process; it allows students to take part in solving problems, participate in innovative processes of organizations, and find the optimal way of professional development.

3 Results

The sustainable development for society and economy enhanced by the changing environment and creative abilities is the main task of smart education development [10]. The smart education initiatives are aimed to help to reach a new level of efficiency in the economy and public administration.

At the same time, smart education should meet the needs of the individual and family [11]. The ICT is the appropriate instrument to connect all stakeholders in a modern environment (Fig. 3). The basis for educational development is the Internet; a great variety of its features and functions enables significant improvement of quality of educational Web services and creation of new types of services for a student.

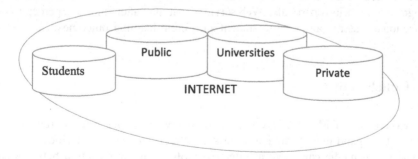

Fig. 3 Internet as a connector of main stakeholders in education development

The new social media and ICT are the core elements of MESI University's most recent innovation – a smart-book. The pilot version of a smart-book is available now for students. The first smart-book from MESI University is the "Marketing in

Social Media" course. It contains a unique learning content, a collection of copy-rights and illustrations which are adapted for a reading using a great variety of mobile electronic devices.

The design of the smart-book was focused on top quality learning outcomes and wide use of various instructional strategies.

Each chapter of a smart-book contains a text block and the author's illustrations – the interactive elements aimed to support the learning process (widgets). It also contains standard features to support learning/training such as cards, glossary, self-tests, etc. At the end of each topic, students/learners are asked to complete an interactive exercise (i.e. to solve a problem or a set of problems) to assess student's knowledge within learned topic.

A set of available widgets allows a student to use various knowledge and/or educational resources without posting those resources into a smart-book. Therefore, learning content designers can use the educational materials according to intellectual property law. Additionally, the widgets are updated automatically.

The smart-book is an innovative product due to implementation of the unique technological solutions in the form of widgets.

4 Discussion

A traditional university usually needs innovations for a sustainable development and to keep students and faculty as most valuable active.

The emergence of MOOCs was one of the most significant recent innovations in e-learning.

With the development of ICT the attention in education development focuses on ICT penetration into a learning process. However, questions of cognitive abilities in e-learning still remain.

A development and provision of educational Web service should be focused on integration of various particular Web services that may alter student's perception of educational material and lead to student's different learning outcomes.

5 Conclusion

The experience of MESI University in innovative activities proves student willingness to expand their participation in the different university activities.

The social media can serve as effective tools for interconnection between students and researches during learning and research processes.

The ICT can significantly enhance the learning outcomes by the innovative technologies like a smart-book.

The development of educational web-service is a new area of competition for online students. The startups now widely use the ICT to compete with traditional universities. However, the same opportunities are available for universities as well.

University should actively use e-business approaches, methods, features, technology, services and functionality to offer students, faculty and Internet users various educational and/or training services to meet their expectations and/or requirements to professional career. However, the development of those innovations take significant resources from main university activities.

MESI University is looking for new ways to provide high quality of online educational services. The MOOCs race imposed by Coursera is less attractive for a traditional university like MESI because of significant difference between resources available at a traditional university and at a successful startup company. However, new ways of offering of educational/training-related services – similar to Web services by e-businesses - may give university a significant advantage in the future.

The developed smart-book methodology seems to be very suitable for the innovative development of the MESI University

References

1. Salins, P.D.: The Smart society: strengthening America's greatest resources, its people, p. 320. New York (2014)
2. European Communities.: Public Strategies for the Information Society in the Member States of the European Union. http://ecdl.com.cy/assets/mainmenu/131/docs/EU-MemberStatesStrategies.pdf (2000)
3. Tikhomirov V.: The Moscow State University of Economics, Statistics and Informatics (MESI) on the way to smart education. In: Proceedings of the 10th International Conference on Intellectual Capital, Knowledge Management and Organizational Learning (2013)
4. Campbell, T.: Beyond smart cities: how cities network, learn and innovate. Earthscan, Abingdon (2012)
5. Clayton, N.: Business Joins the Party, The Wall Street Journal, May 4, 2011. http://www.wsj.com/articles/SB10001424052748703712504576244622146113118 (2011)
6. Tapscott, D.: Macrowikinomics: new solutions for a connected plated. In: Tapscott, D., Williams, A.D. (eds.) Penguin, p. 428 (2012)
7. Dneprovskaya, N., Koretskaya, I., Dik, V., Tiukhmenova, K.: Study of social media implementation for transfer of knowledge within educational milieu, vol. 4, pp. 149–151. Naukovyi Visnyk Natsionalnoho Hirnychoho Univesytetu (2014)
8. Macmanus, R.: The Most Influential Websites in the World: Wikipedia #1, Twitter #4 With a Bullet November 24. http://www.readwriteweb.com/archives/the_most_influential_websites_in_the_world.php (2009)
9. OECD.: Science, Technology and Innovation Indicators in a Changing World, p. 128. Responding to Policy Needs, Paris (2007)
10. Hwang, D.J., Yang, H., Kim, H.: E-Learning in Republic Korea. UNESCO Institute for Information Technologies in Education, Moscow (2010)
11. Dik, V., Urintsov, A., Dneprovskaya, N., Pavlekovskaya, I.: Prospective of e-learning toolkit enhanced by ICT development, vol. 4, pp. 152–156. Naukovyi Visnyk Natsionalnoho Hirnychoho Univesytetu (2014)

Situation Awareness Training in E-Learning

Liubov S. Lisitsyna, Andrey V. Lyamin, Ivan A. Martynikhin
and Elena N. Cherepovskaya

Abstract In this paper the problem of increasing e-learning effectiveness is considered. Students must operate in e-learning environment and sense simultaneously a lot of parameters. This needs particular skills that may be formed by trainings. The paper represents the results of our study proving positive influence of the situation awareness training on e-learning performance. In the randomized controlled study that has been held in ITMO University 104 first-year engineering students participated. The students had been divided into two groups: active and control. All of the participants had to pass an online exam in computer science twice in 28-days interval. The students from active group were asked to pass the situation awareness training within this 28-days interval. At the end of the experiment, necessary parameters had been calculated and analyzed.

Keywords Cognitive training · E-learning performance · Functional state · Online exam

L.S. Lisitsyna (✉) · A.V. Lyamin · E.N. Cherepovskaya
ITMO University, Saint Petersburg, Russia
e-mail: lisizina@mail.ifmo.ru

A.V. Lyamin
e-mail: lyamin@mail.ifmo.ru

E.N. Cherepovskaya
e-mail: cherepovskaya@cde.ifmo.ru

I.A. Martynikhin
Pavlov First Saint Petersburg State Medical University, Saint Petersburg, Russia
e-mail: iam@s-psy.ru

© Springer International Publishing Switzerland 2015
V.L. Uskov et al. (eds.), *Smart Education and Smart e-Learning*,
Smart Innovation, Systems and Technologies 41,
DOI 10.1007/978-3-319-19875-0_25

1   Introduction

Many known computerized cognitive trainings are directed to development of basic cognitive functions [1, 2], e.g. working memory, attention, perception, etc. It's still considered that its real impact on academic performance is hard to estimate [3, 4]. This paper represents the results of our study proving positive influence of the situation awareness training on e-learning performance.

Situation Awareness (SA) is "knowing what's going on" or more formally "the perception of the elements in the environment within a volume of time and space, the comprehension of their meaning and the projection of their status in the near future". According to this SA consists of three levels [5]: (a) perception of the elements in the environment, (b) comprehension of the current situation, (c) anticipation of future status. SA Trainings were used to train pilots, air traffic controllers who interact with complex computers or engineering systems in dynamically changing conditions [6]. In e-learning students deal with different fast changing user interfaces where they must keep an eye on menus, buttons, promptings, etc. Such ability is also necessary during online tests and exams when students are in stressful state. Hence SA Trainings must become necessary elements of e-learning providing readiness of all students to the substantial part of practical exercises.

2   Method

2.1 Participants

For the experimental study of SA Training impact on e-learning performance first-year students of ITMO University were attracted. The study has been conducted at the time, when they had to pass an online exam in computer science. Before starting the experiment, all students gave written consent on participating in it. Independent ethics committee of Bekhterev Saint Petersburg National Research Psycho Neurological Institute had approved realization of this study. 104 engineering students without any significant diseases of cardiovascular system had been selected for participating in the experiment.

2.2 Procedures

Experiment participants were randomly divided into two groups: 67 students in active group and 37 students in control group. Students had to pass an online exam in computer science twice in 28-days intervals (at baseline and at the end of the experiment). A learning of the discipline had not been provided during this period. Between two exams only students from active group were asked to pass SA Training. The study procedures can be seen in Table 1.

Table 1 Study procedures

Visit	Baseline (Exam 1)	Days 1-28	End of Experiment (Exam 2)
Informed consent	X		
Demographic data	X		
Online exam	$X^{(1)}$		$X^{(1)}$
VAS: exam anxiety (5 min before exam)	$X^{(1)}$		$X^{(1)}$
HRV (throughout online exam)	$X^{(1)}$		$X^{(1)}$
PCPT	$X^{(1)}$		$X^{(1)}$
SA Training		$X^{(2)}$	
VAS: like/dislike SA Training		$X^{(2)}$	
VAS: complexity of SA Training		$X^{(2)}$	

VAS – visual analog scale, HRV – heart rate variability, PCPT – PEBL Continuous Performance Test
(1) In a laboratory of ITMO University
(2) Via learning management system, only for active group

2.3 Situation Awareness Training

The training has been developed on basis of SA Test from Psychology Experiment Building Language (PEBL) Test Battery, Version 0.14 [7]. The source code of the test (http://pebl.sf.net) has been revised by adding Russian translations and feedbacks to every task in order to control results. The test was designed to measure the same basic factors as the Situation Awareness Global Assessment Technique [8]. It is framed as a dynamic visual tracking task.

The PEBL SA Test has been used previously to assess the attentional draw that heat strain and encumbrance place on dynamic attention [9], and was shown to detect cognitive impairments from those stressors. It was also used for the patients with OCD [10] demonstrating sensitivity to changes in cognitive processes that are observed in such patients category.

Passing the training, participants have to monitor the locations, identities, and directions of a set of targets. Targets are represented as pictures of bugs and lizards moving on a rectangular panel 700 × 500 pixels (workspace). Targets can change direction of their movement when they reach any of four panel borders or randomly while moving. Two groups of objects are presented in the training: predators (lizards) and preys (bugs). Lizards can eat bugs. After a bug has been eaten by any of the lizards, it appears in different part of the panel in the next moment. Participant had to deal with one of three tasks after targets have been removed from the screen:

- Task 1 – the locations of the tracked targets. A participant was asked to reproduce locations of the targets by clicking on the work-space to set possible locations. After all marks have been placed the average value between participant's marks and actual targets' locations is calculated.

- Task 2 – the identity of tracked targets. One of the targets was randomly chosen for every of two probes of this task. A participant had to choose which target was located in the marked area by clicking on one of the buttons representing target images. After clicking on the button, an accuracy of participant's answer is displayed on the screen.
- Task 3 – the direction, targets were moving in. One of the targets was randomly chosen and its location was marked with the orange circle on the screen. By clicking on the workspace participant placed a target in the direction of the click. After the direction has been set, participant had to submit his answer. Then the difference between participant's and the actual directions is calculated.

The training was divided into three parts: (a) practice block with 3 trials for each of the tasks; (b) pure block with 15 trials per task; (c) mixed block with 35 randomly chosen tasks. In the practice and pure blocks the task had been shown on the screen before targets started moving. In mixed block participant was unaware of which task he or she would pass until the targets had been removed from the screen.

The number of objects on the screen changed for different days (levels) of the training in order to adapt a participant to it. At the same time the number of targets on the first and on the last days of the training was identical in order to obtain commensurable parameters: day 1 – 6 targets, days 2-4 – 3 targets, days 5-8 – 4 targets, days 9-15 – 5 targets, day 16 and further – 6 targets.

The participants had been given the instruction to pass the training at least 16 days from 28 days declared. A participant could pass any number of the trainings during one day. Despite that, he moved to the next level only on the next day. If a participant missed one or more days, in the next session he received the next level of the training after the last level he had passed.

2.4 Online Exam

An online exam in computer science has been held in AcademicNT (learning management system of ITMO University [11, 12]). The exam included 45 questions. Students had been given 45 min to pass the test. All of the presented questions were familiar to students as they participated in similar tests during studying the discipline throughout the semester.

In two tries of the online exam questions' topics had not been changed, though the content of the questions differed. We assumed that experience obtained in first participating in the online exam would slightly affect the result of repeated participating at last day of the experiment.

The online exam has been held in university computer class. Participants were allowed to use only a calculator and a blank of paper. The computation of points for the exam was carried out taking into account the difficulty of every presented question.

2.5 Functional State

The learning load and intensity should not lead to a reduction of student's psychophysiological functional state including learner's mental working capacity. In order to evaluate learner's functional state heart rate variability (HRV) method has been used. HRV method is based on measurement of RR-intervals between the high amplitude peaks of electrocardiogram [13]. The Varikard software/hardware system – the Varikard system – has been used in the experiment in order to calculate RR-intervals, Heart Rate (HR), Index of centralization (IC) and Stress Index (SI) [14]. As the main indexes of HRV IC and SI were calculated by the formulas:

$$IC = \frac{VLF + LF}{HF}, \tag{1}$$

$$SI = \frac{Amo \cdot 100\%}{2 \cdot Mo \cdot MxDMn}, \tag{2}$$

where VLF is spectral density of RR-intervals in a very low frequency range; LF – in the range of low frequency; HF – in the high frequency range; Mo – mode, Amo – mode's amplitude, MxDMn – variation range.

2.6 Continuous Performance Test

Attention indexes have been assessed with PEBL Continuous Performance Test (PCPT) from Psychology Experiment Building Language (PEBL) Test Battery, Version 0.14 [7]. The source code of the test (http://pebl.sf.net) has been revised by adding Russian translation of the task. The PCPT is a faithful implementation of Conners's Continuous Performance task [15]. Series of 20 letters (360 letters in total) gradually appear on the black screen, symbol by symbol. A participant is asked to respond to every letter except the X. The task had been complicated by giving PCPT just after finishing online exam. As the main index of person's attention d-prime index or detectability has been estimated in accordance with approaches accepted in Signal Detection Theory [16].

2.7 Visual Analog Scales

In order to assess participants' subjective attitude to some questions during the experiment visual analog scales (VAS) have been used. They were represented by sliders. Participants were allowed to move slider from central position (0) to any position between minimum (-100) and maximum (100) value which most accurately indicates their attitude to currently displayed question. Three VAS have been used during the experiment:

- Anxiety assessment of the online exam. Before starting the online exam, participants were asked to set the slider in an appropriate position according to their feelings in the current moment between minimum value "Completely calm, relaxed" and maximum value "Extremely strained, anxious".
- Subjective assessment of SA Training complexity. The slider was displayed on the screen after completion of every training level. Participants were asked to set the slider in an appropriate position according to their feelings about complexity of passed training level between minimum value "Extremely simple" and maximum value "Extremely complicated".
- Assessment of person's interest in the training. The slider was displayed on the screen after completion of every training level. Participants were asked to set the slider in an appropriate position according to their interest to passed training level between minimum value"Completely disliked" and maximum value "Extremely liked".

2.8 Data Analysis

Statistical processing of the study results was accomplished with program IBM SPSS Statistics (version 20). Intergroup distinctions were carried out with t-test. For the assessment of intragroup dynamics paired t-test was used. In both cases, the significance level α was 5 %.

3 Results of the Experiment

After the results had been processed, we were able to compare active and control groups in order to confirm our assumption that SA Training influences e-learning performance. At baseline active and control groups of participants were identical in all indexes except the time spent on passing the online exam (Table 2; bold lines in tables represent the most significant parameter values within independent samples). Students of control group spent less amount of time to answer the questions, although the results for both groups are approximately comparable.

Table 2 Comparison of active and control groups at baseline

	Active group (n = 67)	Control group (n = 37)
Number of men (%)	44 (66)	25 (68)
Age (SD)	18,3 (0,6)	18,4 (0,5)
Results of tests in computer science throughout the semester (SD)	65,6 (13,4)	65,8 (15,7)
Exam 1: Average number of correct answers (SD)	22,6 (5,3)	21,2 (7,4)
Exam 1: GPA (SD)	50,8 (10,8)	46,2 (17,6)
Exam 1: Average time in minutes (SD)	**44,35 (8,0)**[*]	**40,3 (8,7)**[*]
Exam 1: Anxiety assessment (SD)	-1,5 (30,8)	-6,0 (32,9)
Exam 1: Average of HRV SI (SD)	74,2 (57,2)	62,2 (42,3)
Exam 1: Average of HRV IC (SD)	4,3 (3,2)	3,3 (2,8)
Exam 1: Average of PCPT d-prime (SD)	2,4 (0,7)	2,3 (0,5)

[*] A rejection of the null hypothesis at the 5 % significance level
GPA – grade point average, HRV – heart rate variability, PCPT - PEBL Continuous Performance Test

3.1 Passing SA Trainings in the Active Group

All 67 students of the active group passed provided number of training levels, so all the results have been considered in the analysis. Average number of training levels passed by participants equals 20 (minimum 16 and maximum 27). Comparison of subjective assessments of training complexity and interest to it did not reveal any considerable differences. During the experiment assessments of training being reasonably complicated and reasonably interesting prevailed.

3.2 Dynamics of SA Trainings

During passing the trainings, the changes in parameters have been observed. Parameters reflected improving of training's performance in every task and stages concerned with changing difficulty of the tasks from level to level. Matching the results of every task of SA Training between the first and the last level, the following changes in parameters have been obtained:

- Task 1: Significant increase of setting correct location of objects has been registered in pure blocks as well as in mixed block. Although there was no essential changes in average time of passing the task.
- Task 2: Significant increases of frequency of giving correct answers and participants' spent time on this type of task have been registered. Amount of correct an-swers increased in pure blocks as well as in mixed block.

- Task 3: Significant increase in choosing the correct direction of an object has not been revealed neither in pure blocks, nor in mixed block. Average time spent by participants on this type of task substantially decreased.

3.3 Results of the Second Online Exam

The most considerable results of the study have been obtained in comparison of calculated parameters for the first and second exam in the active and control groups. The first step was to analyze the second exam results (Table 3). This analysis has revealed that the active group gave larger number of correct answers, so their GPA (Fig. 1B) was higher than in the control group. D-prime parameter (Fig. 1A) value also increased in the active group. No other significant differences have been obtained including the subjective anxiety assessment and HRV indexes.

Table 3 Comparison of active and control groups at the end of experiment

	Active group (n = 67)	Control group (n = 37)
Exam 2: Average number of correct answers (SD)	**25,3 (6,6)**[*]	**22,3 (7,2)**[*]
Exam 2: GPA (SD)	**57,8 (12,9)**[*]	**50,4 (16,3)**[*]
Exam 2: Average time in minutes (SD)	41,7 (8,9)	40,7 (11,2)
Exam 2: Anxiety assessment (SD)	-12,5 (41,2)	-13,5 (40,8)
Exam 2: Average of HRV SI (SD)	83,3 (53,7)	65,4 (50,9)
Exam 2: Average of HRV IC (SD)	5,4 (2,5)	5,0 (2,9)
Exam 2: Average of PCPT d-prime (SD)	**2,6 (0,7)**[*]	**2,3 (0,7)**[*]

[*] A rejection of the null hypothesis at the 5 % significance level
GPA – grade point average, HRV – heart rate variability, PCPT - PEBL Continuous Performance Test

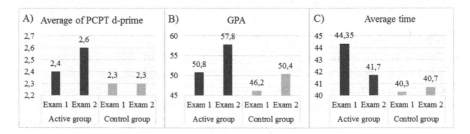

Fig. 1 Comparison of parameter values for active and control groups in the first and the second exams: (A) d-prime, (B) GPA, (C) average time

Comparison of dynamics in both exams for the active and control group separately (Tables 4 and 5) has revealed that during the second online exam the active group gave larger number of correct answers than during the first one, so the GPA increased by 7 points. D-prime parameter value also increased in the active group. In addition to increase of d-prime significant reduction of errors on false alarm type (responding to X symbols) in PCPT has been registered: 50.5 in the first test decreased to 41.9 in the second one.

In the control group only the GPA of the second online exam increased by 4 points compared to the first one. No authentic increase in the number of correct answers and spent time (Fig. 1C) or any changes in other parameters were revealed.

Table 4 Comparison of the parameters at baseline and at the end of experiment for active group

	Baseline (n = 67)	End of Experiment (n = 67)
Average number of correct answers (SD)	**22,6 (5,3)**[*]	**25,3 (6,6)**[*]
GPA (SD)	**50,8 (10,8)**[*]	**57,8 (12,9)**[*]
Average time in minutes (SD)	**44,35 (8,0)**[*]	**41,7 (8,9)**[*]
Anxiety assessment (SD)	-1,5 (30,8)	-12,5 (41,2)
Average of HRV SI (SD)	74,2 (57,2)	83,3 (53,7)
Average of HRV IC (SD)	4,3 (3,2)	5,4 (2,5)
Average of PCPT d-prime (SD)	**2,4 (0,7)**[*]	**2,6 (0,7)**[*]

[*] A rejection of the null hypothesis at the 5 % significance level
GPA – grade point average, HRV – heart rate variability, PCPT - PEBL Continuous Performance Test

Table 5 Comparison of parameters at baseline and at the end of experiment for control group

	Baseline (n = 37)	End of Experiment (n = 37)
Average number of correct answers (SD)	21,2 (7,4)	22,3 (7,2)
GPA (SD)	**46,2 (17,6)**[*]	**50,4 (16,3)**[*]
Average time in minutes (SD)	40,3 (8,7)	40,7 (11,2)
Anxiety assessment (SD)	-6,0 (32,9)	-13,5 (40,8)
Average of HRV SI (SD)	62,2 (42,3)	65,4 (50,9)
Average of HRV IC (SD)	3,3 (2,8)	5,0 (2,9)
Average of PCPT d-prime (SD)	2,3 (0,5)	2,3 (0,7)

[*] A rejection of the null hypothesis at the 5 % significance level
GPA – grade point average, HRV – heart rate variability, PCPT - PEBL Continuous Performance Test

4   Discussion of the Results

Comparing the results of both groups, we may conclude that during the second online exam students from the active group demonstrated better results. The average increase of GPA parameter for the active group was 14 %. Time spent on passing the exam decreased only in the active group. Comparing the baseline and the end of experiment, no significant changes in participants' functional state (HRV indexes) have been revealed.

All participants from the active group of our study assessed SA Training as positive. It is necessary to mention that all of 67 students in the active group had followed given instructions and passed at least 16 levels of SA Training out of possible 28. Although in other similar studies [17] some participants prematurely quitted the experiment.

Positive dynamics of the results in the first and the second tasks of SA Training indicates that these tasks matched the training in the best way. Hereinafter the third task of SA Training which has not revealed any significant changes can be omitted.

Increase of d-prime parameter value indicates that students improved their ability to select necessary signals out of redundant ones. Applying effect of SA Training on PCPT performance, it was confirmed that these trainings develop not only visual working memory, but also lots of other necessary cognitive functions.

5   Conclusion

After the experimental study we found strong evidence confirming our assumption: SA Trainings have positive impact on the online exam results. Comparison of student's results obtained in the two online exams in computer science helped to reveal that during the second online exam without providing any learning GPA of students from the active group has been increased and time on passing the exam has been reduced at the 5 % significance level. Thus SA Training can increase e-learning performance.

Acknowledgments   This paper is supported by Russian Federation Government's grant # 074-U01.

References

1. Jaeggi, S.M., Buschkuehl, M., Jonides, J., Perrig, W.J.: Improving fluid intelligence with training on working memory. Proc. Nat. Acad. Sci. USA. **105**(19), 6829–6833 (2008). doi:10.1073/pnas.0801268105
2. Titz, C., Karbach, J.: Working memory and executive functions: effects of training on academic achievement. Psychol. Res. **78**(6), 852–868 (2014). doi:10.1007/s00426-013-0537-1

3. Shipstead, Z., Redick, T.S., Engle, R.W.: CogMed working memory training: does the evidence support the claims? J. Appl. Res. Mem. Cogn. **1**, 185–193 (2012). doi:10.1016/j.jarmac.2012.06.003

4. Rapport, M.D., Orban, S.A., Kofler, M.J., Friedman, L.M.: Do programs designed to train working memory, other executive functions, and attention benefit children with ADHD? A meta-analytic review of cognitive, academic, and behavioral outcomes. Clin. Psychol. Rev. **33** (8), 1237–1252 (2013). doi:10.1016/j.cpr.2013.08.005

5. Endsley, M.R.: Toward a theory of situation awareness in dynamic systems. Hum. Factors **37** (1), 32–64 (1995)

6. Endsley, M.R., Robertson, M.M.: Training for situation awareness in individuals and teams. In: Endsley, M.R., Garland, D.J. (eds.) Situation Awareness Analysis and Measurement. LEA, Mahwah, NJ (2000)

7. Mueller, S.T.: The Psychology Experiment Building Language, Version 0.14 (2014). Software downloaded from http://pebl.sourceforge.net

8. Endsley, M.R.: Direct measurement of situation awareness: validity and use of SAGAT. In: Endsley, M.R., Garland, D.J. (eds.) Situation Awareness Analysis and Measurement. LEA, Mahwah, New Jersey (2000)

9. Mueller, S.T., Simpkins, B., Price, O.T., Weber, P., McClellan, G.E.: Cognitive performance degradation with the T3 methodology. Final Technical Report. Applied Research Associates Inc., Arlington (2011) (HDTRA-1-08-C-0025)

10. Tumkaya, S., Karadag, F., Mueller, S.T, Ugurlu, T.T., Oguzhanoglu, N.K., Ozdel, O., Atesci, F.C., Bayraktutan, M.: Situation awareness in obsessive-compulsive disorder. Psychiatr. Res. **209**(3), 579–588 (2013). doi:10.1016/j.psychres.2013.02.009

11. Lyamin, A.V., Vashenkov, O.E.: Virtual environment and instruments for student olympiad on cybernetics. In: 8th IFAC Symposium on Advances in Control Education, ACE 2009, pp. 95–100. Kumamoto, Japan (2009)

12. Lyamin, A.V., Efimchik, E.A.: RLCP-compatible virtual laboratories. In: International Conference on E-Learning and E-Technologies in Education, ICEEE 2012, pp. 59–64. Lodz, Poland (2012)

13. Lisitsyna, L., Lyamin, A., Skshidlevsky, A.: Estimation of student functional state in learning management system by heart rate variability method. In: Smart Digital Futures 2014, pp. 726–731. IOS Press, Amsterdam (2014)

14. Uskov, V., Lyamin, A., Lisitsyna, L., Sekar, B.: Smart e-Learning as a student-centered biotechnical system. In: Vincenti, G., Bucciero, A., Vaz de Carvalho, C. (eds.), E-Learning, E-Education, and Online-Training: First International Conference, Eleot 2014, Bethesda, MD, USA, 18–20 Sept 2014. Revised Selected Papers - Lecture Notes of the Institute for Computer Sciences, Social-Informatics and Telecommunications Engineering, vol. 138. Springer (2014)

15. Conners, C.K., Epstein, J.N., Angold, A., Klaric, J.: Continuous performance test performance in a normative epidemiological sample. J. Abnorm. Child Psychol. **31**(5), 555–562 (2003)

16. Stanislaw, H., Todorov, N.: Calculation of signal detection theory measures. Behav. Res. Methods Instrum. Comput. **31**(1), 137–149 (1999)

17. Jaeggi, S.M., Buschkuehl, M., Shah, P., Jonides, J.: The role of individual differences in cognitive training and transfer. Mem. Cognit. **42**(3), 464–480 (2014). doi:10.3758/s13421-013-0364-z

Intra-domain User Model for Content Adaptation

Anwar Hussain, M. Abu Ul Fazal and M. Shuaib Karim

Abstract In learning environment, personalization of contents according to the requirement of an individual student is the most important feature of adaptive educational systems. This process becomes more effective if the system knows the way through which a student learns best. Learning styles are non-stationary and are varied for academic disciplines. Our proposed model considers its non-deterministic nature, effect of the subject domain, and non-stationary aspects during the learning process. Presented approach is novel, simple but more flexible that dynamically and accurately adjusts students learning style variations in a discipline-wise manner. For the evaluation of our proposed model, Visual/Verbal dimension of Felder and Silvermen learning style model is utilized for personalization of Computer Science undergraduate subjects in our experimental prototype. Results show that personalization of contents in a discipline-wise manner is more effective during the learning process of a student.

Keywords e-Learning · User modeling · Personalization · Learning styles · Adaptive educational systems · Content adaptation

1 Introduction

One of the important issues nowadays is to provide contents according to the user requirements. In addition, these contents must be presented according to their preferences. For better adaptation, a system should be aware of a different aspects of a user, e.g. requirements, needs, goals, preferences, learning styles. In educational

A. Hussain (✉) · M.A.U. Fazal · M.S. Karim
Department of Computer Sciences, Quaid-i-Azam University, Islamabad, Pakistan
e-mail: anwarh45@gmail.com

M.A.U. Fazel
e-mail: fazalsidhu@yahoo.com

M.S. Karim
e-mail: skarim@qau.edu.pk

© Springer International Publishing Switzerland 2015
V.L. Uskov et al. (eds.), *Smart Education and Smart e-Learning*,
Smart Innovation, Systems and Technologies 41,
DOI 10.1007/978-3-319-19875-0_26

285

environment, these systems provide educational contents to the students and are known as adaptive educational hypermedia systems [1, 2]. Researchers have utilized different models of human cognition [2, 3] in this field. One of the properties of learning style is its non-stationary nature that can change time to time. For this purpose, researchers are trying to propose models that are capable of updating in terms of learning styles [4, 5]. Keeping user model updated is a complex task because learning is a continuous process and human cognition is not easily quantifiable. To our knowledge, discipline-wise learning style variations are not taken into account for personalization. The following scenario highlights the importance of discipline-wise personalization.

A student uses an adaptive system for his learning that provides course content in the form of text and videos. For example, a student wants to study *Database* course and he prefers written text. The adaptive system will develop a model and will prefer him written text. After some time, the student wants to study *Programming* course and prefers video lectures. Adaptive system will update his model and will prefer video lectures instead of written text. Consider the case if he studies *Database* course again. Adaptive system will again reverse its mechanism. The problem will arise again if he switches once more to *Programming* course again. One of the solutions to this problem is, there should be a model that could handle learning style variations in a discipline-wise manner.

In this study, we have tried to exploit the effect of the subject domain of learning styles too. Our model dynamically and continuously adjusts student learning styles preferences for multiple academic disciplines.

2 Background and Related Work

During the evolution of Computer Science field, interactions between humans and computers have also evolved to increase the usability of computer systems. The first requirement for this is to understand the human. To understand a human, theories of psychology have a vital role that provide information about human's cognitive abilities. These include how a human learns, thinks and processes information. Researchers have come to the conclusion that for better adaptation, along with the physical capabilities, it is more important to consider cognitive abilities of human's [6]. A number of models of human cognition have been developed and used in the field of adaptive education for learning purposes [2, 3, 7]. For example Index of learning style, Field dependent/Field-independent, Verbal-Imager/Holistic-Analytical and Kolb's learning style inventory, which are used most frequently.

In the field of adaptive educational hypermedia systems, learning style models have an important role [2, 3]. The main aim of these studies is to enhance students' learning in a more effective way. These studies give enough information about the learning styles used in the past, modeling approaches, key variables considered and the systems developed for adaptation. Recent work in this field, most of these approaches are in the context of a single domain [4, 5, 8].

Exploiting learning styles in adaptive systems is an active area of research [4, 8]. The main focus of recent studies is to improve user performance and academic achievement. One of the important factors is that a student has variations in difficulty for different academic disciplines [9]. For better adaptation of educational contents, it is important for a user model to have learning style variations according to the requirement of academic disciplines.

3 SWA Model

SWA(Subject-wise Adaptive) is the extension of an existing model presented by [10], for accommodating discipline-wise learning style variation. For personalization of contents and users' learning style preferences, dimensions of Index of learning style model were exploited.

3.1 User's Categories and Their Probabilities

Index of learning style model categorizes users in four dimensions, where each dimension has two further types of user preferences. These preferences are Active (A) vs Reflective (R), Sensing (S) vs Intuitive (I), Visual (Vi) vs Verbal (Ve) and Global (G) vs Sequential (Seq). For each category/dimension, a user will get one of the preferences, i.e. a user will either be Active (A) or Reflective (R) but not both. Learning style of a user is the combination of preferences for dimension a, b, c, and d as shown in Eq. 1. As a result, we have total 16 number of possible learning styles.

$$LSC = \{a \in (A/R), b \in (S/I), c \in (Vi/Ve), d \in (G/Seq)\} \qquad (1)$$

In learning style model, each preference in a dimension has a certain probability. The total probability of a dimension is 1 while preferences in it share part of this. If probability of *Active* preference is w then probability of *Reflective* preference becomes $1 - w$, while the total probability of a dimension remains 1 i.e. $w + (1 - w)$ as shown in Table 1. Where Pr represents probability while subscripts A, R, S, I, Seq, G, Vi and Ve represents the preferences in a dimension.

Selection of a preference in a dimension is based on the number of answers which favor that preference. The general formula, for calculating probability of a preference is obtained from Eq. 2.

$$Pr_i = \frac{A_i}{11} \qquad (2)$$

where Pr_i is the probability of preference in i^{th} dimension. A_i is the number of favorable answers for preference A in a dimension i. Here the number 11 shows the total

Table 1 Probabilities of preferences inside dimensions

Dimension	Preference (A)	Preference (B)	Total Probability (A + B)
a	$Pr_A = w$	$Pr_R = 1 - w$	$Pr_A + Pr_R = 1$
b	$Pr_S = x$	$Pr_I = 1 - x$	$Pr_S + Pr_I = 1$
c	$Pr_{Vi} = y$	$Pr_{Ve} = 1 - y$	$Pr_{Vi} + Pr_{Ve} = 1$
d	$Pr_{Seq} = z$	$Pr_G = 1 - z$	$Pr_{Seq} + Pr_G = 1$

number of questions for a dimension. Through this way, the preferences of a user in dimensions can be found.

3.2 Updating User Model Subject-Wise

The criteria for updating user preferences is the performance of a student in the tests [4, 8]. In the case of lower performance, it is assumed that learning style preferences stored in the user model are not the actual representation of user preferences. Learning style preferences are updated after each unsuccessful learning session. In the beginning, preferences' values are stored same for all disciplines but later these preferences are updated in a discipline-wise manner.

3.3 Adaptation Rules

User model is updated after learning session by a number of rules if desirable performance m is not achieved by a learner in j^{th} subject. When a student fails the test, probability of existing learning style preferences are decremented by factor $R[s_j]$ and probabilities of missing preferences are incremented by the same factor $R[s_j]$. Where value of $R[s_j]$ is determined by Eq. 3. These rules are represented in Table 2. In these rules

Table 2 Rules used in adaptation process

Rule 1

$IF(PFM[s_j] < m)$ AND $(LSC[d_i][s_j] = $ "A") THEN
$SM[d_i][s_j]_A = SM[d_i][s_j]_A - R[s_j]$
$SM[d_i][s_j]_B = SM[d_i][s_j]_B + R[s_j]$

Rule 2

$IF(PFM[s_j] < m)$ AND $(LSC[d_i][s_j] = $ "B") THEN
$SM[d_i][s_j]_A = SM[d_i][s_j]_A + R[s_j]$
$SM[d_i][s_j]_B = SM[d_i][s_j]_B - R[s_j]$

Where m is performance threshold i.e. 60, i represents dimensions of learning style model $[1-4]$, j is the subject numbered from $[1-n]$, performance in a subject is represented by $PFM[s_j]$ in the range $[0-100]$, value of preference for model i dimension for subject j is $LSC[d_i][s_j]$, $SM[d_i][s_j]_A$ is the value of preference A stored in the dimension i for subject j, and $R[s_j]$ is the value of reinforcing for subject j, where $R[s_j]$ is in $[0-1]$ and is not necessarily to be equal to $R[s_k]$ for where $k=1$ ... n and $j \mathrel{!=} k$.

$$R[s_j] = \frac{1}{PFM[s_j] * DLS[s_j]} \tag{3}$$

$DLS[s_j]$ is the distance between preferences in a dimension. DLS can be calculated by the Eq. 4. Reinforcement $R[s_j]$ is inversely proportional $PFM[s_j]$ and $DLS[s_j]$. Inverse relationship between $R[s_j]$ and $DLS[s_j]$ is necessary because when the difference between preferences is too small then these preferences are not considered stronger on either side which become insignificant or undefined preferences. All preferences in four dimensions will be updated for appropriate subject if desirable performance is not achieved in tests.

$$DLS[s_j] = |SM[d_i][s_j]_A - SM[d_i][s_j]_B| \tag{4}$$

The values of R should be in a range for which preferences should not change very rapidly or very slowly. Through simulation, it was identified that maximum value for 0.20 and smallest 0.05 for R is more suitable [10, 11]. We have used these upper and lower limits for the value of R.

4 Experimental Prototype Application

We have developed a web-based application prototype for experimentation and evaluation of proposed model using PHP, MySQL server, JavaScript, CSS and related web-based constructs.

4.1 Dataset and Learning Style Model

We have selected videos and textual documents of two Computer Science subjects from Virtual University Website. Subjects include *Database Management System* and *Introduction to Programming*. Videos and textual documents were divided into small parts i.e. topics. As a result, for each topic we have two forms of representation i.e., visual and written. Index of learning style model is exploited where we have used its *Visual/Verbal* dimension because it is more suitable for coursework hypermedia [12]. For identifying learning style preferences, questions associated with visual and

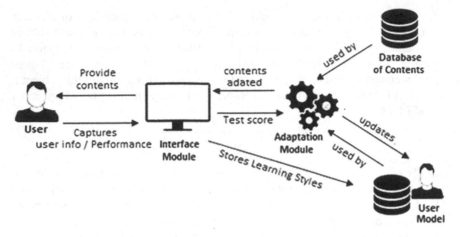

Fig. 1 Architecture of prototype

verbal dimension are selected from the Index of learning style questionnaire.[1] For learning style's dimensions and their associated questions, we can see the work of [13].

4.2 Main Components of Prototype

Our prototype application has main four components. The overall architecture of the experimental prototype is shown in Fig. 1.

- **Database of educational contents:** For each topic, videos and textual documents are stored in a database.
- **User model:** It is the collection of information about students' preferences for each subject.
- **Adaptation Module:** This module updates preferences when a student fails the test at the end of a lecture, using rules as depicted in Table 2.
- **Interface Module:** Interface module provides an interface which captures learning style of a student and presents contents selected by adaptation module. Another responsibility of this module is assessment of student's performance by presenting tests after each lecture. Interface layout for visual preferences and written preferences is shown in Fig. 2.

---

[1] http://www.engr.ncsu.edu/learningstyles/ilsweb.html.

4.3 Adaptation Mechanism

- **Capturing Student Preferences:** For a new user, learning style questionnaire is given. When a student fills and submits it, preferences for the user are stored.
- **Providing Contents:** Contents are given to the student according to his preferences associated with each subject. These preferences could be different for different subjects.
- **Testing User Performance:** At the end of a lecture, the student is asked to give the test. Next lecture will be provided if the student's score on the test is satisfactory otherwise preferences will be updated only for the subject to which the lecture belongs.
- **Updating User Preferences:** Based on the test marks, preferences associated with the current subject are updated.

5 Research Design and Evaluation

Our research design is experimental and followed in recent *learning style based user modeling approaches* [4, 5]. We are using Two-Group Pretest-Posttest Randomized Experimental Design.

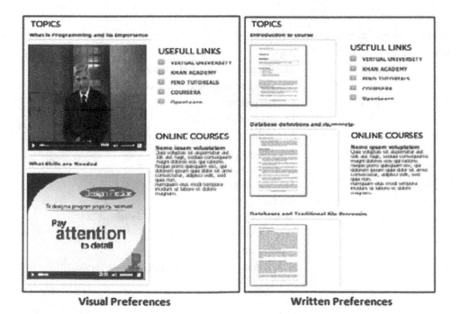

Fig. 2 Interface for visual and written preferences

5.1 Experimental Setup

Evaluation of our model is based upon the comparison between our application prototype and educational website of *Virtual University*[2] that does not have the mechanism for adaptation in discipline-wise manner. Learning style questionnaire was given to the students from *Virtual University of Pakistan* Islamabad campus. These undergraduate *Bachelor* and *Master* students were selected from *Computer Science Department*. Total 30 students were selected, 15 of them were visual while another 15 having preferences for written text. Their preferences were identified through pretest questionnaire, i.e., learning style questionnaire.

Half of visual user were randomly assigned to the control group and other half visual users to the treatment group. The same procedure was used for the written preferences students too. We have also applied independent sample T-test for comparing means of both groups for confidence interval 95 %, for the alpha value 0.05. Value of p was obtained 0.860, i.e., greater than 0.05 which means that both groups are not statistically different. Control group used website of Virtual University for learning 6 lectures, 3 per subject for the duration of one week while Treatment group used the prototype application for learning the same task. At the end of the week, post-survey questionnaire was given to both groups to record their responses to measure efficiency, effectiveness and user satisfaction about these systems. Total 13 questions were used for these variables. The detail of the questions and their purpose is shown in Table 3.

Table 3 Key measures for evaluation

S.No	Measure	Aspects
1	Efficiency	Identification of topics of a lecture
		Switching among topics of a lecture
2	User Satisfaction	Overall perspective
		Presentation of contents
		Scalability
		Future usage
3	Effectiveness	Increasing motivation
		Student assessment
		Usefulness
		Identifying weakness in subjects' lectures
		Help to overcome weaknesses
		Memorization of topics
4	Suggestions	Suggestions to improve the application in future

---

[2]http://www.vu.edu.pk/.

5.2 Results

Comparison between these two groups shows that our proposed model is better and effective for the students during learning. These comparisons are shown in Figs. 3, 4 and 5 where we have plotted *above average* and *very good* responses for evaluation measures.

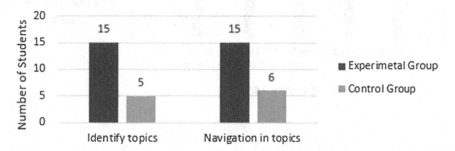

Fig. 3 Efficiency comparison between groups

Figure 3 shows the efficiency comparison. From Fig. 3 we can see that our experimental prototype saves significant time for the tasks *identification of topics* and *navigation among topics*.

Effectiveness of the application prototype is depicted in Fig. 5. We can see that most of the experimental group students agree that our system is better in *pinpointing their weakness* and system presentation help them to *work on their weaknesses*. They believe that *marking is unbiased, helpful and useful* and *increase student's motivation* during the learning process, and system presentation also helps them not to remember topics of a lecture if they need to study these again.

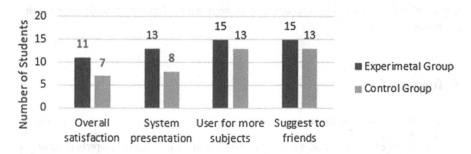

Fig. 4 User satisfaction comparison between groups

Satisfaction of user for their used system are in Fig. 4. We can see from Fig. 4 that most of the students are satisfied with the system's presentation, want to use it for more subjects and to suggest our application to their friends. We can see from these statistics that our proposed model for personalization of contents in term of discipline-wise learning style variations is more effective.

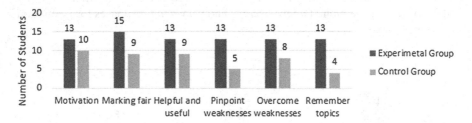

Fig. 5 Effectiveness comparison between groups

6 Conclusion and Future Work

Learning style based user models try to modify/adjust learning style preferences time to time to increase student performance through effective personalization. In our work, we have provided a user model that keeps discipline-wise learning style variations of a student. Our model continuously, dynamically and automatically adjusts learning style preferences for each academic discipline accordingly. Personalization becomes more effective if a user model is aware of discipline-wise requirements of the student. Results show improvements during the learning process. It is important to exploit other dimensions of learning style model for adaptation. We are planning to evaluate the system for a larger number of participants and for a lengthy period of time. It is necessary to investigate the relationship between subject nature and learning style preferences in order to develop more mature adaptive systems for learning purposes.

References

1. Froschl, C.: User modeling and user profiling in adaptive e-learning systems. Graz, Austria: Master Thesis (2005)
2. Knutov, Evgeny, De Bra, Paul, Pechenizkiy, Mykola: Ah 12 years later a comprehensive survey of adaptive hypermedia methods and techniques. New Rev. Hypermedia Multimedia **15**(1), 5–38 (2009)
3. Mulwa, C., Lawless, S., Sharp, M., Arnedillo-Sanchez, I., Wade, V.: Adaptive educational hypermedia systems in technology enhanced learning: a literature review. In: Proceedings of the 2010 ACM Conference on Information Technology Education, p. 73–84. ACM (2010)
4. El-Bakry, H.M., Saleh, A.A.: Adaptive e-learning based on learner's styles. Bull. Electr. Eng. Inf. **2**(4), 240–251 (2013)

5. El-Bakry, H.M., Saleh, A.A., Asfour, T.T., Mastorakis, N.: A new adaptive e-learning model based on learner's styles. In: Proceedings of 13th WSEAS International Conference on Mathematical and Computational Methods In Science and Engineering (MACMESE'11). Catania, Sicily, Italy, pp. 440–448 (2011)
6. Rozanski, E.P., Haake, A.R.: The many facets of hci. In: Proceedings of the 4th conference on Information technology curriculum, pp. 180–185. ACM (2003)
7. Brown, E.J., Brailsford, T.J., Fisher, T., Moore, A.: Evaluating learning style personalization in adaptive systems: quantitative methods and approaches. IEEE Trans. Learn. Technol. 2(1), 10–22 (2009)
8. Fernandes, M.A., Lopes, C.R., Dorca, F.A., Lima, L.V.: A stochastic approach for automatic and dynamic modeling of students learning styles in adaptive educational systems. Inf. Educ. Int. J. (Vol11_2), 191–212 (2012)
9. Jones, Cheryl, Reichard, Carla, Mokhtari, Kouider: Are students'learning styles discipline specific? Commun. Coll. J. Res. Pract. 27(5), 363–375 (2003)
10. Dorça, F.A., Lima, L.V., Fernandes, M.A., Lopes, C.R.: A new approach to discover students learning styles in adaptive educational systems. Rev. Bras. Inf. Educ. 21(01), 76 (2013)
11. Dorça, F.A., Lima, L.V., Fernandes, M.A., Lopes, C.R.: Comparing strategies for modeling students learning styles through reinforcement learning in adaptive and intelligent educational systems: An experimental analysis. Expert Syst. Appl. 40(6), 2092–2101 (2013)
12. Carver, C.A., Howard, R.A., Lane, W.D.: Addressing different learning styles through course hypermedia. IEEE Trans. Educ. 42(1), 33–38 (1999)
13. Graf, S., Viola, S.R., Kinshuk, T.L.: Representative characteristics of felder-silverman learning styles: An empirical model. In Proceedings of the IADIS International Conference on Cognition and Exploratory Learning in Digital Age (CELDA 2006), Barcelona, Spain, pp. 235–242 (2006)

Part IV
Smart Professional Training and Teachers' Education

Preparation and Evaluation of Teachers' Readiness for Creation and Usage of Electronic Educational Resources in School's Educational Environment

Marina V. Lapenok, Olga M. Lapenok and Alevtina A. Simonova

Abstract This paper investigates the teachers training problem in creation of electronic educational resources of education information environment (EER of EIE) for secondary schools, the usage of such resources in the Direct and distant information interaction based on Learning Management System (LMS), as well as the problem of estimating the readiness of teachers to such activities. There has been developed and presented the following: the content and methodological support of teacher training in the development and usage of EER of EIE; multi-criteria assessment of the teachers' readiness in this activity; EER of EIE typing on methodological and functional purpose, the technological implementation; method of pedagogical-quality ergonomic evaluation of EER of EIE; "Integrative organization" of the educational process in the EIE of secondary school that provides the implementation of variant teaching forms and methods.

Keywords Electronic educational resources · Education information environment · Teacher training · Assessment of teachers' readiness

1 Introduction

Modern information society is characterized by the widespread usage of information and communication technologies (ICT), including SMART-technology, distance education technologies, as tools to improve the effectiveness of professional

M.V. Lapenok (✉) · A.A. Simonova
Ural State Pedagogical University, Yekaterinburg, Russia
e-mail: lapyonok@uspu.ru

A.A. Simonova
e-mail: simonova@uspu.ru

O.M. Lapenok
Ural Federal University, Yekaterinburg, Russia
e-mail: olya_lapyonok@mail.ru

© Springer International Publishing Switzerland 2015   299
V.L. Uskov et al. (eds.), *Smart Education and Smart e-Learning*,
Smart Innovation, Systems and Technologies 41,
DOI 10.1007/978-3-319-19875-0_27

and educational activities. The studies by Lozinskaya A.M. [1], Gerova N.V. [2], Rozhina I.V. [3], Starichenko B.E. [4] and others emphasize the reasonability of providing the education information environment (EIE) in school by electronic educational resources (EER).

At present, the development teams and individual authors, teachers have created and placed in the public central storages of EER for students at all levels of education. However, the analysis of the EER for secondary school has showed that their quality does not fully comply with pedagogical and ergonomic requirements.

In this work the education information environment (EIE) refers to a set of conditions for interactive information exchange between teachers, students and electronic educational resource, automation of a control process and organizational management of training activities based on Learning Management System (LMS). The electronic educational resources of education information environment (EER of EIE) refers to the set of educational information presented in electronic formats, the proficiency of which provides the environment for various kinds of learning activities; and the information reflects the certain subject, the study of which, as well as the selection of information from the storage and its delivery to the student, navigation through the content, interactive content services are implemented by using the LMS software.

Certain aspects of the educational process on the basis of EIR in Basic education institutions were studied in the works of many authors (Polat E.S. [5], Chefranova A.O. and others). However, their proposed solutions to the educational process in secondary school by using ICT is not implemented the ability to change from the class and lesson forms to the extracurricular form of education in the case of forced absence from school; but the organization of teaching and research activities of students at the extracurricular form in a distant interaction for training and participating in competitions, certification, training projects is not described teaching methods are not revealed, the use of which contributes to the solution of pedagogical tasks in different modes of training activities and educational interaction of students with the training and the implementation of the EER by realization of LMS' technology opportunity.

To the preparation of future teachers to work in conditions of IT development of education were devoted the works of many modern scholars (Lavina T.A., Robert I.V. [6], Shamalo T.N. and others), which analyzed the components of teacher's work with the usage of ICT and presents the system of teachers training in the conditions of IT development of education.

A number of researchers (Avetisyan D.D., Ivannikov A.D., Osin A.V. [7] and others) noted that the basic bulk of e-learning means is created by professional development teams (for example, closed joint-stock company "1C", closed joint-stock company "Media Education", Limited liability company "Cyril and Methodius" and others) and they cannot be modified by the teacher. The use of such EER limits the implementation of creative approaches to teaching, motivating teachers to develop their own EER, which motivates the preparation of teachers to create e-learning materials copyright by means of instrumented system.

Therefore, in the current works the theoretical thesis in creation of different types of EER of EIE in accordance with the their pedagogical and ergonomic requirements is not fully developed; the issues of the educational process in mass secondary school, providing the use of EER and LMS services at the different types of lessons and extracurricular activities and students' researches, is not fully resolved.

The methodological approaches to teacher training in the field of EER and their systematic use in the educational process organized in EIE are not presented.

The problem of the research is determined by the necessity to train teachers to create EER of EIE for secondary school, to use EER of EIE in the direct and distant information interaction, as well as the necessity for evaluation of teachers' readiness for such activities.

The goal of research is to develop the theoretical basis of creation of EER of EIE, scientific and methodological approaches to the organization of educational process using EER of EIE, as well as training and methodological support of training and assessment of teachers' readiness in this field.

2 EER of EIE's Theoretical Bases of Creation

2.1 Typing of EER of EIE's

As the educational process of a comprehensive school has various training activities (lectures, practical exercises, examinations), to the extent the teacher should be able to create, modify, and use a variety of EER of EIE according to applicable forms and methods of teaching. In the context of this work we have developed a EER of EIE's typology according to various criteria: for methodological purposes, by their functional and technological implementation.

Depending on the methodological purpose were highlighted the following types of EER of EIE: teaching EER designed to let students get some knowledge, formation of skills for training and practice and ensure the necessary level of assimilation are presented in the form of presentations or text with audiovisual insertion with the ability to change the level of difficulty, as well as the sequence and the rate of supply of educational material; training EER are intended for practicing skills in training activities with repetition or consolidation of the material presented in the form of tests of various types and levels of difficulty; controlling EER are designed to control the level of evaluation of learning material presented in the form of creative assignments or tests with the possibility of interactivity.

In addition, in the EER of EIE includes the educational materials presented in the form of downloadable files or web links related to methodological purpose of the following types: reference, designed to organize information; simulation, designed to study the main structural and functional characteristics of an object or a process with a limited number of parameters; modeling, made for creation of models (including interactive) of the studied object or process; demonstrational, intended to visualize the studied phenomena and relationships between objects; educational and entertaining, designed to form the ability to make the best decision.

Depending on the types of functionality were highlighted the types of EER of EIE which provide an automation of processes, design of teaching materials; monitoring of educational achievements; processing of the results of educational experiment; conducting electronic records management.

Depending on the types of technological implementation were highlighted the EER of EIE's types containing not modifiable digitized textual, graphic, audiovisual materials; materials unified modular structure, open for editing and additions; textual materials followed by audiovisual support oriented mainly onto mobile handheld electronic communication device.

2.2 Methodology of Evaluation the Pedagogical and Ergonomic Quality of EER of EIE

Whereas it is not acceptable in the learning process to use EER of EIE of poor quality, the teacher must know the quality criteria and be able to assess the quality of the EER of EIE. We have developed a method for estimating the pedagogical and ergonomic quality of EER of EIE based on marking out the feature of EER of EIE's quality and the formulation of requirements for each characteristic.

We have formulated the requirements for meaningful pedagogical features which include: providing of pedagogical appropriateness (conformity of didactic principles, availability of training materials for all kinds of activities in EER, the availability of the system and links to additional training resources and etc.); according to students' age (EER that contain the educational standards and recommended textbooks; in accordance of EER's interface to students' proficiency in ICT; proving the information security, including the accordance of content to ethical standards and regulations and etc.); providing the implementation of variant forms and methods of training (availability of interactivity in a distant interactive communication, availability of multiple difficulty levels of educational material, the ability to change the sequence and rate of teaching material's presentation and etc.); providing methodological ability (availability of guidelines on the usage of LMS services, at the choice of forms and methods of training in EIE, the availability of the links' system to training resources and etc.).

Requirements to the technical and technological characteristics include the following: maintaining an EER of EIE at the startup of other applications in multi-user mode; matching values of memory amount required for the EER, and download time of the content available in hardware resources; EER content control through various input devices and so forth.

Requirements for the ergonomic design features include the following: ensuring comfort perception of textual and audiovisual information constituting EER content (font quality, non-aggressive visual environment, the ability to adjust the sound scale and so on.); easy usage and convenience of the interface; implementation of multimedia technology in interactive user interaction with the EER (when EER content is in the form of text, audio-visual static and dynamic information, the presence of content with elements of EER selection; navigation and search of learning information; error monitoring diagnostics and feedback, etc.).

To match the EER of EIE to the requirements stated above, you can use the method of group expert assessments of the quality of learning materials presented in electronic form.

While evaluating experts of EER implement the following:

- Definition of actual values of each pedagogical and ergonomic quality characteristic of EER of EIE (in the activation process of content services using LMS and in the process of obtaining the meaningful feedback from the resource);
- Establishment of fact pass/fail of criteria for the obtained values reflected in the state standards;
- Calculation of an avera T_i ge total points accumulated by experts for compliance with performance indicators of the quality of the EER and the establishment of the level of pedagogical and ergonomic quality of assessed EER;
- A statistical analysis of the calculation results on groups of distinguished quality characteristics to ensure the implementation of all groups of requirements for pedagogical and ergonomics of EER of EIE, as well as to identify the consistency of the experts' conclusion.

2.3 Functionality of Modern LMS

Since different schools of the Russian Federation use various LMS (Moodle, NauLearning, Sakai and etc.), it is necessary to teach the teachers to use LMS services for typical pedagogical problems, abstracting from the particularity of LMS. To implement the mentioned above, we distinguished service groups of LMS, reflecting the overall functionality of the most modern LMS: services of educational material presentation; services of learning management; services, implementation of information exchange between the participants of the learning process; services of setup function.

Opportunities of LMS services are implemented in the course of their use in solving of educational problems. Among the typical pedagogical problems we have identified the following: implementation of operational control in the form of massive parallel poll of students (including distant learning conditions) to correct learning process; activation of cognitive activity of students through their involvement in collective learning project; training of learning group for the competition and others. Each of the objectives has been mapped with set of services, the use of which, in our opinion, contributes to its effective solutions. The choice of service is determined by LMS: mode of learning activities and educational interaction (real time or delayed communication); methodological objectives (conducting surveys, consultations, exchange of documents, etc.) of information interaction; methodological objectives application created by EER and representation formats of educational material produced through this service.

3 Scientific and Methodological Approaches to the Organization of Educational Process Using EER of EIE

3.1 Model of "Integrative Organization" of the Educational Process in the EER of Secondary School

The use of EER of EIE allows making the learning process on lessons at school and extracurricular time outside the school. We have developed a model of "integrative organization" of the educational process in the EER, which provides:

- Training of students in groups of constant lineup at school lessons in terms of adaptability of EER of EIE into the traditional class-lesson system;
- Individual students training with the use of EER of EIE in the condition of forced absence from school; group training in non-permanent lineup at the extracurricular time with the automation of individual training plan, monitoring of educational achievements, feedback control, assessment and registration of results of educational activity on the basis of LMS services;
- Group training non-permanent lineup at the extracurricular time with the automation control of extracurricular education and research activities of a student on the basis of LMS services.

3.2 Methodological and Organizational and Instructional Software for "Integrative Organization" of the Educational Process in the EIE of Secondary School

To implement the educational process within its "integrative organization" in EIE of secondary school have been developed methodological and organizational and instructional software, including: guidelines for the choice of form and methods of training in the implementation of the educational process in the EIE; regulations to support the educational process in the EER of secondary school; additions to the job description for the teacher's work in the EIE.

In the developed recommendations were identified and substantiated some teaching methods, the use of which contributes to the solution of pedagogical tasks, depending on: the mode of interaction of students with teacher and EER, teacher's activity and student's activity, organizational forms of learning and technological capabilities of LMS.

Regulations to support the educational process in the EIE of secondary school determines the activities of participants of the educational process and the allocation of resources. The structure of the regulations includes the following documented of procedures: a procedure of planning and carrying out the educational process in the EIE, which determines the order of formation and approval of applications for participation in the learning process; the changeover of a student to learning process with Distant Learning Technologies, which determines the order of formation of student's individual curriculum for a period of his/her absence from school (duration not specified in advance), and the appropriation of network teachers-curators of information exchange for the period of study based on LMS; the procedure of student's individual work planning, which determines the structure and the formation of the individual plan of extracurricular educational and research work of the student; the procedure of reporting the results of learning, determine the conditions of monitoring of educational achievements and etc. additions to the job description for the teacher's work in the EIE in the EIE contain the job description of the teacher in the conditions of implementation of e-learning.

4 Theoretical and Methodological Teachers' Training to Create and Use the EER of EIE

4.1 The Content of Teacher Training in the Development and Use of EER of EIE

The main topics of teacher training in this area are: general issues of IT development in education; typing of EER of EIE; opportunities of LMS services for creating and using EER of EIE; evaluation of pedagogical and ergonomic quality of EER of EIE; methodological aspects of the educational process in the EIE of secondary school in the implementation of variant forms and methods of teaching.

We have also identified the types of learning sessions undertaken in the preparation of teachers in the said area. The lectures form the teacher's theoretical knowledge and skills to ensure the creation of the EER of EIE of high pedagogical and ergonomic qualities, as well as the use of EER of EIE in terms of educational communication and learning management service based on LMS. At the workshops, realized in the form of role-playing games, scenarios that simulate the pedagogical situations, typical of "integrative organization" of the educational process, forming practical skills in the use of services for effective LMS implementation in EIE. To carry out a laboratory work teachers create an e-learning material of various types based on EER of EIE.

To train teachers in the development and use of EER of EIE was developed methodological support, including: working curriculum of the discipline; textbook "technologies of distance learning courses"; guidelines on the use of LMS services for creating and using EER of EIR; role-play scenarios, describing the learning process in the EIE of secondary school; measurement and control of materials for evaluation of learning outcomes.

4.2 Assessment of Teacher's Readiness to Create and Use EER of EIE in Secondary School

Methodology of teacher training level evaluation in the development and use of EER of EIE is based on multi-criteria approach to the assessment of knowledge and skills. The level of training in this area, made as a result of the teacher training program we can estimated of the measure of T_i:

$$T_i = T_{i1} + T_{i2} + T_{i3}, \ i = 1, \ 2, \ \dots, \ m, \tag{1}$$

where: m – the number of teachers participating in the training;

i – the teacher's serial number;
T_{i1} – indicator of the training level of i-th teacher in the field of theoretical knowledge and skills in creating and usage of EER of EIE;
T_{i2} – indicator of the training level of the i-th teacher in the field of practical skills for using the services of distance learning in the educational process;
T_{i3} – indicator of the training level of the i-th teacher in the field of practical skills to create the EER of EIE.

The indicators values T_{i1}, T_{i2}, T_{i3} of the index T_i can be estimated from the results of the teachers' final certification works (final diagnostic work tasks in the test form, the final diagnostic role-playing game; defense of training project in creation of the EER of EIE), bringing the obtained evaluation to a single measuring scale according to the following formula:

$$T_{ij} = \frac{X_{ij} - \overline{X}_j}{\overline{\sigma}_j} \times 10 + 50 \quad j = 1, \ 2, \ 3, \tag{2}$$

where: X_{ij} is a mark received by the i-th teacher on the results of the j-th final certification work; $\overline{X}_j$ is the sample mean of the random value (marks estimated by the i-th teacher on the results of the j-th final certification work); $\overline{\sigma}_j$ is a sample standard deviation of the random value.

The scales of measuring the T_i and its components T_{i1}, T_{i2}, T_{i3} can be divided into four areas: [0; 0,7); [0,7; 0,8); [0,8; 0,9); [0,9; 1,0] which quantitatively corresponds to a low, medium, and high level of basic teacher training.

5 Pedagogical Experiment

The Pedagogical experiment on proper teacher training in the development and use of EER of EIE was held at the Ural State Pedagogical University during 2007–2012. In pedagogical experiment were involved high school teachers of various subjects (algebra, history, social and economic geography, Russian language and literature, chemistry, biology, physics) from Yekaterinburg city, as well as teachers and students of the university.

On ascertaining stage of the experiment (during 2007–2010) were successively formed four experimental groups, which are typical for the qualitative composition series (based on academic subjects, age, and employment history in the school): in 2007 there was 1st group - 53 teachers; in 2008 there was the 2nd group - 52 teachers; in 2009 there was the 3rd group - 54 teachers; in 2010 there was the 4th group - 51 teachers.

During the formation of the experimental groups was evaluated an initial level of knowledge and skills of every teacher in the field of searching and processing information using ICT. The results of the input diagnostic work showed that all teachers have the necessary level of basic knowledge and skills in this area, so their additional training wasn't conducted.

On the forming stage of the experiment (during 2007–2011.) was carried out teachers' training in the development and use of EER of EIE in the program that implements the theoretical and methodological approaches presented above. The duration of teacher training is one academic year. Upon completion of a course, each teacher has fulfilled the final diagnostic work with tasks in the test form; took part in the final diagnostic role-playing game, and has developed and defended an educational project in creating of the EER of EIE.

At the final stage of the experiment (during 2011–2012) was carried out a statistical processing of empirical evidence obtained in the course of pedagogical experiment.

The analysis of the results of assessing the training level of teachers in the development and use of EER of EIE showed that 14 % of teachers reached high level, 56 % of teachers - baseline, 26 % of teachers - the average level of training in this field. However, only 4 % of teachers have demonstrated a low level of training in the development and use of EER of EIE. Thus, the results of the experiment showed that the majority of teachers (70 %) reached base and high levels of education in the development and use of EER of EIE, which confirms the effectiveness of our developed theoretical and methodological approaches to teacher training.

6 Key Findings

Scientific novelty of the research lies in: the grounding and development of "integrative organization" of the educational process in the EIE of secondary school providing the implementation of variant forms and methods of training; methods of evaluation of pedagogical and ergonomic quality of EER of EIE; to develop the content and educational provision of teacher training in the development and use of EER of EIE; in the formulation of recommendations for selecting LMS services to present the content of EIE, implementation of-students interaction with the EER, management training and research activities of the students.

The theoretical significance of the study is theoretical grounding and development of: EER of EIE typing in methodological and functional purpose, the technological implementation; performance of requirements for pedagogical and ergonomic quality of EER of EIE.

The practical significance of the study is to develop: a working curriculum of a subject; a textbook for teachers' technology of distance learning courses "; guidelines for choosing the forms and methods of training in the implementation of the educational process in the EIE; guidelines on the use of LMS services in the process of creating and using EER of EIE; role-playing games as means of formation of practical skills in the use of LMS services in the conditions of realization of variant forms and methods of training; test materials for evaluating the level of teachers' training in the development and use of EER of EIE, regulations to support the educational process in the EIE of secondary school.

References

1. Lapenok, M.V., Lozinskaya, A.M.: Formation of pedagogical competence of interactive communication in the information environment of distance learning. J. Teacher Educ. Russia **5**, 78–82 (2012). (in Russian)
2. Gerova, N.: Methodical system of future teachers' information training in higher education. In: SGEM Conference on Psychology and Psychiatry, Sociology and Healthcare Education, vol. 3, p. 541–547. Bulgaria (2014)
3. Lapenok, M., Rozhina, I.: Distance learning system: from designing to introduction at schools. In: 11th IASTED International Conference on Computers and Advanced Technology in Education, pp. 100–108. Crete (2008)
4. Starichenko, B., Korotayeva, E., Semenova, I., Slepukhin, A., Mamontova, M., Sardak, L., Egorov, A.: The method of use of information and communication technologies in educational process. Manual in 4 parts. Ural State Pedagogical University, Yekaterinburg (2013) (in Russian)
5. Polat, Y.S.: Educational Technologies in Distance Learning. Academy, Moscow (2008) (in Russian)
6. Robert, I.V.: The development of didactics in conditions of informatization of education. In: International Scientific and Practical Conference on informatization of Education, pp. 3–9. Omsk (2012) (in Russian)
7. Osin, A.V.: Open Educational Modular Multimedia System. Publishing Service, Moscow (2010) (in Russian)

ICT Proficiency Measurement While Realizing Information Activity of Students Majoring in Pedagogical Education

Natalya Gerova

Abstract This paper deals with the components of information activity of students majoring in pedagogics and their correlation to the future profession. The author suggests structuring the disciplines in computer science, information and communication technologies for students specializing in pedagogical education. In accordance with the current state of higher education in the Russian Federation, there are two levels of bachelor and master's programs. The author offers a set of disciplines, each containing from 2 to 4 modules, for both levels. A module has an invariant and variable part and includes theoretical questions, practical assignments, laboratory exercises, requirements for independent work, presentations and measuring materials. The article also contains proficiency measurement of students in computer science and ICT due to the modular presentation of the academic programs.

Keywords Bachelor · ICT · Information activity · Master · Math statistics

1 Introduction

The actual level of ICT development dictates the transfer of accumulated knowledge in different spheres of human activity as a priority task. This knowledge reflects vital needs of the information-oriented society.

The framework of education is the activity as a generalized structure of human activity in a definite sphere. The information activity presupposes the acquisition of knowledge and skills through a reasoned and well-targeted task solution.

N. Gerova (✉)
Informatisation of Education and Methods of Teaching Department, Ryazan
State University n.a. S.A. Esenin, Ryazan, Russian Federation
e-mail: nat.gerova@gmail.com

© Springer International Publishing Switzerland 2015
V.L. Uskov et al. (eds.), *Smart Education and Smart e-Learning*,
Smart Innovation, Systems and Technologies 41,
DOI 10.1007/978-3-319-19875-0_28

The development of inborn cognitive abilities and their perfection in the information-oriented society are necessary for the realization of one's potential and self-improvement in the professional activity on the basis of ICT.

ICT create unlimited opportunities for intellectualization of various kinds of activity. They ensure both the international cognition principles in different subject areas and the individualization and differentiation of education.

The most important requirement for carrying out the professional activity of graduates is not the acquisition of some abstract knowledge but the skill of useful and effective work under definite circumstances while performing certain duties.

If we examine pedagogical skills as mastering means and techniques of education and upbringing, we should mention that continuous information training at the level of higher education is the basis of ICT competence in educational and future professional activity of students-humanitarians majoring in pedagogics.

This means that the acquisition of skills for carrying out various kinds of educational and future professional activity on the basis of ICT is an integral part of pedagogical thinking and professional training of students. The components of information activity of future bachelors and masters don't depend on the specialization. They are based on the common types of information activity. It's evident that large-scale implementation of ICT means in academic activities at a university influences all types of educational and future professional activity considerably.

2 Components of Student Informational Activity

The structure of the teachers' professional activities in the use of ICT was investigated in works by different authors [1, 3, 6–9]. Each type of bachelors' (pedagogical, cultural-educational, etc.) and masters' (teaching, research, management, design, methodological and cultural-educational) professional activities is closely related to the components of informational activity, the content of which is to be implemented at the programs.

If we try to single out relatively independent functional types of pedagogical activity, we should take into consideration the informational nature of the education process in the course of which a teacher:

- Collects information, uses available knowledge, elicits new information from different sources (books, magazines, databases, the Internet, etc.), communicating with colleagues by means of conferences, methodological associations, postgraduate courses, etc.
- Works out the information model of the educational process, which includes reasoned goals, the algorithm of the educational process, selected teaching material, forms, means and methods of teaching the discipline.
- Transfers information and stimulates the feedback from students to evaluate the results.

- Analyses the efficiency of the algorithm of the educational process used for its further correction.

To define the components of students' information activity and the interrelation we should single out:

- The structure of the information activity of students-humanitarians (bachelors and masters) according to the types of a professional activity, this structure being invariant for various specializations.
- The content of the chief components of the students' information activity as related to ICT application in the field of the educational and future professional activity.

Meanwhile, the organization of a certain kind of future professional activity on the basis of ICT is focused on the information culture acquisition in the course of academic activities. Nowadays education milieu of a university combines educational, scientific and industrial areas. The latter ones include different interrelated information departments, organizational and legal structures, means and technologies of data processing, knowledge databases, ensuring information processes. Higher vocational education brings about competent qualified specialists possessing information culture and ready to use ICT in the educational and professional activity. The information culture is formed through various types of educational activity.

In the course of education students master chief educational programs, attend different classes, do assessment tasks. Carrying out various types of educational activity (academic, research, creative, social, independent), a student achieves definite results, which are performance indicators. The indicators of academic activity are: attendance, the accomplishment of laboratory reports, fulfillment of term and graduation papers, the accomplishment of reports, colloquia, summaries, essays, etc., passing mid-term and final assessment. The indicators of research activity are: participation in grants, conferences, symposia, Olympiads, academic competitions, innovation projects, published articles, received patents and certificates.

The indicators of creative activity are: participation in cultural events of the university, city, region, etc. The indicators of sport activity are: participation in sport competitions at the university, city, region, country level or international ones, fulfillment of training standards, receiving titles.

The indicators of social activity are the participation in the life of the group, department, university, city, etc. Taking into consideration the notions stated above, we represent the structure of students' educational activity as possessing academic, research, creative, sport and social components.

Bachelors of Education prepare themselves for the pedagogical and cultural professional activity with specific kinds of professional activity selected by the university. A bachelor, working as a teacher, is supposed to solve specific professional problems. He is supposed to study the abilities and interests of the students, to scheme the results of their learning, education, development, to use ICT and other technologies typical of this or that subject area in teaching, to be ready for self-instruction and personal development. Cultural activity requires the creation of a cultural environment on the basis of ICT and the elaboration and realization of cultural and educational programs for different social groups.

Masters of Education are to solve professional problems according to their specialization. The types of professional activity include not only the pedagogical and cultural activity but also the research activity, the management activity, the design and methodological activity. As far as the pedagogical activity is concerned, both masters and bachelors are to mark the guidelines for their professional career, self-education and perfection.

Te research activity requires the analysis, systematization and generalization of the results of the investigation due to a number of research methods, the evaluation of the results by means of modern scientific and ICT methods, communication with Russian and foreign colleagues, the usage of existing educational milieu together with the formation of a new one to solve various problems, participation in testing activity, profession and personal development.

The management activity includes the examination of the conditions and the potential of the educational system, organization a management process with the usage of innovative management methods, cooperation with the colleagues and social partners, the maximum usage of the opportunities of educational organizations, working out the guidelines of development to guarantee the management quality.

The design (project) activity combines the elaboration of various components of educational milieu and programs, including individualized ones, new subjects and elective courses, the forms and methods of control, especially those dealing with ICT.

The methodological activity consists of the examination and analysis of teachers' professional needs, working out individual methodical routes, the usage of innovations for planning and the realization of methodical support for teachers who design new educational milieu on the basis of ICT and distance learning.

The cultural activity presupposes working out strategies for the realization of the educational activity, the elaboration of educational programs and their realization through the transfer of cultural traditions and scientific knowledge, the usage of ICT and mass media means to solve cultural problems.

Teachers help students to create a systematized information picture of the world. Thus, the humanistic function is fulfilled. We suppose that this task may be solved by means of correcting the content of informatics and ICT courses at the level of higher education while preparing humanitarians. New teaching forms and methods will also be of use.

According to the structure and types of professional activity, the chief components of the information activity are the following: constitutive, gnostic, organizational, designing, communicative and motivational. Each of the kinds of professional activity of bachelors (pedagogical and cultural) and masters (pedagogical, research, management, design, methodological, cultural-educational) is tightly connected with the components of the information activity, the content of which is to be taken into account in the course of teaching bachelors and masters. We'll define each component of the information activity of students.

The constitutive component presupposes the elaboration and realization of methodic models, pedagogical technologies and teaching methods, the analysis of the results of their usage in different educational organizations, the systematization, generalization and interchange of Russian and foreign experience.

The constitutive component includes collecting information from teaching guides, electronic media and information educational resources as well as planning independent work in the course of preparation for classes.

In the future professional activity the constitutive component presupposes: planning and preparation of professionally important data for different events on the basis of ICT didactic opportunities, searching for necessary didactic and methodic information in local and global nets, producing and publishing genuine study guides in nets and on the basis of standard applications and device tools, the automatization of introduction, storage, accumulation and analysis of professionally important data and the informational and methodical support of the educational process, presentation of information in the form available to every user.

3 Modular Principle of Discipline Formation

The researches in the theory of modular training [2, 5, 11] identify its following features: the development of the didactic system's components, structuring learning content, providing structural, organizational and methodological units.

The transition of the Russian education system to the modular one has led to the fundamental changes in the educational process of universities. The reformation has triggered off the following changes in the educational process: the introduction of undergraduate and graduate training of highly qualified personnel, the introduction of the credit system, the introduction of a points-rating system to evaluate students' knowledge, providing opportunities for personal students' participation in the formation of the individual curriculum by selecting courses, the reorganization of training programs for bachelors and masters, developing the mechanisms to evaluate the competence of the graduates, the promotion of the mobility of students and staff. The system is based on the integration of the competency and modular approaches to the organization of the education process. The implementation of educational programs is based on the modular principle of the curriculum development, the curriculum content, teaching practices, etc. Such an implementation provides the formation of professional competencies, the transparency of vocational training, the introduction of individual educational paths.

From the prospective of the competency approach, the purpose is to develop the student's competence, the identification of the expected results, the definition of the structure and content as a module and its components (as a part of the cycle of discipline, practice, etc.) [3, 5]. It can be observed from simple to complex ones, from the methodological disciplines to the application. Thus, the module can contain both basic and applied disciplines. This leads to the creation of open and flexible educational structures, the integration of different methods, tools and training forms depending on the target audience.

The analysis of the research in modular training shows that the characteristics of modern teaching theories and concepts are reflected in the modular technology of training, combining different approaches to the selection of the content and methods. The authors note that the purpose of modular training is to adapt the material to the individual interests of students and create the conditions for the personal development by providing training content flexibility. Thus, the modular training allows to implement a comprehensive training program for the visual, problem-oriented content, to adapt the educational programs to the target audience, to use the forms and methods of active learning, to motivate the students' cognitive activity, to form communication skills, to encourage students to develop and improve themselves in various subject fields.

The analysis allows to identify the following features of the modular training: the need for detailed study of each component, the structuring of the learning content, the necessity to provide the methodological materials and technologies for evaluation, the need in providing the organizational and methodological units which can be adapted to the learning process, individual abilities and interests of students.

The goal of the modular concept is to provide the flexibility of educational programs. There are different definitions of the module. On the one hand, it is an independent entity in a set of learning activities, a conceptual unit of educational material and prescribed actions for students, the overall theme of the course, an information block containing a logically complete unit of study materials. On the other hand, the module is a organizational and methodological structure of the interdisciplinary teaching material, the amount of material that provides the acquisition of theoretical and practical skills for a particular job. All the points stated above lead to the conclusion that, at present, there is no single approach to the definition of the term "module". The notion of "unit" will be considered from several positions: as a structural unit within the same discipline; as an interdisciplinary structural unit; as a unit of the curriculum. In the context of our study the module will be understood as the organizational and methodical block of the discipline, which contains didactic purpose, a complete unit of teaching material, learning materials and the evaluations. The module is characterized by an isolated learning content, educational material's variability, etc. The organization of the educational process on the modular principle includes the presence of various educational paths.

To structure the educational material different methods we used: the integration of didactic units, the allocation of structural and systematic knowledge units, the modular construction method, the graph theory. To develop the block of disciplines "Computer Science and ICT" for students majoring in pedagogical education one should use the modular approach for the planning and organization of the educational process as well as the construction of training programs. The development of training and the curricula content is a complex process. The content should provide interdisciplinary connections, as well as the formation of common cultural, professional, special information competencies and the ability to self-development.

The structure has two major modules in curricula for bachelors and masters. Each of the blocks includes a set of disciplines for Bachelor (Fig. 1) and masters (Fig. 2). Each discipline contains a summary, the purpose and the objectives of the study subjects, the requirements for the results, the modular presentation of the educational material, reports or essays, questions, the list of recommended resources.

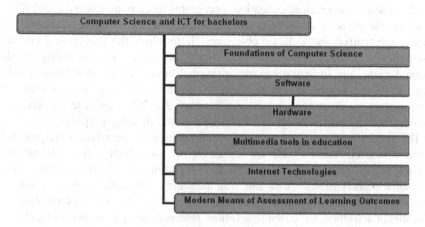

Fig. 1 The structure of disciplines in the field of computer science and ICT for bachelors

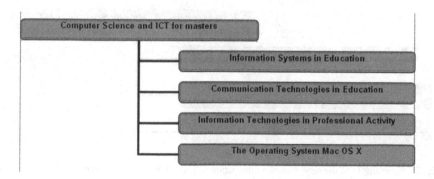

Fig. 2 The structure of disciplines in the field of computer science and ICT for masters

The purpose of the modular organization is: to ensure the learning content, the construction in relation to other learning methods and theories, aimed at the keeping individual characteristics and needs, the formation of information competencies, the possibility of organizing nonlinear educational paths.

The learning material in each discipline is presented as a module that has an invariant and variable parts, and includes theoretical questions, practical assignments, laboratory exercises, assignments and the requirements for independent work, presentation and test materials. The invariant component reflects the current state of scientific and technical progress in the field of computer science and ICT, the general requirements to the training level of students majoring in pedagogical education, etc. The variable component in this structure includes questions, teaching methods, themes and reflects the use of computer science and ICT in students' future professional and educational activities.

The modular structure of educational programs in computer science and ICT is based on the division of the educational material on the structural modules with certain substantive and didactic objectives. It provides the ability to form new disciplines in accordance with the educational objectives, the variability of educational forms, the structural and organizational flexibility, the non-linear training (Fig. 3). The content of the module within the discipline should meet the requirements of consistency, integrity, autonomy, etc. It is a guide to mastering the training material and is designed to achieve specific didactic purposes.

The special features of the modular representation of the educational programs are: a structured learning content, the rating system which allows students and teachers to adjust to the learning process, the adaptation of the learning process to the individual characteristics and capabilities, the training variability. The structure and content provides in-depth knowledge, skills and experience in the implementation of educational activities, including innovation, research, design, organization, etc.

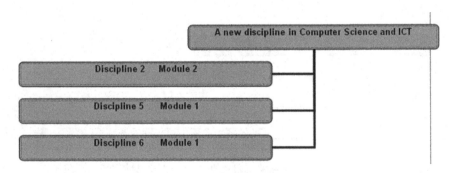

Fig. 3 An example of the new discipline formation in accordance with the learning objectives and audience

Currently, universities organize the educational process (form the curricula, regulations, etc.). This organization is based on the credit system. The credit system is an integral part of the reformation process. It leads to significant changes in the educational process and the restructuring of its content. The introduction of the credit system is directed to ensuring the transparency of the educational process, including the results of the final certification, providing the possibility of educational process independent planning, improving and diversification of educational technology, the usage of new educational methods, forms and means, etc.

The labor intensity of the educational activity is a set of 1 credit units, each being equal to 36 academic hours. The semester contains 18 study weeks, so the modular structure of the course will be based on the set of disciplines 18 weeks long and 2 or more credits value. Each module is an autonomous and logically completed part of the discipline, but has a close relation to the other disciplines of the block "Computer Science and ICT", as well as a part of the educational program.

4 Proficiency Measurement of Students in Computer Science and ICT

We use the mathematical statistics methods to determine the level of the information activities implementation in accordance with the modular principle [4, 10, 12]. The assessment of the students' training level can take only integer values and is a random process with discrete frequency distribution. To evaluate this level we can use an indicator:

$$Z_i = \sum_{k=1}^{l} \sum_{j=1}^{m_k} z_{ijk} \qquad (1)$$

where: l- the number of academic disciplines;

m_k- the number of training modules in the k-th discipline;

Z_{ijk}- the training level indicator of the i-th student within the j-th unit k-th discipline;

$$z_{ijk} = \frac{X_{ijk} - \overline{X}_{jk}}{\overline{\sigma}_{jk}} \qquad (2)$$

where: X_{ijk}- the number of points obtained by the i-th student in passing the final control on the j-th unit k-th discipline;

$\overline{X}_{jk}$- the average value of the random variable (the number of points earned by students during the passage of the final control on the j-th unit k-th discipline), which is defined by the formula

$$\overline{X}_{jk} = \frac{1}{n} \sum_{i=1}^{n} X_{ijk} \qquad (3)$$

(n - number of students); $\overline{\sigma}_{jk}$- the average standard deviation of the random variable, which is defined by the formula

$$\overline{\sigma}_{jk} = \frac{1}{n-1} \sum_{i=1}^{n} (X_{ijk} - \overline{X}_{jk})^2 \tag{4}$$

The condition of the calculation of the formula (1) is that the final results of the control for each module are normally distributed. Then we check that the experimental groups are homogeneous in training levels on the j-th unit k-th discipline. Thereafter, the values of parameters are calculated by the formula (2) and the formula (1). Further, the values of the distribution function corresponding to the training level of students majoring in pedagogical education are calculated.

5 Conclusion

At present the nature of students' future professional activity changes its character and is closely related to the information activity's components, the content of which must be considered in the implementation of training programs. ICT provide the opportunities for the innovative activities development and offers the innovative potential formation.

References

1. Gerova, N.: Methodical system of future teachers' information training in higher education. In: Gerova, N. (ed.) SGEM Conference on Psychology and Psychiatry, Sociology and Healthcare, Education. vol. 3, pp. 541–547. Bulgaria (2014)
2. Gerova, N.: The requirements to information competence of future russian teachers of the humanities. In: Gerova, N., Rogovaya, O. (eds.) SGEM Conference on Psychology and Psychiatry, Sociology and Healthcare, Education. vol. 1, pp. 256–261. Bulgaria (2014)
3. Gerova, N.: Use of IT in formation of personal achievements in educational activity of university students. In: Gerova, N. (ed.) The International Conference on the Transformation of Education. pp. 107–115. Acessed 22–23 April 2013
4. Gmurman, V.E.: Probability Theory and Mathematical Statistics: A Manual for Students of Technical Colleges. Higher School, Moscow (1977)
5. Gordon, J., Halasz, G., Krawczyk, M., Leney, T., Michel, A., Pepper, D., Putkiewicz, E. Wisniewski, J.: Key Competences in Europe: Opening Doors for Lifelong Learners Across the School Curriculum and Teacher Education, Study Undertaken for the Directorate General Education and Culture of the European Commission. URL: http://ec.europa.eu/education/more-information/reports-and-studies_en.htm
6. Kuzmina, N.V.: Methods of Teaching. Leningrad State University (1970)
7. Lapenok, M.: Teachers Training for Using of Distance Learning System at Schools. Lapenok, M. (ed.) 9th IEEE Conference on Advanced Learning Technologies (Riga, Latvia, July 15–17, 2009), pp. 602–603 (2009)
8. Slastenin, V.A.: Pedagogy: Textbook for Students of Higher Educational Universities. Academy, Moscow (2002)

9. Talyzina, N.F.: Managing the Process of Learning: Psychological Foundations. Moscow University, Moscow (1984)
10. Venttsel, E.S.: Probability Theory. Nauka, Moscow (1969)
11. Yutsyavichene, P.: Theory and Practice of Modular Training. Kaunas (1989)
12. Zaks, L.: Statistical Estimation. Statistics, Moscow (1976)

Training of Future Teachers in Development and Application of Computer Tools for Evaluation of Student Academic Progress

Marina Mamontova and Petr Zuev

Abstract Evaluation of academic progress is an important stage in academic activity. The educational program for training of future teachers at the Ural State Pedagogical University includes the course "Modern Tools for Academic Progress Evaluation". Mastering this course the undergraduates become familiar with the theory and methodology for development and application of evaluation tools using computer technologies. The paper presents an attempt to integrate the contextual learning technology with WEB 2.0 (World Wide Web) and SaaS (Software as a Service) Internet technologies for training undergraduates in joint design and application of computer assisted tools for academic progress evaluation - educational tests, mind maps, memorization maps, etc.

Keywords Higher school training · Contextual learning · Saas business model · Academic progress evaluation tools · Pedagogical test · Mind map

1 Introduction

Informatization and computerization in the contemporary educational practice set new requirements to professional competence of future teachers in using information and communication technologies for training and evaluation of students" academic progress. Many researchers and practitioners both in Russia and abroad have been looking into modernization of educational process at higher school based on modern information and communication technologies (ICT). The authors suggest several options for combining traditional educational technologies with

M. Mamontova (✉) · P. Zuev
Ural State Pedagogical University, Yekaterinburg, Russia
e-mail: mari-mamontova@yandex.ru

P. Zuev
e-mail: Zyuew@yandex.ru

© Springer International Publishing Switzerland 2015
V.L. Uskov et al. (eds.), *Smart Education and Smart e-Learning*,
Smart Innovation, Systems and Technologies 41,
DOI 10.1007/978-3-319-19875-0_29

321

modern ICT. That said when training future teachers, the dynamics and prospects of the development of both information and educational technologies should be taken into account.

The paper presents a methodology for training of future teachers in the development and application of computer tools for academic progress evaluation. The methodology is based on the synthesis of the contextual learning technology with network Internet technologies - WEB 2.0 (World Wide Web) and SaaS (Software as a Service business model). The proposed methodology in the course "Modern Tools for Academic Progress Evaluation" taught to the undergraduates of the Institute for Mathematics, Informatics and Information Technologies of the Ural State Pedagogical University.

2 Integration of Networking Technologies with Learning Technologies

2.1 Rationale for Integration of Networking Technologies with Contextual Learning Technology

Why do we believe that integration of network technologies with the contextual learning technology is promising?

The advent of new IT-technologies brings about changes in the minds of users. Young people for whom Internet, services offered through Internet and usage of social networks have become customary are coming to work in the education system. In the future professional network communities will actively develop.

That is why young teachers should bring their network experience in the school corporate practice. Each Member of the professional network community makes his contribution and the whole community uses the results of their activity.

The Bologna process resulted in the gradual change in the training paradigm "from teaching to learning."

In the course of higher school training the new paradigm is implemented through student-centered [8] and competence-oriented approach to training [11].

Competence-oriented education programs are designed and implemented at higher educational establishments. Higher school didactics is gradually acquiring new features. Those include orientation of education on objectives and results, promotion of independently organized and active learning, focus on motivational, volitional and social aspects of education». The focus is on "... learners and their learning process, and on changing of the teacher role (guidance and counseling on the learning environment or learning situations)" [10].

2.2 Limitations of Traditional Technologies

Traditional higher school educational technologies have limitations in terms of simulation and modeling of a variety of real professional situations. This disadvantage can be compensated by introduction of the contextual learning technology in the educational process.

The concept and method of the contextual learning have been developed by A.A. Verbitsky. The theory and methods of the contextual learning are presented in details in [9].

Let us give a brief description of the key idea and characteristics of the contextual learning.

The idea of the contextual learning is "systemic use of the professional context i.e. a set of tasks, organizational and technological forms and methods of activities, and situations of social and psychological interaction specific for certain professional environment" [9].

While developing the methodology of the contextual learning various types of subject and social contexts and their combinations can be used.

The undergraduates master profession in the course of training; however its forms are not relevant to professional activity.

Knowledge and skills acquired in the course of training, in professional activities are applied in the context of real professional situations and processes. Mastering professional knowledge in the course of training is discrete in nature (disciplinary approach) which conflicts with the systemic use of knowledge in professional activities. The learning process is based on the mental process of the undergraduates (memory, attention, perception, etc.). In professional activities labor process involves personality as a whole, with all peculiarities of thinking, motivation, and social activity.

In traditional education the undergraduates often play a passive role, completing tasks as directed by the teacher. By contrast, in professional activities independent decisions and responsibilities for the result of the activity are to be taken. A person is to be proactive and interact with colleagues and students.

The educational activities are focused on assimilation of past social experiences. Knowledge and ways of activity acquired during training become ways of professional activity. When switching from one type of activity (cognitive) to another one (professional), the needs, goals, motivation, tools, objects and results of operations are changing. With traditional his transition takes place outside the educational establishment, with the contextual learning - as part of the learning process.

The contextual learning considers learning and professional activities as successive stages of the development of one and the same activity – the professional activity.

2.3 Modeling of Professional Context in Learning

Professional context is reconstructed in the educational process reflecting the content and technology of a real working process, relations and their system of values.

Modeling of subject and social content of professional activity in the educational process involves reflection of content and context of the professional activity in a variety of undergraduate studies. Among those we reckon the below: selection of learning forms and methods activating the undergraduate learning activities, enhancement of motivational, volitional and social aspects of studies; establishment and gradual complication of relations between forms and models of learning at different stages of specialist training.

The contextual learning of future teachers is based on phased transition of the undergraduates from academic activities (lectures, seminars, laboratory works) through quasi professional activities (implementation of individual and group projects) to educational professional activities (school practice).

2.4 Characteristics of Educational Activities Stage in Contextual Learning

At the stage of educational activities new and traditional forms and methods of learning are applied, subject and social content of the future professional activity is modeled using a variety of symbolic aids. At this stage semiotic training models are mainly used (work with texts, processing of basic information, creation of the information fund). Training at this stage should result in systemic professional knowledge and mastered working methods.

2.5 Characteristics of Quasi Professional Activities Stage in Contextual Learning

At the stage of quasi professional activities real professional situations are modeled when the formed knowledge system is to be applied.

At this stage, simulation and social learning models are appropriate.

Training assignments go beyond the symbolic information; the information is correlated with the future professional activities; to do so the undergraduates are to be involved in the studied situation. Professional situation characterized by content and social ambiguity and contradictions becomes the main structural element of the learning process. The tasks are performed both in groups and collectively. While analyzing situations an undergraduate becomes a specialist and a member of the collective, gains experience of collective work in the future professional environment, and forms his attitude to work and profession.

Fund of didactic methods in the sphere of higher education, which in our opinion is appropriate to use at the quasi professional stage is presented in the «New guide on didactics of higher education»: coaching, small group training, problem training, self-managed learning, project-based learning and other [6].

2.6 Characteristics of Professional Teaching Activities Stage in Contextual Learning

In professional teaching activities (school practice) the undergraduates can jointly use the products created at the previous step.

In traditional educational process the three mentioned stages are normally spread over different courses and periods of time. The stages are implemented by different teachers, which, in our view, does not contribute to logical transition from stage to stage. A peculiarity of the proposed methodology is that all three stages are implemented within single discipline and used models of teaching (semiotic, simulation and social) are integrated with modern information and communication technologies.

3 Characteristics of Professional Context of Future Teachers' Training

3.1 Content and Conditions of Training of Future Teachers in Modern Professional Context

The professional context of the training of future teachers is formed by specific social and historic conditions (goals, content of education) and instrumental and technological conditions (goal achievement methods, tools, and technology) of school. These conditions have been rapidly changing over the past two decades.

Russia, like other countries, is upgrading the education quality assessment system. Training of specialists in this field of activity is conducted. The theory of educational measurements is developed. Mass standardized procedures are being applied in education to evaluate the quality of training of educational institutions graduates. The number of quality monitoring programs at different levels of the education system is growing.

Russian students and undergraduates participate in international comparative studies. Some of them are assessing the quality of general education (Trends in International Mathematics and Science Study, TIMSS; Programme for International Student Assessment, PISA; Progress in International Reading Literacy Study, PIRLS). Others are assessing information and computer literacy (International Computer and Information Literacy Study, ICILS) [5].

The role of computer technology in the development and application of various evaluation tools has increased. Electronic educational resources and systems of distance learning are being developed [3].

The new purpose of education is "self-aware, internally motivated, quickly thinking, problem-solving and ready to risk person acting together with others, and sufficiently armed with knowledge, tolerant and socially oriented" [7].

New goals and needs are reflected in the educational standards. These standards are based on competence approach to content, methods and teaching aids, and to evaluation of academic progress. In this regard, along with standardized procedures, tools and methods of self-evaluation of academic progress are increasingly in demand in the educational process.

The modern professional context where teachers operate today sets, on the one hand, the new requirements to the teachers' professional competence and, on the other hand, involves changes in the content and the conditions in which future teachers are trained.

3.2 Appropriateness of Application of SAAS Business Technologies in Contextual Training

The teacher must be able to develop, independently and in a team, computer tools for evaluation and self-evaluation of academic progress, to use the results of external and internal (in the first place, classroom) evaluation for adjustment of the educational process.

What technologies should be used? Do we need to teach future teacher programming and development of automated control software?

In practice, the teacher would rather use off-the-shelf software. These challenges can be met using the potential of SaaS business technologies that are currently underutilized in schools. The future teachers need to have an idea about these technologies and their potential uses for development of evaluation tools, and to learn how to use them in practice.

In accordance with the standard requirements, in the course of study the undergraduates shall master the theoretical basis and methodology for the development and application of various evaluation tools (including computer programs), be able to use the testing results to make decisions about changes in the educational process and adjustment of the students knowledge.

In our opinion, the set goals can be achieved through the establishment within the framework of the academic discipline of the virtual educational information and communication environment based on modern network and computer technologies with consideration of the professional context.

4   Brief Description of the Teaching Method

4.1 Content and Teaching Methods at Learning Activities Stage

Let us give a brief description of the methodology of the future teachers training in the development and application of computer-assisted evaluation of academic progress. This methodology is based on the synthesis of the contextual learning technology and modern information and communication technologies.

At the first stage of the contextual learning the undergraduates study the basics of classical and modern theories of educational measurements, familiarize themselves with development and application of various tools for evaluation and self-evaluation of academic progress – educational tests, concept maps, memorization matrixes, mind maps, complex tasks, rating scores, portfolios, etc.

Study of various evaluation programs (classroom evaluation, the final state examination, large-scale comparative studies to assess the quality of education, the education quality monitoring programs, etc.) give the undergraduates the opportunity to see the application of various evaluation tools and methods.

In the course of the study of theoretical material the undergraduates are encouraged to use the information retrieval systems which contributes to the development of their information search, structuring and storage skills. Among those are: the sites of the Federal Institute of Pedagogical Measurements [1], the National Center for Educational Quality Assessment (Institute of Tools and Methods of Education of the Russian Academy of Education [4]), and the Federal Testing Centre [2]. They provide to the undergraduates the idea of the spheres of application of measuring materials in education, of the use of modern information and communication technologies for the purpose of education quality assessment and quality management.

This is how the undergraduates form the basis of the information fund on evaluation and management of education quality. It will be used in the second and third stages of the contextual learning and in their future work at school.

An important element in the training of a future teacher is the ability to choose the automated systems to develop evaluation tools.

Automated knowledge control systems should satisfy the requirements of comprehensiveness (automation of the development of individual assignments and their banks, test generation, test distribution through communication channels, data storage, and aggregation of test results), of high performance (transmission of large volumes of information through communication channels), reliability, ability to use both individual computers and networks - LANs and the Internet), responsiveness (while developing metrics, and processing of results). During learning process the undergraduates are encouraged to become familiar with the SaaS business technology and with different online services to develop tests, questionnaires. They can carry out their comparative analysis based on the characteristics provided by the lecturer. For this work the undergraduates are divided into micro groups, each analyzing one system. The results of the analysis are presented in tables. In the result of this work the undergraduates information fund is replenished with comparative characteristics of the services. They can be used by a teacher in developing evaluation tools in a particular educational institution.

Along with specialized knowledge the undergraduates are encouraged to study and master a series of tools for analysis, processing, structuring and presenting the course material studied during its perception, memorization and practical use.

In our work we use the concept of self-directed learning. The undergraduates develop self-learning skills, learn to evaluate their own academic progress as well as to plan their own academic activity (work with new educational material).

The process of acquiring new knowledge is accompanied by creation of various visual and conceptual knowledge representation models – memorization matrices, structural and logical schemes of educational material, concept maps, and mind maps which at the stage of acquiring new knowledge serve as self-evaluation tools.

To create mind maps, concept maps, logical and semantic models of educational material our methodology also uses the SaaS technology. The undergraduates familiarize themselves with online services for mind map creation, and conduct comparative analysis and choose service that best suit their work and correspond to their capabilities (in terms of access, language, payment, the need for registration, technological capabilities, a possibility to save created files in different formats).

The result of an undergraduate" work with the educational material is a set of mind maps, structural and logical schemes of educational material, and concept maps.

In order to improve the efficiency of the learning process the undergraduates get acquainted with the patterns of remembering information by humans, with mnemotechnics, and define the individual features and technical capabilities of personal work with information (in the first place, presence and characteristics of mobile devices).

Special attention is given to technique of meaningful, systematic, "cumulative repetition" of the studied material, which, in our opinion, is consistent with the information and communications technologies used by modern man. The undergraduates are encouraged to draw up individual timetable for work with the created models of educational material - to identify specific points in time for repeated systematic conscientious perception of the created schemes, to register these moments in time using the "calendar" service with a reminder feature, for example, available on a mobile device and to follow this timetable. The undergraduates following these recommendations mention the positive effects of the use of the "cumulative repetition" technology and demonstrate a higher level of academic motivation and, consequently, a higher level of knowledge of the discipline in question.

In our opinion, the most convenient for individual and group development of visual and descriptive models of educational material are the Google services.

The result of the first (academic) stage of the contextual learning is as follows: theoretical knowledge of the undergraduates about various tools and methods of evaluation and self-evaluation of academic progress, knowledge of possibilities for automation of evaluation procedures using modern online services, and practical mastery of the technologies for the development of various tools for knowledge evaluation.

The use of these tools not only allows the undergraduates to master the necessary professional knowledge. It also contributes to development of conscious approach to the process of self-learning, to see the progress in the development of their own knowledge, and to manage their knowledge.

Thus, the undergraduates create individual virtual information and communication environment that allows them not only to create and store information on external media, but also to plan the work with it both individual and in a group.

4.2 Content and Methods of Training at Quasi Professional Activities Stage in Contextual Learning

At the second stage (quasi professional) of the contextual learning the undergraduates master methods of individual and group design combining them with the pedagogical and network technologies mastered at the first stage of training.

In the course of the project implementation the undergraduates learn self-organization and organization (planning of time and activities, selection of tools and technologies), setting goals, coordination of work on the project, preparation of material, obtaining and evaluation of results, preparation and delivery of presentations.

A modern teacher should be able to use the project method in teaching. To teach students project activities a teacher is to be well trained in planning and organization of instructional design, creation of didactic, methodological and material and technical support. The teachers must be able to develop their pedagogical projects.

They have such a possibility at the quasi professional stage of the contextual learning. While implementing projects, the undergraduates by their own experience learn the methodology of project and conversion activities. They implement a group project "Development of a Set of Computer Tools of Evaluation of Academic Progress of Computer Science Students".

At this point they need to apply in practice the tools and technologies mastered at the first phase. The undergraduates jointly choose pick a school class (a group of classes) and a teaching kit used to teach a specific discipline. They use the information fund created at the first stage of training (textbooks, programs, manuals, standard texts, etc.). A set of computer tools for evaluation and self-evaluation of academic progress for different stages of learning of one topic from the school course in informatics are being designed and developed.

Why did we choose a topic as a structural and meaningful unit of educational material? There are several reasons for that. A topic is a minimum amount of logically related educational material, to which specific requirements in terms of results are imposed in the discipline program (what students are supposed to learn). It is important to ensure that the undergraduates see a complete cycle of studying a topic - from goal setting to final checking of academic progress of specific content. The tools for evaluation and self-evaluation are used at all stages of the cycle which helps the undergraduates to form a systemic view of the evaluation toolkit as opposed to the traditional teaching when students learn each tool separately, in fact in isolation from the "context". The learning takes place in a limited time period, so most optimal would be to choose a topic of limited volume.

Development of a set of evaluation tools is carried out in several stages. Project work planning is carried out using a collective general mind map (performing the "Road Atlas" function) that reflects the main stages of the set development, specifies the content of work, tools and methods of their implementation, and indicates project coordinators and implementers. The collective map is being developed with the help of a network service selected by the undergraduates.

During the first stage the goal of the topic study and planned results are defined. Based on their analysis a sheet of goals is drawn. In the future the list of goals can be personalized and converted into a progress evaluation list where students would mark their achievements and their quality.

Next, the undergraduates determine which evaluation tools should be used at the stage of mastering by the students of the new material, at the stage of subject revision, individual work on the material and at the final test, and discuss the possibility of using computer technologies to develop each of the selected tools.

Also, a network service is defined that will create a generic product. A common "cloud" for work of the established professional community is created. After that the undergraduates are put in several micro groups (pairs) according to the number of the tools to be developed. Those tools are concept maps, memorizations matrices, structural and logical schemes of educational material, mind maps, tests, assignments/units (integrated), and series of tasks to test the formation of skills. Each micro group develops evaluation and self-evaluation tool and chooses appropriate online services. Inside the micro group works are distributed and work execution plan is developed, the responsibility for results is distributed among the implementers. The teacher acts as a consultant.

4.3 Content and Learning Methods of Training-Professional Stage in Contextual Learning

The final point is presentation of the set of developed tools. The undergraduates (from different micro groups) act as experts and evaluate the developed tools (in terms of content and technologies), completeness of the developed set, and contribution of each micro group in the overall result. The result of the quasi professional stage is a set of tools for academic progress evaluation on one of the topics.

At the third stage of the contextual learning the undergraduates use the developed sets in the course of their school practice. During school practice they are faced with the need to develop and use available computer tools for academic progress evaluation. Many schools use automated learning process management systems and systems of distance learning. The developed set of evaluation tools can be used directly or be regarded as "standard" for development (or selection) of control and measurement materials.

5 Conclusions

The proposed methodology of training future teachers in development of computer tools for evaluation of academic progress of students through combination of the contextual learning technology with information and communication technologies enables the undergraduates in the course of mastering one academic discipline to transfer from academic to professional activity - from mastering the theoretical basis of a discipline to active joint use of knowledge obtained in the course of studies to solving practical professional tasks.

Knowledge and mastered methods acquire personal and socially significant meaning. A special effect is achieved when using capabilities of the network professional community when the activity of an undergraduate is transferred from the contextual learning to the context of professional interaction.

References

1. Federal Institute of Pedagogical Measurements. http://www.fipi.ru
2. Federal Testing Center. http://www.fct.ru
3. Lapenok, M., Rozhina, I.: Distance learning system: from designing to introduction at schools. In: 11th IASTED International Conference on Computers and Advanced Technology in Education, pp. 100–108. Crete (2008)
4. National Center for Educational Quality Assessment (Institute of Tools and Methods of Education of the Russian Academy of Education). http://www.centeroko.ru
5. Naumann, J.: TIMSS, PISA, PIRLS, and low educational achievement in world society. Prospects **35**(2), 229–248 (2005)
6. Berendt, B.: Studentische Literaturarbeit in Zeiten des World Wide Web. In: Berendt, B., Voss, H.-P., Wildt, J. (eds.) Neues Handbuch Hochschullehre (G3.3), pp. 1–30. Berlin (2003)
7. Mortimore, P.: The Road to Improvement. Reflections on School Effectiveness, p. 384. Swets and Zeitliner Publishers, Lisse (1998)
8. Student-Centred Learning: Toolkit for students, staff and higher education institutions of ESU. Brussels (2010)
9. Verbitsky, A.A.: Active learning at higher school: contextual approach: textbook of methodics. Vysshaya Shkola, Moscow (1991) (in Russian)
10. Wildt, J.: Vom Lehren zum Lernen. Zum Wandel der Lernkultur in modularisierten Studienstrukturen. In.: Berendt, B, Voss, H.-P., Wildt, J. Neues Handbuch Hochschullehre. Berlin (2004)
11. Zürcher, R.: Informelles Lernen und der Erwerb von Kompetenzen: Theoretische, didaktische und politische Aspekte. In: Materialien zur Erwachsenenbildung. Nr. 2, Bundesministerium für Unterricht, Kunst und Kultur, Wien (2007)

The Diagnostics' Methods of Students' Readiness for Professional Pedagogical Activity Within Information Educational Environment

Alexander V. Slepukhin and Natalia N. Sergeeva

Abstract Based on the analysis of the concept essence of "students' readiness for professional pedagogical activity" and the problems of organizing the diagnostics of the level of its formation, that are prevailed in the Russian educational system, we propose a model of diagnostics techniques of students' readiness and specify the sequence of actions for its implementation in the educational process. In addition, the competence and system-activity approaches are selected as the major approaches in constructing of diagnostics' methods, and the means of Information and Communication Technology are considered as the main tools for implementing actions of a teacher. This paper was written as a part of the state task of the Ministry of Education and Science of Russian Federation 2014/392, project № 2039.

Keywords Readiness for activity · Diagnostics of readiness · Information technology

1 Introduction

The main focus of modern education is associated with the training of a competitive labor market graduate who is able to adapt to the changing conditions of the life and acquire new technologies. The readiness to be included in further life activity, the ability to solve practically the arising life and professional problems depends not only on obtained knowledge and skills, but also on some additional qualities that should be formed among students. In that aspect the ideas of competence-oriented education, and also the importance of formation, and so, the diagnostics of level of formation of students' readiness for implementation of professional activity, the ability of self-control and self-development implementation within professional activity are the answer of education system to the changing requirements to

A.V. Slepukhin (✉) · N.N. Sergeeva
Ural State Pedagogical University, Ekaterinburg, Russia
e-mail: srbrd@mail.ru

© Springer International Publishing Switzerland 2015   333
V.L. Uskov et al. (eds.), *Smart Education and Smart e-Learning*,
Smart Innovation, Systems and Technologies 41,
DOI 10.1007/978-3-319-19875-0_30

university graduates. Thus, the setting up the activity, methodological, psychological components of competences as the components of readiness for professional activity define new understanding of the purpose of education and mean the use of the technologies allowing to create another pedagogical space with changed evaluation system of education results.

Above all, in order to construct such evaluation system it is necessary to understand the essence of methods of competencies and competence formation as the components' parts of readiness for professional pedagogical activity. In foreign pedagogical literature these problems are actively discussed (particularly, [20, 21]), however, we should pay special attention to the results of Russian research on various aspects of professional specialists' training. So a number of important aspects of professional training and, in particular, formation of a professional culture are analyzed in studies of V.G. Bocharov, V.A. Slastenina, E.I. Kholostova. The Psychologists I.A. Zimnyaya and B.Y. Shapiro disclose an individual-oriented nature of the professional activity. A.A. Verbitsky, Ts. Jotov, N.V. Kulyutkin, J. Sheylz engaged in learning of formatting tools of particular components of high school students' readiness for the future professional activity. The analysis of pedagogical conditions of formation of the particular components of high school students' readiness for their future professional activity is reflected in the works of A.A. Verbitsky, V.V. Gorshkova, Ts. Jotov, M.V. Clarin, M.I. Lisina, V.J. Lyaudis. We studied the formation of the particular components (psychological, communicative, creative) of high school students' readiness for the future professional activity, but these processes have been observed separately from each other (A.A. Verbitsky, M.V. Clarin, E.A. Klimov etc.).

The process of professional formation of the teacher and formation of readiness for professional activity is considered by these authors (in particular, [2, 8, 10, 12, 15]) from different points of view. From the point of view of activity approach the more successfully rules and norms of activity are studied, the level of professional development is higher (A.G. Asmolov, A.N. Leontyev, V.A. Petrovsky, G. P. Schedrovitsky, etc.). From the point of view of personal approach the professional development is the development of self-concept in the direction of spiritual values. From the point of view of personal-activity approach (E.N. Volkova, L.M. Mitina, V.A. Slastenin, V.E. Tamarin, etc.) the professional development is connected with the change of a personal position from object to subject. Within functional approach the readiness is considered as a mental state. The competence approach, aimed at the training of a student to solve independently subject and professional-oriented tasks, is one of the main approaches in the national education (in particular, [5, 6, 9, 16, 18, 19]). Not assimilation of the sum of data, but development of such abilities by students, which would allow them to define the purposes, to make decisions and to work in typical and non-standard situations, becomes the main value of training within the specified approach.

Despite the theoretical results, received in the presented researches, many pedagogues (in particular, A.I. Lyashenko, V.A. Slastenin, N.B. Shmelyova) note that graduates of the higher school experience difficulties in implementing the professional kinds of activity. From their point of view there is a mismatch of "model of professional activity" formed in students' consciousness during the years of high school education, with that real situation in which they have to begin professional activity. Methodists and teachers see the reasons of such situation in prevalence of the informative side of education in professional training of students of higher education institution; the operating "knowledge" approach in education conflicts to nature of the follow-up professional activity of the graduate. The contradiction between the organization of educational and professional activity is expressed in the fact that the content of training is divided in different subjects, and students receive static information. Activity of students is considered as response to the operating influence of teachers, and the professional pedagogical activity requires initiative, skillful applying of system knowledge, abilities for the solution of professional tasks, rather high level of formation of psycho-physiological qualities of future teacher as organizer of educational interaction, which, unfortunately, quite often drops out of a field of vision of pedagogical collective.

The allocated problems, and also the interrelation of such types of pedagogical activity as training, formation of competencies, formation of competences, readiness formation, diagnostics of level of formation (development) of competences and competencies, diagnostics of readiness cause the need of implementation of the system-activity approach within complex diagnostics of readiness assuming, first of all, a specification of the formed readiness components to the level of reliably recognizable, and so diagnosable kinds of activity.

A variety of the allocated approaches to the construction of methods of formation of students' readiness for pedagogical activity defines ambiguity of approaches to the diagnostics of level of student readiness for the professional pedagogical activity (PPA). Their implementation in practice (for example, [1, 3, 4, 7, 11, 13, 17, 22]) is characterized by a number of contradictions, in particular, between: requirements to the level of students' preparedness and modern system of diagnostics of professional competence of future teacher; the need for systematic complex diagnostics of successful student advance in educational route, giving accumulation of information diagnostic base, and lack of methodological and methodical bases for carrying out such work; the need to process a large volume of information about students' readiness for professional activity received during pedagogical diagnostics and impossibility of its implementation only by traditional means.

Therefore, the use of information and communication technologies (ICT) which are capable to cope with the forming and diagnostic function of educational process in the conditions of clear ratio understanding of component structure of competencies, competences and readiness has to become one of the main links of methods of formation and diagnostics of competences formation, and also students' readiness.

The need to resolve these specified contradictions actualizes a research problem which essence consists in development of scientifically based methods of complex diagnostics of competencies, competences and students' readiness for professional activity, and also features studying of their formations for the correct management of educational process. In this case, the specificity of the implementation of

A.V. Slepukhin and N.N. Sergeeva

diagnostic methods of activity, psychological, personal components of competences and readiness is that its results have to be taken into account by all pedagogical staff of educational institution.

In order to construct the methods of complex diagnostics let's analyze, first of all, the ratio of concepts such as "competency", "competence" and "readiness", having allocated the structural components of readiness for professional activity.

2 The Ratio of Concepts

In the concept of the competence-oriented education competency is understood as ability of a student to effectively self-organize internal and external resources for achieving educational goals. A student, mastering any way of activity, gets experience of integrating various results of education and setting (or assigning) the goal, and therefore, it occurs an understanding of management process of own activity.

In contemporary research competence is often understood as the integral person quality, which is shown in the general ability and readiness for the activity based on knowledge and experience that are obtained during the training and are focused on independent and successful participation in activity. Currently in the literature there are following key competences such as: the readiness to resolve the problems, technological competence, readiness for self-education, readiness for using the information resources, readiness for social interaction, communicative competence and others.

Readiness for professional activity considered in the works of I.I. Agapov, O.V. Akulova, V.I. Baidenko, N.F. Efremova, E.S. Zair-Beck, L.S. Lisitsyna, A.M. Mityaev, S.A. Pisarev, E.V. Piskunova, D.A. Pogonysheva, N.F. Radionova, A.P. Tryapitsyna, A.V. Khutorskoy, A.I. Chuchalin, O.V. Shemet, S.E. Shishov, etc., assumes free possession of the profession and orientation in adjacent areas, the competitiveness in the labor market, readiness for professional growth, social and professional mobility, the ability to adapt in the changing external conditions.

The ratio of the main approaches to the treatment of the considered concepts in the context of features of pedagogical activity allows to draw a conclusion on interdependence of these concepts. Synthesis of approaches gives the opportunity to specify the definition of readiness as follows: "The readiness for implementation of professional pedagogical activity is the complex state of the subject which is characterized by formation of set of such qualitative characteristics of the person as motives of activity, formation of the actions included in professional pedagogical activity, a sufficient level of development of professionally significant mental processes, methodological literacy, and also a positive psycho-physiological tune on a certain kind of activity. That is the student readiness for professional activity will be considered as a result of accumulating the high-quality personal changes and finding the professional competence by the student.

According to the given definition for implementing complex diagnostics of readiness we set up the component structure of readiness. The obtained solution of the set task is based on the analysis of existing approaches and integration of structure components of readiness, which were selected from these approaches. All this allowed us to make a conclusion about sustainable unity of the following components: motivational component, which includes a positive attitude to pedagogical activity, an interest in it and other stable enough professional motives; cognitive (orientation) component, which provides a high enough level of formed knowledge of subject, methods and methodology, the understanding characteristics and conditions of professional activity, its requirements for the personal characteristics of a teacher; technological (the operational) component, which means the possessing methods and techniques of professional activity; psycho-physiological component, which requires from the future teacher to have a high enough level of formation of intellectual, research, creative abilities; social (volition) component, which includes the self-control and self-diagnostics abilities, the other regulatory skills, as well as the strong-willed qualities, communication skills; reflexive (evaluative) component, which provides the self-assessment ability of own professional preparedness and compliance of solutions' process of professional tasks to required (specified) conditions.

Depending on a certain type of readiness, for example, psychological, moral, practical, etc. one of the structural components from selected set can dominate. Besides, prevalence of a component(s) can be caused by specifics of educational institution, professional interests and student needs. Selected components will be considered as the content base of the activity for the constructing the process of diagnostics of students' readiness for professional pedagogical activity.

3 The Diagnostics Methods of Students' Readiness for PPA

We interpret the diagnostics of readiness for PPA as the process of result determining of the formation of all parts of readiness' components, that allows to detect, evaluate, analyze, correct and predict their further development. Let's point out the basic activity components of the diagnostics methods of students' readiness for professional pedagogical activity: the definition of the diagnostics purpose and objectives, the content selection of diagnostic materials and indicators of readiness for professional activity, the choice of diagnostics forms and methods, corresponding to the tasks of certain stage of diagnostics, the system choice of formation levels and the criteria that defines the student's identity to each of the levels, the focus on the pedagogical diagnostics principles, the principles of using ICT for diagnostics, the formulation of conditions of using ICT for pedagogical diagnostics, the implementation of diagnostic procedures, the organization of reflexive-evaluative activity.

Let's comment on the essence of the main components of the methodology.

The purpose of diagnostics is to determine the level of students' readiness for professional work. The objective of diagnostics from positions of competence-based approach is to carry out diagnostic procedures for each of the selected components and to develop correcting influences. Competence-based approach in the organization of diagnostics assumes the transfer in emphases from the controlling function of diagnostics to the training, developing and focusing functions.

The content of diagnostic actions assumes the check of not only subject knowledge, abilities, but also the types of professional activity formed within subject domain. Diagnostic tasks are directed on identifying the level of formation of each of the selected components of readiness. Thus we will consider that specification of diagnostics subject to the level of reliably recognized and therefore, diagnosed actions (operations) is the basis of any diagnostic activity. Indicators of students' readiness for future professional activity correspond to the selected groups of its structural components. Besides, let's point out the following feature of the content of diagnostics: it is impossible to form and therefore to diagnose the readiness within only one academic discipline. Therefore it is expedient to compile a matrix of compliance of the formed components of readiness for subject disciplines and units (sections) of subject disciplines. Within the framework of one academic discipline it can be only concluded whether the process of forming (developing) the components of readiness in this discipline is going or not (whether the dynamics of changes is being observed). The total diagnostic conclusion can be drawn only on the basis of results' comparison of implementing the educational tasks within studying the subject material of all academic disciplines.

Methods of pedagogical diagnostics must be connected with training methods with using the information and communication technologies, methods of use of ICT and methods of management of educational students' activity (for example, according to [14]), and results of their use must be taken into account by all pedagogical collective. Choosing the methods of diagnostics it is necessary to keep in mind that readiness for PPA is a set of the complex integrating qualities of the person which cannot be diagnosed only by means of one method of diagnostics (for example, testing). That's why the combination of methods of pedagogical diagnostics (poll, monitoring, role and projective methods, etc.), a set of various didactic tasks for realizing the set of methods during the implementation of diagnostics of all components of readiness is necessary. Processing the results of applying the specified methods it is expedient to use the method of element-by-element analysis during all period of training and the method of rating assessment for collecting final statistical data on all components of readiness.

The analysis of approaches to system construction of levels of formation of readiness components allows to draw a conclusion about their variety both in the process of allocating the levels and in the process of allocating the criteria by which it is possible to define a certain level of a student. From our point of view it is expedient to use a combination of two main approaches to allocation of levels of formation (development) of the main readiness components. Within one subject discipline: for initial, current and thematic diagnostics the levels are determined by a dichotomizing scale: it is formed − it isn't formed, there is a development − development is absent. Within final diagnostics by comparing the diagnostics' results of formation of readiness for PPA in all subject domains it is the offered to use the following scale of levels: minimal, basic, increased and high readiness.

Considering that the level of formation reflects the degree of expressiveness of certain properties and also the dynamics degree of characteristics formation on each component of readiness, let's concretize the allocated levels: the levels of minimal, basic, increased and high readiness, characterized by a row defined a component.

Defining the rules of fixing the diagnostic judgment, let's point out the expediency of use of the following criteria corresponding to the allocated approaches to construction of system of levels of diagnosed components formation (development): for one subject discipline within initial and current diagnostics – if the change process is observed (to the best) at least for one of components of readiness, then the component is developing; if all components of readiness are formed, then the readiness is formed; if at least one of components isn't formed, then the readiness isn't formed.

In order to compare the diagnostic data on all subject domains within the final diagnostics the diagnostic conclusion is similarly formulated only on the basis of comparison of diagnostic data on all academic disciplines: if readiness is not formed in a single academic discipline, then the result output contains the conclusion about unformed readiness for PPA.

The principles of pedagogical appropriateness, didactic and cognitive compatibility, methodical significance, not inconsistent with the classical principles of the system, consistency, objectivity, systematic principle and others are the fundamental principles of pedagogical diagnostics. Diagnostics of readiness for professional activity organized on the basis of competence approach must to promote the forming of reflexive and estimated activity, adequate self-assessment that allows each student to correct the individual educational route.

4 Technologization of Teachers' Activity in Constructing Diagnostic Methods of Students' Readiness for PPA

According to the activity approach taken in the development of diagnostics methods of the level of formation of all components of readiness for PPA, let's present a sequence of teacher's actions, the implementation of which involves the use of ICT, particularly, the spreadsheets.

1. Analyzing the components of readiness, providing specification of the activity set, bringing the component to the level, when the student actions (operations) can be reliably recognized and therefore, diagnosed. Let's give the examples of specifying the activity of the particular components of student's readiness: the ability to analyze: a student can allocate (declare) the essence (purpose) of the partition, is able to allocate all the component parts (elements) of the whole; the ability to synthesize: a student can allocate (declare) the purpose(s) of the compound, is able to allocate the elements of the compound, is able to establish connections between elements as components of a whole, is able to combine elements into a whole, is able to combine the elements to another (modified) whole declaring (formulating)

new target of compound; the ability to generalize: a student can allocate a set of elements for generalization, is able to establish a common, is able to identify common in relation to each unit of knowledge; the ability to perceive information: a student is able to concentrate at perception of educational material, is able to perceive the information provided in a different form, is able to choose the manner of presentation of information, that facilitates the best perception of the main part of information; the ability to set a goal and to choose the ways of its achievement: a student knows the difference between goals and objectives (tasks), is able to formulate goals, knows how to formulate objectives, differed from the goal, is able to build the minimum set of tasks for achievement the goals, etc.

2. Supplementing (making adjustment of) the set of elements of knowledge, skills, that were selected during preparing the tasks for the diagnostics of assimilating the subject didactic units, by the set of elements of knowledge, skills, possessions to diagnose the level of formation (development) of the readiness components, that formed within a specific section of academic discipline.

3. Matrix compiling of compliance between forming (diagnosable) didactic units (DU) of subject material of educational discipline and activity components of readiness. Beforehand it is necessary to analyze the activity set of forming each didactic unit of subject domain. Then the proposed comparison can be made only on the basis of such analysis. The result of the comparison will be the basis for meaningful content of studied themes (sections) by educational tasks aimed at the forming, developing, diagnosing of developing the components of readiness not only within one discipline, but also for another disciplines. In this case, one of the rational variants of filling such tables is to use cloud technologies, in particular, Google services or saving/uploading (developing) the specialized files in the resources of the information environment of the educational institution.

4. Developing and using the tasks constructors and action verbs for the constructional designing and formulating the set of diagnostic tasks that can be used for estimating the level of formation of didactic units and components of readiness. Here is an example of the set of action verbs and constructors of educational tasks for the component of readiness such as the ability to analyze. The verbs of action: analyze ...; divide into parts ...; classify ...; compare ...; differentiate .... Constructors of tasks: analyze the structure ... in the view of ...; make a list, characterizing ... in terms of ...; build a classification... on the basis of ...; compare the point of view ... and ...; identify the principles underlying ...; specify the purpose of the separation ....

5. Selecting educational (cognitive), diagnostic tasks at which performance the content and activity components assume demonstration, and, so the opportunities for diagnostics of certain components of readiness, supplementing (specifying, enriching) thereby a set of diagnostic tasks. Let's give the examples of formulation of similar diagnostic tasks, referring to the use of different diagnostic methods (not just computer-oriented): select the interrelation between the basic concepts of the course theme (the result can be presented in the form of drawings, charts, etc.); classify the found in the literature points of view on the issue ...; make annotated catalog of information resources (with the release of printed sources and electronic resources) for studying the topic ..., etc.

6. Evaluating the results of implying the educational and diagnostic tasks by method of element-by-element analysis and formulating the diagnostic judgment (getting the formulation by means of spreadsheets) based on the selected rules of fixing the diagnostic output and criteria (indicators) for which the student can be referred to a concrete level of development of readiness component. In addition, let's point out the expediency of using a three-point evaluating scale: 0 points – a student has not coped with the task element, 1 point – a student has made insignificant errors (had the understanding the essence of a kind of activity), 2 points – a student completely coped with an element of a diagnostic task.

The final conclusion about dynamics of formation of readiness component is formulated only on the basis of comparison of received results with previous results of control arrangements containing information about results of diagnostic tasks implementation directed on evaluating the same components of readiness.

During the carrying out the final diagnostics there is the accumulation of obtained points (method of accumulated mark), the share of performance of all educational diagnostic tasks in relation to the most possible number of points that can be given for the full performance of these tasks is also counted.

7. Fixing the final diagnostic judgment based on a comparison of the diagnostic data of all academic disciplines. In order to carry out the statistical analysis of all received results it is necessary to count a share of performance of all educational, cognitive and diagnostic tasks offered on all academic disciplines in relation to the most possible number of points.

The final data are represented in the scoring system (if necessary it can be easily converted in percentage representation) based on the summing the points on all subject domains. The final data can be also presented by charts and graphs showing the development dynamics of readiness components during the training periods (for example, semester) in each subject domains. The teaching staff formulates the final diagnostic judgment based on the chosen approach of the level selection of components' formation of readiness for PPA.

5 Conclusion

The diagnostics' methods of students' readiness for professional pedagogical activity (PPA) should be based on competence, system-activity approaches and on the use of ICT in the information environment of the educational institution. Use of system and activity approach assumes the conclusion about possibility of application of a technique not only to graduates of educational institution, but also in the course of all training, that allows teaching collective to draw conclusions about the level of formation of the students formed a component of readiness.

Taking into account the possibilities of ICT for collecting, storing, processing and accumulating information about the formation (development) level of all readiness' components we can formulate the statement about pedagogical expedience and didactic significance of the spreadsheets' use, primarily in the cloud services in information systems of the educational institution for implementing the diagnostics of the components of students' readiness for professional activity.

Use of the specified means allows to allocate a role of the information educational environment of educational institution (including, including, and cloudy services) as the defining implementer of the offered diagnostics technique access to which have to have all participants of educational process.

References

1. Velichko, E.V.: Psychological students' readiness of teacher-training colleges for the professional activity and its diagnostics. Pedagog. Educ. Russ. **3**, 203–207 (2011) (in Russian)
2. Dyachenko, M.I., Kandybovich, L.A.: Problems of Readiness for Activity. Publishing House of the Belarusian State University, Minsk (2008). (in Russian)
3. Efremova, N.F.: Approaches to the Competencies Evaluation in Higher Education: Educational Manual. M: Research Center challenges the quality of training (2010) (in Russian)
4. Kirkpatrick, D., Kirkpatrick, J.: Four Steps to Successful Training. Practical Guidance on Evaluating the Effectiveness of Training. HR Media, Moscow (2008) (in Russian)
5. Akulova, O.V., Zair-Beck, E.S., Pisarev, S.A., Piskunova, E.V., Radionova, N.F., Tryapitsyna, A.P.: Competence Model of the Modern Teacher: Educational and Methodical Manual. Izd RSPU. Herzen, St. Petersburg (2007) (in Russian)
6. Kurneshova L.E. (ed.): Competencies and Competence Approach in Modern Education. Quaity Assessment of education. Moscow Center of Education Quality (2008) (in Russian)
7. Solomin, V.P. (ed.): A Set of Test Materials for Assessment of Students' Competences. Publishing house of RGPU of A. I. Herzen, St. Petersburg (2008) (in Russian)
8. Koksheneva, E.A.: Formation of students' readiness of higher education institution for the future professional activity. Dissertation of the candidate of pedagogical sciences, Kemerovo, 226 p (2010) (in Russian)
9. Lisitsyna, L.S.: Methodology of Designing Modular Competence-Oriented Education Programs: Method. Manual. SPbSU of ITMO, St. Petersburg (2009) (in Russian)
10. Mityaeva, A.M.: The content of multilevel higher education in the implementation of competence model. Pedagogy **8**, 57–64 (2008) (in Russian)
11. Mikhailova, N.S, Minin, M.G., Muratova, E.A.: Development of Fund of Assessment Tools in the Design of Educational Programs. Publishing House of TPU, Tomsk (2008) (in Russian)
12. Motsar, L.S., Nekrasov, S.D.: About student readiness of higher education institution for professional activity. Person. Commun. Manage. **1**, 110–118 (2011) (in Russian)
13. Orudzhalieva, E.R.: Diagnostics of readiness for professional work of undergraduate students of teacher education. Author's abstract of dissertation of the candidate of pedagogical sciences (2011) (in Russian)
14. Semenova, I.N.: Methodology of teaching mathematics methods designing in the modern educational paradigm: monograph. Science Book Publishing House, Yelm (2014) (in Russian)
15. Sidneva, I.E.: The level characteristic of readiness of students of higher educational institution for organizational and administrative activity. The Theory and Practice of Education in the Modern World: Materials of the III International Scientific Conference, pp. 157–159. Renome, St. Petersburg (2013) (in Russian)
16. The Structure of Professional Competence of Bachelors and Masters of Education in the Field of Humanitarian Technologies: Method. Manual. Publishing house of RGPU of A. I. Herzen, St. Petersburg (2008) (in Russian)
17. Tatur, Y.G.: How to improve the objectivity of measurement and evaluation of education results. High. Educ. Russ. **5**, 22–31 (2010) (in Russian)

18. Chuchalin, A.: Educational programs design based on the credit assessment of graduates' competencies. High. Educ. Russ. **10**, 72–81 (2008) (in Russian)
19. Shemet, O.V.: Didactic bases of competence approach in higher professional education. Pedagogy **10**, 16–22 (2009). (in Russian)
20. Baltusite, R., Katane, I.: The structural model of the pedagogy students' readiness for professional activities in the educational environment. Rural Environ. Educ. Personal. 29–41 (2014)
21. Kocor, M.: Teachers' professional competences in theory and practice. Pract. Theory Syst. Educ. **7**(2), 175–187 (2012)
22. Saudabayeva, G., Alnazarova, G., Aitbayeva, M.: Educational diagnosis in modern education: a systems approach to cognitive-converting activity teacher. World Appl. Sci. J. **29**(9), 1183–1186 (2014)

The Diagnostics of Well-Formed Ability of Students and Teachers to Make and to Evaluate the System of Modern Methods of Teaching Mathematics

Irina N. Semenova and Sergey A. Novoselov

Abstract Taking into account the constructed structure of knowledge and actions the approach for technologization of diagnostics process of well-formed ability to make and evaluate the systems of modern methods of teaching mathematics is offered in this article, the examples of tasks which can be used for constructing a diagnostic conclusion during the training of students and expertise of the mathematics teacher' activity are given. This paper was written as a part of a state task of the Ministry of Education and Science of Russian Federation 2014/392, the project № 2039.

Keywords Training methods · Complexity of action · Difficulty of action · Didactic system · Pedagogy · "participation" of a computer · "help" of a computer

1 Introduction

Professional competence of the modern mathematics teacher assumes well-formed ability of choosing effective methods of training. This specified ability plays the dominating role in ensuring of education quality when using new means, forms and types of training. It has composite set nature and contains, according to our own developed ideology [6], the following knowledge and actions on two different levels.

Methodological level (knowledge and understanding): knowledge of cause and effect relationships between elements of methodical system and conditionality of a choice of training methods in modern didactic system; knowledge of methods classifications of training on different bases, advantages and disadvantages of existing classifications of general methods of training and methods of training in mathematics, their correlation; knowledge of efficiency criteria of training methods;

I.N. Semenova (✉) · S.A. Novoselov
Ural State Pedagogical University, Ekaterinburg, Russia
e-mail: semenova_i_n@mail.ru

© Springer International Publishing Switzerland 2015
V.L. Uskov et al. (eds.), *Smart Education and Smart e-Learning*,
Smart Innovation, Systems and Technologies 41,
DOI 10.1007/978-3-319-19875-0_31

345

knowledge of software tools allowing to model and design educational objects and educational processes (in particular, diagnostics), and also the tools for "participation" and "help" of the computer (in terminology V.P. Bespal'ko [2]); the understanding the essence of aggregation and disaggregation methods for designing of training methods; the understanding of possibility of using the modern information and communication technologies (ICT) for enrichment of methods of organizing activity and training in mathematics in various forms of training and communication modes described, for example, in [11];

Practical level (readiness and abilities): ability to allocate the dominating conditions defining a choice of training methods for concrete educational process; ability to choose methods of organizing the educational cognitive activity and methods of training in mathematics in specific conditions of educational process; readiness to prove a choice of methods of training in mathematics in methodical system for concrete educational process; ability to construct the sets and systems of methods of training in mathematics and methods of organizing the educational and informative process, using one and several classifications of training methods; ability to choose criteria, to estimate literacy and readiness to optimize the sets of methods of training in mathematics and methods of organizing the educational cognitive activity of students in specific conditions of educational process.

2 Theoretical Foundations for Diagnosing the Formation of Ability to Make and Evaluate the Systems of Methods of Training in Mathematics in the Conditions of Using the ICT

Identifying each action with the solution of the task that we understand as an initial ratio of conditions and requirements based on the works of V.I. Krupich, A.V. Brushlinsky, K.A. Slavskaya, etc., let us agree that the actions included in ability to make and evaluate the set of modern methods of training in mathematics are characterized by "complexity" and "difficulty" (as in [3]). In this case:

complexity of action is defined as the sum of quantity of the main relations which are determined in the process of intellectual operations, namely: analysis and classification through synthesis based on generalization and restriction, number of explicit links between the main relations and quantities of types of links (which can be explicit and implicit), that is the last summand can accept value 0 – at existence of only one main relation, and also value 1 and 2; in general complexity is established by the following rule: complexity from 1 to 2 – the general complexity 1, complexity from 2 to 4 – the general complexity 2, complexity higher than 5 – general complexity 3, difficulty of action is defined by degree of problematical character which depends on what components of information structure of action (task), allocated by Y.M. Kolyagin and called A (an initial state), C (basis of the decision), R (transformation), B (a final state), are unknown; the level of difficulty corresponds to the serial number of component of information structure in the given sequence.

Within the adopted agreements let's represent hierarchical structure of actions composing the ability to make and evaluate systems of modern methods of training in mathematics when using ICT in the pedagogical field integrated in ICE. In this structure let's specify complexity and difficulty for each action (a special symbol (*) is used for marking out the actions which are shown in the research and creative activity having special character, for example, when writing the master thesis):

- action 1_1: declaring of cause and effect relationships between two elements of methodical system at a choice of methods of training in mathematics (complexity 1, difficulty 1),
- action 1_2: declaring of cause and effect relationships between three (and more) elements of methodical system at a choice of methods of training in mathematics in didactic system of the pedagogical field integrated in ICE (complexity depends on quantity of elements of methodical system between which connection is established, and is defined by number of combinations of the main relations from the chosen quantity of elements of methodical system (n) taken two at a time, C_n^2, difficulty 1),
- action 1_3: allocating from the specified literature the justifications of cause and effect relationships between elements of methodical system at a choice of methods of training in mathematics in didactic system of the pedagogical field integrated in ICE (complexity depends on quantity of elements of methodical system between which connection is established, difficulty 2),
- action 1_4*: independent carrying out justification(s) of cause and effect relationships between elements of methodical system when choosing the methods of training in mathematics in didactic system of the pedagogical field integrated in ICE (complexity depends on quantity of elements of methodical system between which connection is established, difficulty is established by an expert assessment);
- action 2_1: classifying the methods of training in mathematics on one set basis (complexity 1, difficulty 1),
- action 2_2: classifying the methods of training in mathematics on the different bases (complexity is defined by quantity of the allocated bases, difficulty 1),
- action 2_3: correlating the classifications of the training methods which are carried out on the different bases (complexity is defined by number of the carried-out correlation, difficulty 2, if there is the allocation of various correlation between the chosen classifications – difficulty 3),
- action 2_4 * (research level): allocating the basis of classifications, classifying on the allocated bases, correlating the classifications of methods of training in mathematics in the certain conditions of educational cognitive activity of students (complexity is defined by number of the executed classifications and their correlation, difficulty is defined by an expert assessment of individual work of student);

- action 3_1: based on information sources giving the examples of using the modern ICT for enrichment of methods of organizing activity in concrete form of education and mode of communication (complexity 1, difficulty 1),
- action 3_2: with use of information sources giving the examples of using the modern ICT for enrichment of methods of training in mathematics in concrete form of education and mode of communication (complexity 1, difficulty 1),
- action 3_3: with use of information sources giving the examples of using the modern ICT for enrichment of methods of organizing activity and training in mathematics in concrete form of education and mode of communication (complexity 2, difficulty 1),
- action 3_4: with use of information sources giving the examples of using the modern ICT for enrichment of methods of organizing activity and training in mathematics in various forms of education and modes of communication (complexity 2, difficulty 2),
- action 3_5: giving the independently made examples of using the modern ICT for enrichment of methods of organizing activity when training in mathematics in various forms of education and modes of communication (complexity 1, difficulty 2),
- action 3_6: giving the independently made examples of using the modern ICT for training in mathematics in various forms of education and modes of communication (complexity 1, difficulty 2),
- action 3_7: giving the examples of using the modern ICT for enrichment of methods of organizing activity and training in mathematics in various forms of education and modes of communication (complexity 2, difficulty 3),
- action 3_8* (the creative level assuming scientific research): developing the methods of training in mathematics with using the modern ICT at the allocated subject and research goal (complexity and difficulty is defined by an expert assessment);
- action 4_1: from given set choosing the software tools allowing to model and design educational objects and educational processes (in particular, diagnostics) for the computer "participation" (complexity 1, difficulty 1),
- action 4_2: from available information resources choosing the software tools allowing to model and design educational objects and educational processes (in particular, diagnostics) for the computer "participation" (complexity 1, difficulty 2),
- action 4_3: from given set choosing the software tools allowing to model and design educational objects and educational processes (in particular, diagnostics), and also tools for the computer "participation" (complexity 2, difficulty 1),
- action 4_4: from available information resources choosing the software tools allowing to model and design educational objects and educational processes (in particular, diagnostics), and also tools for the computer "participation" (complexity 2, difficulty 2),
- action 4_5: from given set choosing the software tools allowing to model and design educational objects and educational processes (in particular, diagnostics) for the computer "help" (complexity 1, difficulty 1),

- action 4_6: from available information resources choosing the software tools allowing to model and design educational objects and educational processes (in particular, diagnostics) for the computer "help" (complexity 1, difficulty 2),
- action 4_7: from allocated (given) information resources choosing the software tools allowing to model and design educational objects and educational processes (in particular, diagnostics), and also tools for the computer "help" (complexity 2, difficulty 2),
- action 4_8: from available information resources choosing the software tools allowing to model and design educational objects and educational processes (in particular, diagnostics), and also tools of the computer "help" (complexity 2, difficulty 3),
- action 4_9: from specified information resources choosing the software tools allowing to model and design educational objects and educational processes (in particular, diagnostics), and also tools for the computer "participation" and "help" (complexity 3, difficulty 3),
- action 4_{10}: from available information resources choosing the software tools allowing to model and design educational objects and educational processes (in particular, diagnostics), and also tools for the computer "participation" and "help" (complexity 3, difficulty 4),
- action 4_{11}*: (the creative level assuming scientific research) creating the software tools for the computer "participation" and "help" in the process of organizing the educational cognitive activity od students in subject domain of "Mathematics" (complexity and difficulty are established individually based on expert assessment);
- action 5_1: from given set choosing the methods of the organizing the educational cognitive activity in specific conditions of educational process (complexity 1, difficulty 1),
- action 5_2: from information sources independent choosing the methods organizing the educational cognitive activity in specific conditions of educational process (complexity 1, difficulty 2),
- action 5_3: changing the set of methods of organizing the educational cognitive activity of students when changing the conditions of educational process (complexity 2, difficulty 3),
- action 5_4: from given set choosing the methods of training in mathematics in specific conditions of educational process (complexity 1, difficulty 1),
- action 5_5: from information sources an independent choosing the methods of training in mathematics in specific conditions of educational process (complexity 1, difficulty 2),
- action 5_6: changing the set of methods of training in mathematics when changing the conditions of educational process (complexity 2, difficulty 3),
- action 5_7: from given set choosing the methods of organizing the educational cognitive activity and methods of training in mathematics in specific conditions of educational process (complexity 2, difficulty 2),
- action 5_8: from information sources choosing the methods of organizing the educational cognitive activity and methods of training in mathematics in specific conditions of educational process (complexity 2, difficulty 2),

- action 5_9: changing the set of methods of organizing the educational cognitive activity of students and methods of training in mathematics when changing the conditions of educational process (complexity 3, difficulty 3),
- action 5_{10}* (research level): describing the adaptation of the available methods in concretely chosen conditions of educational process (complexity and difficulty are defined by an expert assessment);
- action 6_1: from given set (or system) of techniques and methods of training composing the aggregate (set) of methods of training in mathematics for specific conditions of educational process (complexity 1, difficulty 1);
- action 6_2: based on information sources independent choosing and composing the set of techniques and methods of training in mathematics for specific conditions of educational process (complexity 1, difficulty 2),
- action 6_3: changing the set of techniques and methods of training in mathematics and justifying this changing when varying specific conditions of educational and cognitive process (complexity depends on quantity of setting variants of changing conditions, difficulty 3),
- action 6_4* (the creative level assuming scientific research): independent designing of method of training in mathematics (in specific conditions of educational cognitive activity of students) (complexity and difficulty are marked out on the basis of expert assessment),
- action 6_5* (the creative level assuming scientific research): independent designing of system of methods of training in mathematics as the element of a certain methodical system or methods (in the specific conditions of educational cognitive activity of students) (complexity and difficulty are marked out on the basis of expert assessment);
- action 7_1: from given set choosing the criteria for assessing the set of methods of training in mathematics in specific conditions of educational process (complexity 1, difficulty 1),
- action 7_2: from information sources choosing the criteria for assessing the set of methods of training in mathematics in specific conditions of course of educational process (complexity 1, difficulty 2),
- action 7_3: from given list choosing the criteria for assessing the set of methods of organizing the educational cognitive activity of pupils in specific conditions of educational process (complexity 1, difficulty 1),
- action 7_4: from information sources choosing the criteria for assessing the set of methods of organizing the educational cognitive activity of pupils in specific conditions of educational process (complexity 1, difficulty 2),
- action 7_5: from given list choosing the criteria for assessing the set of methods of training in mathematics and methods of the organizing the educational cognitive activity of pupils in specific conditions of educational process (complexity 2, difficulty 1),
- action 7_6: with use of information sources independent choosing the criteria for assessing the set of methods of training in mathematics and methods of organizing the educational cognitive activity of pupils in specific conditions of educational process (complexity 2, difficulty 2),

● action 7_7 * (the research level, creative level assuming scientific research): allocating the criteria of assessing the methods of training and (or) methods of the organizing the educational cognitive activity of pupils for specific conditions of modern educational process (complexity and difficulty are defined by expert assessment).

The number of levels of formation of ability to make and evaluate the sets of modern methods of training in mathematics can be established under the agreement of experts based on the analysis of operational structure of each specified action and research of correlation between actions. Thus, for the description of the maintenance of level the configuration (set) of the actions entering in each level can be constructed. It is convenient to carry out the creation of such configurations in the base of a profile composed of the actions entering in ability with detailed disclosure of methodology of determination of complexity and difficulty of action. In the base of a profile (Fig. 1) constructed by us in the context of the formulated theoretical points, actions are connected between themselves by obvious communication, or don't contain obvious communication.

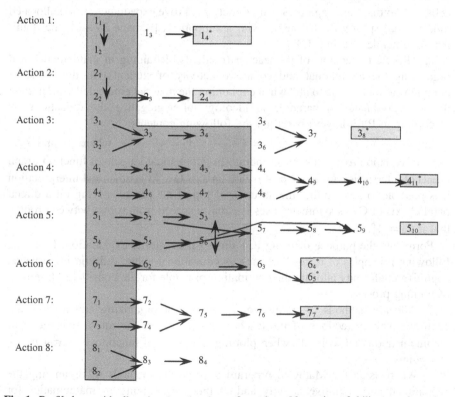

Fig. 1 Profile base with allocating three levels configuration of formation of ability to make and to evaluate the system of modern methods of teaching mathematics and methods of organizing educational and cognitive activity of students

Obvious communication means the need of the result accounting of the previous action when performing the following (in Fig. 1 such communication is designated by an arrow), implicit communication means autonomous performance of action – the solution of an informative (educational and informative) task (designation in drawing – lack of an arrow). Thus in figure we allocated with color three levels (offered by us) of formation of ability to make and evaluate the sets of methods of training in mathematics and methods of organizing the educational and cognitive activity of students (in certain cases, to design methods of training in mathematics) which make the hierarchical sequence: sufficient, high and advanced level.

3 The Examples of the Tasks for Students' and Teachers' Diagnostics of Ability to Make and Evaluate the System of Modern Methods of Training in Mathematics

1. Allocate and, according to [11], name the mode(s) of possible communication for organizing the cognitive activity of students when studying new material on the subject "Plotting the Graph of Square Function". Prove expediency of the allocated mode(s) and specify for it (them) (with a support on [8, page 39–40]) the dominating principle of using ICT.

2. Allocate the actions of the teacher (methods of training in mathematics and organizing the educational and cognitive activity of students) in the mode of communication by "one to all" when explaining the material connected with finding the top coordinates of parabola in ideology of aggregating the didactic units (according to P. Erdniyev) based on the following contents:

$$x_1 = \frac{-b + \sqrt{D}}{2a}, \quad x_1 = \frac{-b}{2a} + \frac{\sqrt{D}}{2a}; \quad x_2 = \frac{-b - \sqrt{D}}{2a}, \quad x_2 = \frac{-b}{2a} - \frac{\sqrt{D}}{2a},$$ where x_1 and x_2 – roots of equation $ax^2 + bx + c = 0$, that is (in geometrical language) function graph points of intersection $y = ax^2 + bx + c$ with an axis Ox. At geometrical interpretation it is possible to assume that the abscissa of top of the parabola lying on a direct, parallel axis of Oy as symmetry axes is equally spaced from roots (between roots), that is $x_B = \frac{-b}{2a}$.

Formulate the purpose of using ICT during the material explanation. From the following principles of using ICT such as educational value, didactic importance, cognitive conformity allocate the dominating principle for the considered fragment of training process.

3. Allocate methods of training and methods of organizing educational and cognitive activity (actions of the teacher and pupils) in the interactive mode of communication "all-with-all" when plotting the graph of function $y = ax^2 + bx + c$ "by points".

4. When using the MathCad program allocate the methods of organizing the educational and cognitive activity and methods of training in mathematics for formation of students' knowledge and abilities which are necessary during the plotting the graph of function $y = ax^2 + bx + c$.

Specify the purposes of training in mathematics and the purpose of using ICT.

Allocate methods of training in mathematics for formation of students' knowledge about symmetry of the graph about the line passing through parabola top and parallel axis of Oy.

5. Describe the scenario of educational presentation taking into account ICT advantages for use at a lesson of the subject "Plotting the graph of the function $y = ax^2 + bx + c$".

Develop the educational presentation on the specified subject for using it in the mode of distance learning. Mark out the differences of the developed presentations. Allocate the tools of ICT in that and other case. For each presentation specify groups (classes) of the used methods of training in mathematics. Answer the questions: "Is there a difference in classes of the specified methods? Why?".

With use of results [4] for studying the specified subject allocate the techniques connected with applying the computer graphics in the associative-synektics technology developing the direction specified by K. Byuler, who considered drawing as the graphic story about action.

6. Based on the principle of pedagogical expediency of using the tools of ICT in the communication mode "all to one", choose the tool and develop the test tasks providing control of the knowledge and abilities of pupils, diagnostics of the knowledge and abilities of pupils for any chosen by you subject of a school mathematics course.

Mark out the advantages or prove the expediency of using ICT for the considered stage of formation of the knowledge, understanding and abilities of pupils.

Formulate the requirements to the test tasks for ensuring reliability of control at the chosen tools. Formulate requirements to the tool and allocate conditions for carrying out the diagnostics with using ICT.

From the point of view of V.S. Avanesov ([1]), B.E. Starichenko [9] that the test checks the level of mastering the material, but not a level of thinking development, that the purpose of testing is receiving the students distribution on levels of material mastering, evaluate the practical importance of the materials developed by you.

From a position of the specified point of view characterize the efficiency of tests on the subject of other authors chosen and found by you in Internet resources.

7. Describe psychology and pedagogical features of some contingent trained in mathematics. In system of the principles of availability (classical didactics), multimedia, information humanity (information didactics [10]) and cognitive conformity (the principle of using ICT) when classifying the methods of training by information source (verbal, evident, practical) in the mode "one to all" allocate the methods of training in mathematics on the subject "Proof of Pythagorean Theorem".

Make the detailed answer to the following questions: Will the constructed set of training methods change if:

(1) educational and cognitive activity will be organized without using the tools of ICT; (2) classification of "interring" will be accepted as classification of training

methods in the conditions of using the tools of ICT (according to definitions [7, page 101–104]); (3) there will be the changes in the psychology and pedagogical characteristic of the pupils? And how?

8. Make the version of the short psychology and pedagogical characteristic of some group of pupils, containing the 2–3rd characteristics, for example, a modality, degree of interest in training in subject domain, mentality or another. Describe features of the content and allocate methods of a statement of training material for the pupils' contingent characterized by you trained on (a) lesson of studying of new material, (b) lesson of formation of skills.

Give an example (prove efficiency) means of the information technologies allowing the teacher to solve the educational (developing) lesson problems proceeding from the accounting of the specified characteristic of the trained.

Offer options of activity of the teacher on condition of change in the characteristic of pupils of any component. Whether the choice of means of ICT will change thus?

9. Allocate the role of ICT when carrying out the set tasks.

Read the text [5, page 5–20] and taking into account the work done by you express your point of view on philosophy of science of Feyerabend.

4 Conclusion

The presented approach worked out for developing and diagnosing the formation of ability to make and estimate systems of modern methods of training in mathematics, contains also the technologization possibility of designing new tasks, and also increasing "complicity" or "difficulty" of already composed tasks. The ideology of such technologization consists in educational and cognitive realization of the actions verbs allocated in the description of methodological and practical levels of developing (diagnosing) ability of teachers and students.

References

1. Avanesov, V.S. : www.testolog.narod.ru (in Russian)
2. Bespal'ko, V.P.: Education and Training with Participation of Computers (pedagogy of the third millennium). Publishing House of The Moscow Psychologist-Social Institute, Moscow; Publishing House of NGO "MODEK", Voronezh (2002) (in Russian)
3. Krupich, V.I.: Theoretical Bases of Training in the Solution of School Mathematical Tasks. Prometheus, Moscow (1995) (in Russian)
4. Novoselov, S.A., Ivanova, N.P.: Methods of Applying Computer Graphics in an Associative and Synektics Technology: Educational and Methodical Manual. Urals State Pedagogical University. Ekaterinburg (2012) (in Russian)
5. Sviridyuk, G.A., Monakova, N.A.: Concept of modern natural sciences, Part 2: Chemistry, Biology, Humanitarian and Social Sciences: Manual. Chelyabinsk Publishing Center SUSU, (2014) (in Russian)

6. Semenova, I.N.: Formation and Assessment of Level of Proficiency to a Choice and Compose The System of Methods of Training in Mathematics of Future Masters of Education for Realization of Electronic Training. Teacher education in Russia, vol. 11, pp. 190–194 (2014) (in Russian)
7. Semenova, I.N.: Methodology of Teaching Mathematics Methods Designing in the Modern Educational Paradigm: Monograph/Semenova//Yelm, WA. Science Book Publishing House, USA (2014). (in Russian)
8. Semenova, I.N., Slepukhin, A.V.: Methods of using information and communication technologies in educational process, Part 2. In: Starichenko, B.E.(ed.) Methodology of Using Information Educational Technologies: Manual. Urals State Pedagogical University. Ekaterinburg (2013) (in Russian)
9. Starichenko, B.E., Mamontova, M.Y., Slepukhin, A.V.: Methods of using information and communication technologies in educational process, Part 3. In: Starichenko, B.E. (ed.) Computer Technologies of Diagnostics of Educational Achievements: Manual. Urals State Pedagogical University. Ekaterinburg (2014) (in Russian)
10. Starichenko, B.E.: Is it the right time for a new didactics? Educ. Sci. **4**, 117–126 (2008) (in Russian)
11. Yavich, R.P.: Management of Mathematical Training of Students of Technical College on the Basis of Telecommunication Technologies: Thesis of the Candidate of Pedagogical Sciences. Urals State Pedagogical University, Ekaterinburg (2008). (in Russian)

The Word Cloud Illustration of the Cognitive Structures of Teacher Candidates About Education Concept

Nuray Zan, Burcu Umut Zan and F. İnci Morgil

Abstract Graduates of Arts and Sciences Faculties attend Pedagogical Formation Certificate Programme as Pedagogical Formation Education is a necessity to be a teacher. This study was conducted for searching the views of teacher candidates attending the Pedagogical Formation Certificate Programme about education concept. The data of the study were the replies of 234 participants with a bachelor's degree. The reply words of the teacher candidates for "education" concept in the Word Association Test were analyzed by word cloud technique. "Preparation for life", "examination", "course", "education", "stress", "civilization" and "acquisition" words proved to be the mostly preferred words. It can be said that the students' word associations with education concept were valid. However, some of these associations were far from the scientific information and the teacher candidates gave superficial answers more so. According to the findings of the study, Word Association Test was affirmed to be an effective technique to manifest the cognitive structure, and the results of the Word Association Test were illustrated by Word Cloud technique in order the cognitive structure manifested in the study to be comprehensible.

Keywords Word association test · Word cloud · Education · Pedagogical formation

N. Zan (✉)
Faculty of Arts and Science, Department of Education Science, Çankırı Karatekin University, Çankırı, Turkey
e-mail: nurayyoruk@gmail.com

B.U. Zan
Faculty of Arts and Science, Department of Information and Records Management, Çankırı Karatekin University, Çankırı, Turkey
e-mail: burcumut@gmail.com

F. İnci Morgil
Secondary School Maths and Science Department, Hacettepe University, Ankara, Turkey
e-mail: incimorgil@gmail.com

© Springer International Publishing Switzerland 2015   357
V.L. Uskov et al. (eds.), *Smart Education and Smart e-Learning*,
Smart Innovation, Systems and Technologies 41,
DOI 10.1007/978-3-319-19875-0_32

1   Introduction

Many countries have implemented different reforms to increase the quality of their education systems since 2000's. These reforms include a wide range from education programmes and teacher training models to finance and management of the education. Teacher training is one of the prominent education subjects in the country. In Turkey, different practices were carried out in distinctive ways for teacher training after the Republic [1]. From Republic to date, the level of teacher training in Turkey has increased permanently. However, as Özoğlu states in his study [2] it cannot be said that historically, the need for teacher positions in the country was catered in a planned and qualified way. The mission of training teachers for secondary and high schools was assigned to 4-year education institutes at first, and then to the universities after converting institutes to education faculties. Teacher candidates graduated from education faculties take the major area courses from the academicians of the related departments of arts and sciences faculties and take the major education courses from the academicians of the education faculties. The criteria of Ministry of National Education require teacher candidates having a satisfactory liberal education, specific major education and pedagogical formation to be teachers [3]. The pedagogical education which the graduates of arts and sciences faculty prefer to be teachers is one of the practices in teacher training. Thus, to train branch teachers for secondary schools, some regulations for the graduates of arts and sciences to become teachers such as practical teaching training under the control of experienced teachers and giving two-term pedagogical formation courses including necessary teaching courses were made [4]. This age is information age and it needs educated labor force in education. Therefore, well trained, sophisticated, modern educators are needed. First of all, to provide a quality education, teachers who gained education concept should be employed. Teacher candidates' view of education has utmost importance in forming a qualified education understanding. In this study, views of teacher candidates who are bachelors of arts and sciences about education concept were searched at the beginning of Pedagogical Formation Education, their words to explain education concept were found and the cognitive structures were studied. This study is a part of comprehensive study which is planned to be repeated with the same group when the Pedagogical Formation Education is over to monitor the alteration of the views of teacher candidates about education concept.

The main aim of Educational Science Department is to raise teachers and educational specialists, who are competent in their professions, to conduct professional teaching knowledge courses, which are necessary for teaching and up skilling in teaching, and to conduct activities, increasing quality of education.

Making research in education field, arranging in service training for the administrators and teachers of primary and secondary educational institutions, providing basic education courses to students, graduated from various faculties and departments, and providing consultancy service to educational institutions takes place among the foundation aim of our department, right along with the main aim of the department,

The main goal of Pedagogical Formation Certificate Programme Department is to raise teachers that competent national, individual and Professional norms, like their profession and productive in their profession. Our principle for making this goal real is to raise education volunteers, who are interrogator, researcher, solution oriented and have skill of analytical thinking, universally appropriate understanding of science and ethical values.

The program offers formal teaching education to the students who aim to teach in various subjects. This program also provides methodological and practical information to the students and guides them in their future teaching profession position. Courses have been conducted interactively in classrooms on campus. The students, who are volunteered to study in Pedagogical Formation Certificate Program, are in the mandatory to study in two terms. In the first term, theoretical courses are given to students and second term continues with the applied courses. There are seven courses in the first term. Five of them is mandatory and two of them is selective. Defined selective courses were selected from the selective course pool by the administrative court of the faculty. These courses are called "Program Development in Education" and "Turkish Education System and School Management".

The contents of courses which are mentioned above is given below.[1]

Table 1 Education of pedagogical formation certificate program courses[2]	Theoretic courses
	Introduction to educational sciences
	Principle and methods of teaching
	Assessment and evaluation in education
	Guidance
	Classroom management
	Selective course I (program development in education)
	Selective course II (turkish education system and school management)
	Applied courses
	Special teaching methods
	Instructional technologies and material design
	Teaching internship

Introduction to Educational Science: Basic concepts, principles and characteristics of education; social, political, philosophical, and historical foundations of education in general and the Turkish educational system in particular.

---

[1]Information is accessed on 04.01.2015 from http://dwww.boun.edu.tr/en_US/Content/Academic/Undergraduate_Catalogue/Faculty_of_Education/Department_of_Educational_Sciences.

[2]Information is accessed on 01.01.2015 from the http://www.yok.gov.tr/web/guest/icerik/-/journal_content/56_INSTANCE_rEHF8BIsfYRx/10279/7052802

Table 2 Proposed elective courses list[3]

Elective course I group	Elective course II group
Research in education	Educational psychology
Program development in education	Usage of technology in education
History of education	Philosophy of education
Sociology of education	History of turkish education
Developmental psychology	Turkish education system and school management
Ethics of teaching profession	Character and values education
Lifelong learning	Special education
Individualized teaching	Computer assisted instruction

Principles and Methods of Instruction: Theoretical approaches to curriculum development and instruction. Concepts and principles of instructional processes. Methods and techniques of instructional planning. Designing, developing, and producing lesson plans on learning units. Development of formative and summative evaluation instruments.

Assessment and Evaluation in Education: Theories of measurement, evaluation and assessment and their historical development. Concepts of objectivity, reliability and validity. Theories and concepts related to different types of tests. Statistical analyses of test scores and the interpretation of test results.

Guidance: Introduction into the field of guidance and counseling; its conceptualization, historical background, nature, professional roles, direct and indirect functions. A brief review of career counseling, counseling with special groups, basic assessment techniques, and development of these services in Turkey.

Classroom Management: Development of effective classroom management skills. Understanding and handling factors influencing student behavior, student motivation, communication and group interaction. Establishing and maintaining standards of behavior. Planning and modification of classroom tasks, materials, time and environment to facilitate learning.

Program Development and Evaluation: Processes and elements of program development and evaluation. Survey of different theoretical approaches and models. Program development under supervision.

Turkish Education System and School Management: Theories and approaches in educational administration. Administrative functions and processes, organizational behavior, leadership, culture and climate in educational organizations. Projects in organizational analysis.

Instructional Technologies and Material Design: Selection of an instructional problem and development of a product. Demonstration of competence in analysis (problem, learner, task, content). Designing (objectives, criterion tests, instructional strategies, media and format selection, content design, and alternative scripts),

---

[3]Information is accessed on 01.01.2015 from the http://www.yok.gov.tr/web/guest/icerik/-/journal_content/56_INSTANCE_rEHF8BIsfYRx/10279/7052802

developing (layout, composition, production), and evaluating (field test, feedback, revision) of instructional products.

Teaching Internship: Supervised practice in schools or other educational settings. Development, implementation and evaluation of group guidance.

Special Teaching Methods: The study of pedagogical and policy issues in secondary education. Examining these issues in the Turkish social and political context. Review of research in the field.

This study is applied before the students take Pedagogical Formation Certificate Program Courses. At the end of these courses, it is thought to provide the development of cognitive structures of the students' which is related with education sciences.

Aim of the Study. The aim of this study is to reveal the opinions of the students attending pedagogical formation training about education concept. In order to fulfill this aim, the students attending the Pedagogical Formation Certificate Programme in Çankırı Karatekin University were called for an explanation of the "education" concept using Word Association Test (WBT). Study group was composed of 234 teacher candidates who have a bachelor's degree and registered the certificate programme in 2014, August. WBT is one of the techniques which show up the students' cognitive structures and the relationships between the concepts in these structures which means the information network, and is used for detecting the sufficiency and significance of the relations between the concepts [5]. The participants were to write the words evoked by a key concept given in a time about any concept. Those reply words are evaluated and the structural relations are set up. Word Cloud was used to reveal the relations between the reply words given by the teacher candidates about education concept.

2 Method

Since it is a quite convenient method to find out the perceptions of the students, qualitative research design was used in this study. In a qualitative research, data collecting techniques such as observation, interview, and document analysis are used and it enables the perceptions and cases to be manifested in a realistic and holistic way in their habitats [6].

Table 3 Number of participants

Gender		Field of study				
Female	Male	Geography	Philosophy	Biology	Mathematics	History
132	102	24	32	22	77	79
		10 Male	11 Male	4 Male	30 Male	47 Male
		14 Female	21 Female	18 Female	47 Female	32 Female

2.1 Study Group

Sampling of the study was the graduated students attending the Pedagogical Formation Certificate Programme in Çankırı Karatekin University in 2014–2015 academic year. Sampling was 234 volunteer students from this programme.

As it is seen in Table 1, the total number of the participant students was 234. 132 of them were female and 102 of them were male students and they were graduated from Geography, Philosophy, Biology, Mathematics and History departments.

2.2 Data Collecting Tool

This study was conducted with the volunteer students in the Pedagogical Formation Certificate Programme. Before the students took Education Sciences courses, WBT, as a data collecting tool, was applied to find out how they define education. The students were given papers to write their definitions on and the statements on these "documents" were the main data source of the study. Completing this process, the data were ready to be analyzed. Based on these definitions, the students were to write their statements for each concept. To indicate the frequency of the words and show the concepts visually, word cloud was created with WordItOut programme. The frequency of the preferred words for each concept was shown in parenthesis numerically, and the concepts were plotted in word cloud map based on their sizes designed by using these numbers. The validity and reliability of the study; reporting the collected data in details and explaining the ways of finding the results are the significant validity criteria of a qualitative research [6]. Assuring the validity of this study, data analysis process was explained and all the concepts found out after students' feedbacks were presented in findings part.

Education is written 10 times on a sheet of paper given to the students. A sample page layout is given below.

> Education......................................
> Education......................................
> Education......................................
> Education......................................
> Education......................................
> Education......................................
> Education......................................
> Education......................................
> Related   statement   ....
> ....

Before the application, the teacher candidates were informed about WBT and some practices were done related to the up to date concepts.

Key concept was written on a page one under the other. This aimed the students to write their opinions about just the given concept. At the end, there is "related statement" part. Considering the studies in the literature, each concept in the Word Association Test was given 30 s in average [7, 8]. The participant teacher candidates could not complete the application in that duration, thus, the duration was extended to two minutes. During the application, the students wrote the words which they thought related with education concept, and were to write the sentences came to their minds about education. Also, since the related statement is more complicated and at a higher level than a word, such cases as whether the statement is scientific or includes misconceptions in different features affects the education process [9]. Test was comprised of two parts. In the first part the demographic information of the students were taken and in the second part the students would write words and a sentence related to key word "education".

Figure 1 is the 15th student's paper. According to our education system, our lecture language is in Turkish. Translated version of Fig. 1 can be find is Table 2.

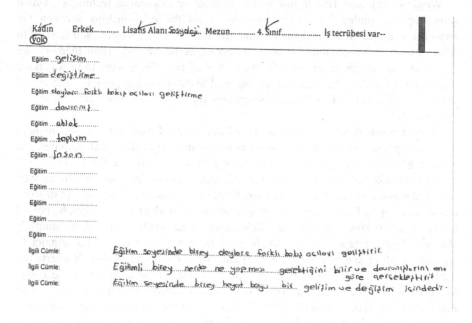

Fig. 1 Test sample applied to the teacher candidates attending to the pedagogical formation certificate programme (15th papers)

Data Analysis. Papers of the participant students were numbered 1 to 234. All the words written on the papers were taken carefully and each word was saved in an Excel table. Preferred words for key concept were loaded to WordItOut programme in order to find out the frequency of the words and create the word cloud (Tables 3, 4 and 5).

Table 4 Tranlated text of Fig. 1

Female	Department: sociology	4th Class	Work experience: no
Education...	Develop		
Education...	Changing		
Education...	Behavior		
Education...	Ethics/Moral		
Education...	Society		
Education...	Human		
Education...			
Education...			
Related statement 1	Through education systems, individuals are able to develops different perspectives to the events		
Related statement 2	Educated individuals know what to do		
Related statement 3	Education can be able to change and develop humans		

Word Association Test is one of the alternative evaluation techniques which shows up the students' cognitive structures and the relationships between the concepts in these structures which means the information network, and is used to detect whether the relations between the concepts in the long term memory are sufficient and significant or not [7]. Word Association Tests were used for many different objectives in the literature. It was used to expose the cognitive structures of the students [7, 11], define the misconceptions [8, 9] and determine the conceptual changes [12].

Word Cloud. A word cloud is a "visual depiction of words. The more frequent the word appears within the text being analyzed the larger the word becomes." In essence a word cloud "plots" word frequency by the size of the Word [13].

Word clouds developed from web based social networking sites, which are web sites that allow a group of common users to share information. The concept of word clouds developed from the tags or descriptors used to identify photographs posted to social networking sites such as Flickr, a site specifically designed for multiple sharing of photographs. Word clouds have been viewed as a useful adjunct to teaching reading and writing skills [14] and for summarizing research interviews [15] but there is a dearth of research into their use to enhance student learning. Word Cloud can be defined as visual depiction of words. In other words, it is an illustration programme based on the frequencies of the preferred words. The programme defines the sizes of the concepts on the map according to the frequency. Word clouds are usually the textual images with different colors and designs after the analysis of the frequencies. There are some programmes to create a word cloud[4] one of which is Wordle.

The reply words taken after the Word Association Test were analyzed; then, the word cloud was created and illustrated by Wordle programme. These kinds of

---

[4]http://wordle.net, http://worditout.com/, http://www.tadgcrowd.com.

studies were applied in different fields. Word cloud is used in classroom discussions to provide students with visual perception, in the studies about key concepts, mostly preferred words were illustrated visually as well. In the study of Miley and Read in 2011 [16], they illustrated the students' key words on a word cloud and shared it with students instantly concluding that it increased the success level of the students.

Total number of the reply words was 1826. On the graphic below, mostly preferred words of the teacher candidates to explain education were given. There are 46 words and they were used 1082 times. Out of these, there are 744 words excluded from the list since they were used only once to 5 times. Some examples are stair, method, complex, full, mother, pleasure, temperament, permanent, place,

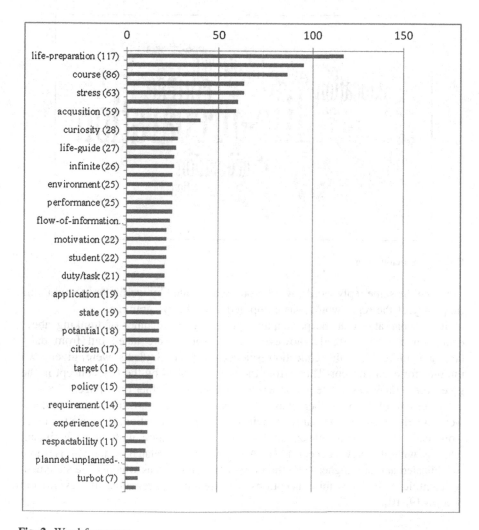

Fig. 2 Word frequency

general, exclusive, impossible, absolute, friend, cinema, different, integration, once, ministry, government, alphabet, roof, border, ambition, depression....

3 Findings

Reply words for the education concept are given in Fig. 2. The words with a frequency of 5 and below were excluded from the list. The total frequency is 406, including the words out of the list. Reply words are shown in the table below. The word cloud created after Word Association Test is seen below (Fig. 3).

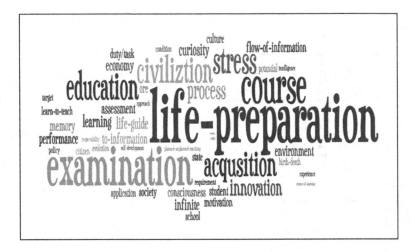

Fig. 3 Word cloud map

Using the same reply words, word cloud was created by wordle programme, the frequency of the reply words was computed with this programme.

It is remarkable that the participants perceived the manner of thinking about education concept correctly, however, reply words were taken out from daily language rather than the education sciences jargon. Students were given two minutes for each concept. They wrote the words evoked by the key concept in the given time. Each key concept was written one under another on a page.

At the end of the page, there was a "related statement" part. In this part students were to write the sentences came to their minds about education. Because the reply word can only be at an evocation level or a connotation without a significant relation with the key concept [17]. Also, since the related statement is more complicated and at a higher level than a word, such cases as whether the statement is scientific or includes misconceptions in different features affect the evaluation process [9, 10].

The sentences of the teacher candidates were categorized according to the meanings they encapsulate. The categorization was done considering the table developed by Ercan and Taşdere [9–11]. This table categorizes the sentences as the statements including scientific information, the statements including unscientific and superficial information and the statements including misconceptions.

Table 5 Frequency table of the related statements about key concept

Key concepts	The statements including scientific information	The statements including unscientific and superficial information	The statements including misconceptions	None
Education	60	66	47	61

According to the study conducted by Ercan and Taşdere [9], the students who did not write a sentence or were not able to write a meaningful sentence could not learn the related concepts at cognitive level. Analyzing the number of the sentences, most of the teacher candidates attending the Pedagogical Formation Certificate Programme could not make a meaningful sentence about education and teaching concepts. The sentence for learning concept was blank. Teacher candidates wrote statements including unscientific and superficial information mostly. These statements were generally from daily language with superficial meanings, unrelated to the key concept, up to date expressions of momentary emotions or opinions rather than scientific ones.

Some samples of those related statements are seen below.

The statements including scientific information

(Paper 14) Desired behaviors occurring with behavioral alteration.

(Paper 36) Education is a lifelong process.

(Paper 58) It is the permanent behavioral alteration, changes the mistakes.

The statements including unscientific and superficial information

(Paper 24) Getting as much information as he wishes while sitting in front of a computer.

(Paper 35) Education is individual's directing his own life.

(Paper 43) Education is an adventure starting from the family going on at school.

The statements including misconceptions

(Paper 37) Teacher obeys the programme in a planned way in education.

(Paper 73) Education has a curriculum.

(Paper 154) Education should include students' interests and skills.

4   Discussion and Conclusion

Main component of education is teacher and qualified person which the teacher aims to train. In this context, training the teacher and the qualifications of the teacher candidates are significant. In this study, views of teacher candidates about education concept were searched.

Cognitive structures and preferred words or word groups about education concept of teacher candidates attending the Pedagogical Formation Certificate Programme were illustrated by using Word Association Test and word cloud. The most preferred word group was preparation for life with a frequency of 117. Second word was course (86), third word was stress (63) and teaching (63), fourth word was civilization (60) and the fifth word was acquisition (59). Also, information flow was one of the most preferred words to explain education concept. The mostly preferred reply words and word groups of the teacher candidates were illustrated on a word cloud created by Tag Crowd and WordItOut programmes. On the other hand, this study claims that while the graduates of faculties of arts and sciences specialize in their fields of study, they are not at a scientific level to explain education concept before attending the Pedagogical Formation Certificate Programme.

5   Suggestions

In this study, cognitive concept structures for education concept of teacher candidates were searched by WBT, their word or word group preferences were illustrated using word cloud. A quality education system brings the students in the up to date information and skills is a common belief all over the world. To fulfil this aim, meaningful learning targets should be defined and appropriate environments for students and teachers to learn should be set. By means of these studies, following the improvements in the education about training, observing the individual's progress and the conceptual improvement will be possible. If this study is repeated after the completion of the Pedagogical Formation training with the same group, the alteration in the views of the teacher candidates about concepts will be seen explicitly. Furthermore, while it is used in the evaluation of the home works at elementary and secondary school levels, analyzing the students' preferred concepts by only using word cloud may be a better option to monitor conceptual improvement. According to the results of Polat's [18] study it is shown that the reasons for the tendency in choosing teaching as a profession were revealed as loving the teaching profession, comfortable work conditions and obligations.

This study is planned according to scanning model. That is why it is planned to define the current cognitive situations of the student, which is related with education. This study includes only one application. But in the different WAT study models, it is planned as a two test application which are called pretest and post test. Hereby conceptual change of the students can be identified. That can be reveal the cognitive differences of the Pedagogical Formation Certificate Programme student's situations with the comparison of before the certificate programme and after the certificate programme. Two test model, which includes pre-test and post test, is going to be the continuity of this study. It is suggested to apply the post test to the same group of students at the end of the Pedagogical Formation Certificate Programme, to expose the cognitive developments of the students.

References

1. Akyol, M.: Gayri Resmi Yakın Tarih (Informal Recent History). Nesil Basım Yayın, İstanbul (2011)
2. Özoğlu, M.: Türkiye'de Öğretmen Yetiştirme Sisteminin Sorunları (Problems of Teacher Education System). Seta Analiz, Ankara (2010)
3. MEB, http://ttkb.meb.gov.tr/meb_iys_dosyalar/2014_09/10093949_9_cizelgeveesaslar.pdf. (Accessed on 8.12.2014)
4. Gürüz, K.: Yirmi Birinci Yüzyılın Başında Türk Milli Eğitim Sistemi (Turkish National Education System in Beginning of the Twenty First Century). Türkiye İş Bankası; Kültür Yayınları, İstanbul (2008)
5. Bahar, M. (ed.): Geleneksel- Allternatif Ölçme ve Değerlendirme (Traditional - Alternative Measurement and Assesment). Pegema Yayıncılık, Ankara (2006)
6. Yıldırım, A., Şimşek, H.: Sosyal Bilimlerde Nitel Araştırma Yöntemleri. (Qualitative Research Methods in Social Sciences). Seçkin Yayıncılık, Ankara (1999)
7. Bahar, M., Johnstone, A.H., Sutcliffe, R.G.: Investigation of students' cognitive structure in elementary genetics through word association tests. J. Biol. Educ. **33**, 134–141 (1999)
8. Bahar, M., Özatlı, N.S.: Kelime İlişkilendirme Yöntemi ile lise 1. Sınıf öğrencilerinin canlıların temel bileşenleri konusundaki bilişsel yapılarının araştırılması (Examining the Cognitive Structure of 9th Class Students in the subject of Basic Components of Living Through Word Associate Test). Balıkesir Üniversitesi, Fen Bilimleri Dergisi **5**(1), (2003)
9. Ercan, F., Taşdere, A., Ercan, N.: Kelime İlişkilendirme Testi Aracılığıyla Bilişsel Yapının ve Kavramsal Değişimin Gözlenmesi (Observing Conceptual and Cognitive Changes Through the Word Associate Test), Türk Fen Eğitimi Dergisi (TÜFED), **7**(2), (2010)
10. Işıklı, M., Taşdere, A., Göz, N.L.: Kelime İlişkilendirme Testi Aracılığıyla Öğretmen Adaylarının Atatürk İlkelerine Yönelik Bilişsel Yapılarının İncelenmesi (Examining the Cognitive Structures of Teacher Candidates' towards Atatürk's Principles Through Word Associate Test). Uşak Üniversitesi Sosyal Bilimler Dergisi **4**(1), 50–72 (2011)
11. Cardellini, L., Bahar, M.: Monitoring the learning of chemistry through Word association tests. Australian Chemistry Resource Book **19**, 50–69 (2000)
12. Nakiboğlu, C.: Using Word Associations for Assesing Nonmajor Science Students Knowledge Structure Before and After General Chemistry Instruction: The Case of Atomic Structure. Chem. Educ. Res. Pract. **9**, 309–322 (2008)
13. Ramsden, A., Bate, A.: Using Word Clouds in Teaching and Learning. University of Bath. (Unpublished) (2008)
14. Hayes, S.: Toolkit: Wordle. Voices from the Middle **16**(2), 66–68 (2008)
15. McNaught, C., Lam, P.: Using Wordle as a supplementary research tool. Qual. Rep. **15**(3), 630–643 (2010)
16. Miley, F., Read, A.: Using Word Clouds To Develop Proactive Learners, Journal Of The Scholarship of Teaching and Learning, Vol. 11, No. 2, April 2011, pp. 91–110 (2011)
17. Nartgün, Z.: Fen ve Teknoloji Öğretiminde Ölçme ve Değerlendirme (Measurement and Assesment in Science Technology Teaching), Bahar, M. (ed.), Fen ve Teknoloji Öğretimi. Pegema Yayıncılık, Ankara (2006)

18. Polat, S.: The reasons for the pedagogical formation traınıng certificate program students'-who have jobs-tending to choose the teaching professıon Bir mesleğe/işe sahip olan pedagojik formasyon eğitimi sertifika programı öğrencilerinin öğretmenlik mesleğine yönelme nedenleri. Int. J. Human Sci. **11**(1), 128–144 (2014)

Formalization of Knowledge Systems on the Basis of System Approach

Natalia A. Serdyukova, Vladimir I. Serdyukov
and Vladimir A. Slepov

Abstract In recent times amid deep mathematization of science, including its humanitarian areas, all the more evident becomes the importance of system approach as the most powerful tool for qualitative analysis of the regularities of the word. The link between science and society is education. Education is the process of transmission, as a rule, subsequent generations of accumulated information, followed by quality control. The important part in the education is the organization of the process of information transmission. The system approach is of the paramount importance in learning processes as the basis of methodologies that allow one to shed light upon the patterns of learning processes, as they provide feedback between an information transmission and a control of her perception. The goal of this paper is to build and formalize the structural scheme of the system of knowledge with the help of a new formalization of the system approach.

Keywords Formalization · Knowledge system · System approach

1 Introduction and Key Definitions

The conceptual strategy of this paper is based on using a new formalization of the system approach developed in [4, 5, 6] and obtaining and justification on this basis, the substantive results of General systems theory, and then, as an application of this, solutions of specific problems in the field of General learning theory. The proposed

N.A. Serdyukova (✉) · V.A. Slepov
Plekhanov Russian University of Economics, Moscow, Russia
e-mail: nsns25@yandex.ru

V.A. Slepov
e-mail: vlalslepov@yandex.ru

V.I. Serdyukov
Moscow State Technical University n.a. N.E. Bauman, Moscow, Russia
e-mail: wis24@yandex.ru

© Springer International Publishing Switzerland 2015
V.L. Uskov et al. (eds.), *Smart Education and Smart e-Learning*,
Smart Innovation, Systems and Technologies 41,
DOI 10.1007/978-3-319-19875-0_33

methodology allows, in contrast with previously used ones [1] to characterize the examined properties in a single, integrated complex as numerical indicators and synchronized with them relationships, in dynamics and in statics, with the help of static predicates, the theory of which was proposed by A.I. Maltcev, [2]. For the general theory of systems the system's dynamics may be reflected by the complexity of relationships, gradually emerging in the process of functioning of the system. For educational systems dynamics may be reflected by the difficulty levels of the tests, aimed at controlling of the learning process.

Practice has shown that it is impossible to limit the group theory by the questions of its use only in the nearest areas. The development of group theory, believed A.G. Kurosh, [3], should have an increasingly significant impact in various fields of science, and even outside the area of mathematics. We believe that this statement applies not only to the mathematical fields, but to the scientific research in general, because the group theory is well-formalized and reflects a fundamental property of things – symmetry and therefore is associated with notions of appropriateness, proportionality, optimality. Application of group theory, for example, in physics, is possible because the theory of groups is well formalized and allows one to explore the invariance principle, according to which physical law with the need is invariant in any inertial system and, therefore, retains its shape under all transformations of four-dimensional space-time transforms the inertial system into inertial ones. The group of all transformations under which Maxwell's equations for electromagnetic fields remain invariant, described by H. A. Lorentz, led to the formation of the special theory of relativity. We propose to develop applications of the theory of algebraic systems, model theory and group theory to formalize General systems theory. The proposed formalization may be used to obtain and prove the main results in General Systems' Theory, and, as a consequence, in General Theory of Education.

Let's formulate all needed for understanding the text main notions and after that we shall introduce the notions of P - innovative system and dual notion of P-pseudo-innovative system on the basis of which we shall give answers to formulated questions. Let's consider key definitions, [6].

Definition 1. Under the algebra of factors a system will be understood algebra $\overline{A} = \langle A | \{f_\alpha^{n_\alpha} | \alpha \in \Gamma\} \rangle$ with the fundamental set of factors A and the set of operations $\{f_\alpha^{n_\alpha} | \alpha \in \Gamma\}$ describing the interaction of the factors.

Definition 2. Sub-algebra $\overline{B} = \langle B | \{f_\alpha^{n_\alpha} | \alpha \in \Gamma\} \rangle$ of an algebra $\overline{A} = \langle A | \{f_\alpha^{n_\alpha} | \alpha \in \Gamma\} \rangle$ is called P - pure (P - clean) in $\overline{A}$ if every homomorphism $\overline{B} \to {}^\varphi \overline{C}$ of the subalgebra $\overline{B}$ into $\overline{C}$, where $\overline{C}$ is an algebra of the signature $\{f_\alpha^{n_\alpha} | \alpha \in \Gamma\}$ of the of $\overline{A}$ and $P(\overline{C})$ is true and P is a predicate on the class of algebras of the signature $\{f_\alpha^{n_\alpha} | \alpha \in \Gamma\}$ closed under taking sub-algebras and factor algebras, can be continued to a homomorphism $\overline{A} = \langle A | \{f_\alpha^{n_\alpha} | \alpha \in \Gamma\} \rangle$ into $\overline{C} = C | \{f_\alpha^{n_\alpha} | \alpha \in \Gamma\}$ i.e. the following diagram is commutative:

$$0 \to \bar{B} = \langle B | \{f_\alpha^{n_\alpha} | \alpha \in \Gamma\} \rangle \xrightarrow{\alpha} \bar{A} = \langle A | \{f_\alpha^{n_\alpha} | \alpha \in \Gamma\} \rangle$$

$$\varphi \searrow \qquad \bar{C} = \langle C | \{f_\alpha^{n_\alpha} | \alpha \in \Gamma\} \rangle \nwarrow \beta$$

that is $\beta\alpha = \varphi$

2 Main Examples of P – Purities

Let's consider the meaning of purities and examples of P – purities in the class of all groups. Chart (1) has the following meaning in the class of groups: epimorphic images of B and A in the class of all finite groups are one and the same. For P –purities the meaning runs as follows: B and A have the same epimorphic images in the class of all groups satisfying the condition P.

Examples:

- P allocates the class of all finite groups in the class of Abelian groups, get the usual purity in the class of all Abelian groups;
- P allocates the class of all Abelian groups in the class of all groups;
- P allocates the class of all finite groups in the class of all groups;
- P highlights the diversity in the class of all groups i.e. the class of groups closed under subgroups, homomorphic images and Cartesian products, such as Burnside's variety of all groups of the exponent (indicator) n defined by the identity $x^n = 1$, the variety of nilpotent groups of class of nilpotent is not more than n, soluble groups of length not exceeding the number l, etc.

Now consider a system with full implementation of all links that satisfy the predicate P. An algebraic system of the signature Ω which is injective with regard to all P-pure sequences in the class of all algebraic systems of the signature Ω is a system with full realization of all links that satisfy the predicate P.

Let's concern inverse limits of systems and the embedding of a system into a system with a full implementation of the R relations. Generalizations of theorems on the structure of algebraically compact groups and the theorem that any reduced abelian group can be embedded as a pure subgroup into an algebraically compact abelian group (inverse limit of cyclic groups) are effective to build the P-pure embedding of a system into a system with full realization of the P - links. In addition, for General Systems Theory, this approach provides an opportunity to prove the analogue of the theorem that an algebraically compact abelian group is allocated as a direct summand of the group containing it as a pure subgroup: P – pure subsystem with a full implementation of the P - links in it is a retract of a system containing it. So we obtain the following theorem.

Theorem 1. Any system with a full realization of the P - links can be offline. Purities are in fact the fractality of links. P – purities are the fractality of links with the property P.

Let's begin with formalization of an axiomatic description of a system.

Definition 3. By the system we should understand a two- dimensional vector $S = \langle \{ \langle S_\alpha, Q_\alpha, U_\alpha \rangle | \alpha \in A \}, I(S) = \langle \{ a_\beta | \beta \varepsilon B \} | \Omega_S = \{ f_\gamma^{n_\gamma} | \gamma \varepsilon \Gamma \} \rangle \rangle$,

where $\{ S_\alpha | \alpha \varepsilon A \}$ is the set of all system S statuses which are possible as a result of a system S functioning, $\{ Q_\alpha | \alpha \varepsilon A \}$ is the set of all statuses of a system Q upon which system S affects, $\{ U_\alpha | \alpha \varepsilon A \}$ is the set of all statuses of a external environment which are possible as a result of system S functioning, $\{ a_\beta | \beta \varepsilon B \}$ is the set of all inner factors acting on system S that is determining its behavior, if the composition of factors $a_1^\circ a_2^\circ \dots ^\circ a_{n_\gamma} = a$, than let $f_\gamma^{n_\gamma}(a_1, a_2, \dots, a_{n_\gamma}) = a$ where $\{ f_\gamma^{n_\gamma} | \gamma \varepsilon \Gamma \}$ is a set of operations on the set of factors $\{ a_\beta | \beta \varepsilon B \}$, $f_\gamma^{n_\gamma}$ is n_γ -argument operation, $I(S) = \langle \{ a_\beta | \beta \varepsilon B \} |_S = \{ f_\gamma^{n_\gamma} | \gamma \varepsilon \Gamma \} \rangle$ is an algebraic system of inner factors of a system S. Let $\{ P_i | i \in I \}$ be a set of all properties of a system S which it holds as a result of it's functioning, $\{ B_j | j \in J \}$ – is a set of all subsystems of a system S, $\{ v_m^n | m \in M, n \in N \}$ is a set of all connections of a system S, $*$ is the operation of composition, $G(S)$ is a goal of a system S.

So we possess an algebra $< \{ P_i | i \in I \} \mid * >$ in conjunction that the set of all properties of a system S is closed under composition $*$ that is that we have full description of a system S.

3 Background

3.1 Axiomatic System's Description

To build a formalization of the innovation system one must set the axiomatic system's description. At first let's consider the formalization of a system's goals. The system's goals can runs as follows:

1. New status S_α of a system S,
2. New status Q_α of a system Q, on which system S affects.
3. New status U_α as a result of an external agency of a system S upon outdoor environment U, $\alpha \in A$.

As a result, we have the target vectors $\langle S_\alpha, Q_\alpha, U_\alpha \rangle$, $\alpha \in A$, and a final status S_f of a system S, corresponding to the vector $\langle S_f, Q_f, U_f \rangle$, in which the system S moved as a result of it's functioning. The following cases are possible:

(a) $2 \Rightarrow 3$,   (b) $3 \Rightarrow 2$,   (c) $3 \Rightarrow 1$,   (d) $1 \Rightarrow 3$.

The graph $G(S)$ – the impact of targets of a system S upon the environment U – runs as follows (Fig. 1):

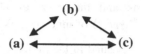

Fig. 1 The graph $G(S)$

Based on the introduced notations and concepts one can go to the formalization of an axiomatic description of the purposes and principles of the system approach. Generally adopted axiomatic (even verbal one) of the system approach or of the theory of systems does not exist. Usually the following main principles are chose while describing the system: the presence of a target (set of targets), integrity (wholeness), hierarchy, good structure. Basic system's principles are opened in the following way. Integrity means fundamental irreducibility properties back to the sum of the properties of its constituent elements and revivalist of the last properties of the whole, the dependence of each element, properties and relationships of the system from his place, functions, etc. within the whole.

Good structure means the ability to describe the system through the establishment of its structure, i.e., network connections and relations of the system, the dependence of the system behavior from the behavior of its individual elements and the properties of its structure, the interdependence of the system and environment.

Hierarchy means that each component of the system can be considered as a system and the analyzed system represents one component of a broader system in this case.

The adequate knowledge of the system requires building many different models, each of which describes some aspect of the system because of the principal complexity of each system.

Let's describe main system's principles using narrow predicate calculus; it runs as follows.

The principle of integrity is set to items 1 and 2:

1. $\vDash \bigwedge_{i \in I} P_i(\langle S_\alpha, Q_\alpha, U_\alpha \rangle, \alpha \in A) \bigwedge (\exists P)(P(\langle S_\alpha, Q_\alpha, U_\alpha \rangle,$

   $\alpha \in A) \bigwedge (\lnot P(\langle S'_\delta, Q'_\delta, U'_\delta \rangle, \delta \in \Delta)$

2. - system S possesses all the properties $\{P_i | i \in I\}$, and there exists at least one property P, such that no one of own subsystems S' of the system S does not possessed P. Thus, the property of integrity allocates system as one having a synergistic effect (at least one).

3. The graph of all relations $\Gamma(S)$ of the system S is isomorphic to the graph $\Gamma(S_f)$ of all links of the system S_f – the final status of the system S.

4. The system structure is understood to be the lattice of its subsystems. Good structure principle is set to item 3:

5. Let $\{a_\beta | \beta \varepsilon B\}$ be a set of all inner factors acting on the system S that is determine its behavior. If $a_1 * a_2 * \ldots * a_{n_\gamma} = a$, then $f_\gamma^{n_\gamma}(a_1, a_2, \ldots, a_{n_\gamma}) = a$, where $\{f_\gamma^{n_\gamma} | \gamma \varepsilon \Gamma\}$ is the set of operations on the set of factors $\{a_\beta | \beta \varepsilon B\}$, $f_\gamma^{n_\gamma}$ is n_γ-dimensional operation. There exists one to one correspondence between the

class of all real systems and the class of all finite algebraic systems $I(S) = \langle \{a_\beta|\beta\varepsilon B\}|\Omega_S = \{f_\gamma^{n_\gamma}|\gamma\varepsilon\Gamma\}\rangle$ of inner factors of systems.

The principle of hierarchy revealed in Definition 3.

3.2 Formalization of the Goal of a System

Let's formalize the goal of a system. The goal of the system S can be specified by a single or n-dimensional predicate Q in the following way:

$$\bigwedge_{i\in I} P_i \Rightarrow Q$$

or

$$\bigwedge_{i\in I} P_i(S) \Rightarrow Q(A_1, \ldots, A_n)$$

where $A_1, \ldots, A_n$ belong to the domain of predicate Q. Numerical characteristics of a system S or structural characteristics of a system S can be changed but the resulting modified system S' may satisfied P which is the property of the integrity of the system S. That is $P(S) \Rightarrow P(S')$, and therefore, to perform almost all the same functions as the system S, and, ultimately, to achieve the goal Q. Thus the numerical change or structural change of the system S within certain limits does not violate its integrity. This property will be call quasi sustainable one.

Definition 4. The system S is quasi sustainable one in relation to integrity property P if there exists the system S' and the congruence $\equiv_P$ on the system S' such that factor- model $S'/\equiv_P \cong S$.

To determine the class of innovation systems and to study it one should make following definitions. Now let's define the attribute characteristics of the system.

Definition 5. Let Q be a target of a system S. The set of predicates $\{P_\alpha|\alpha \in \Lambda\}$ is called internal attribute signs of a system S if the formula

$$\bigwedge_{\alpha\in\Lambda} P_\alpha \Rightarrow Q$$

is an identically true formula for S.

Definition 6. Let Q be a target of a system S. The set of predicates $\{P_\beta|\beta \in B\}$ is called external attribute signs of a system S if the formula

$$Q \Rightarrow \bigwedge_{\beta \in B} P_\beta$$

is an identically true formula for S.

Definition 7. Let Q be a target of a system S. The set of predicates $\{P_\gamma | \gamma \in \Gamma\}$ is called attribute signs of a system S if the formula

$$Q \Leftrightarrow \bigwedge_{\gamma \in \Gamma} P_\gamma$$

is an identically true formula for S.

4 Main Results: P – Innovative System and P-Pseudo-Innovative Systems

In [8] the notion of P-innovative system was introduced; it runs as follows.

Definition 8. Let S be a system with integrity property P. Then S is called an innovative system with deciding integrity property P if S can be off line in every super system S' such that S is P - pure in S', that is containing it in a way that is not distorted P - connections.

An innovative system S with integrity property P differs from a system with integrity property P higher (not lower than in super system containing it) implementation performance numeric innovative properties, or the fact that analogues of a system with property P does not exist.

The main indicators of innovation are: newness, the degree or level of newness, consumer value, degree of implementation in practice, effectiveness, the presence of a single indicator of efficiency, the phenomenon of "flash" (a synergistic effect) characterizing the beginning of the autonomous work of the innovation system, the graph of all links of all innovations.

Let's consider examples of innovative systems:

IT – technologies, autonomous work in digital libraries, expert systems and so on.

Now let's formulate and prove some theorems concerning innovative systems.

Let P be a predicate defining the integrity property of a system S. We shall consider P-pure embeddings in the class of all systems. From [8] we have the following main theorem.

Theorem 2. An innovative system S with full realizations of P –connections not distorting P –connections of containing it super system is off line.

Definition 9. Let $S = \{S_i | i \varepsilon I\}$ – be a split of a main set of a system $\hat{A}$ with integrity property P, and $P(S_i)$ is true for every $i \varepsilon I$. The intersection $\bigcap_{i \varepsilon I} S_i = K$ is called the P–kernel of a system $\hat{A}$. As P is closed under intersections then $P(K)$ is true.

In practice $\{S_i|i\varepsilon I\}$ are different spheres of functioning operating system $\hat{A}$. For example, in the field of learning processes (learning processes) one can distinguish the following areas: - the range of subject areas, in which, for example, are studied discipline with interdisciplinary linkages, and the like; - the field of learning technologies, in which, for example, the following technology and generalized educational technologies are included: problematic instruction, concentrated training, developing training, modular training, differentiated instruction, active learning, 's training, and so on; - the field of exit (field performance).

So the P-kernel of a system is essentially its infrastructure.[1]

In practice the spheres that are affected when the system $\hat{A}$ is functioning have as a rule non-empty intersection, and thus $\{S_i \cap A|i\varepsilon I\}$ is not a split of A. So one can consider $S' = \{S_i'|i\varepsilon I\}$, where $S_i' = S_i \setminus \bigcap_{i\varepsilon I} S_i$, $i\varepsilon I$, instead of $S = \{S_i|i\varepsilon I\}$. Then $\bigcap_{i\varepsilon I} S'_i = \varnothing$ and $\bigcup_{i\varepsilon I} S'_i = A$ and $S' = \{S_i'|i\varepsilon I\}$ is a split of A.

Theorem 3. For a subsystem $\hat{B}$ of a system $\hat{A}$ to be an innovative one it is necessary that the main set B of the system $\hat{B}$ contains P-kernel of every its supersystem.

Let's introduce the notion of P-pseudo- innovative system following the principle of duality.

Definition 10. Algebra $\bar{G} = \langle G|\{f_\alpha^{n_\alpha}|\alpha \in \Gamma\}\rangle$ is called P – pure projective (P-pseudo- innovative system) if every diagram with exact P - pure string

$$\bar{G} = \langle G|\{f_\alpha^{n_\alpha}|\alpha \in \Gamma\}\rangle$$

$$0 \to \bar{B} = \langle B|\{f_\alpha^{n_\alpha}|\alpha \in \Gamma\}\rangle \xrightarrow{\alpha} \bar{A} = \langle A|\{f_\alpha^{n_\alpha}|\alpha \in \Gamma\}\rangle \xrightarrow{\pi} \bar{C} = \langle C|\{f_\alpha^{n_\alpha}|\alpha \in \Gamma\}\rangle \to 0$$

with arrows φ and ψ

that is Im α is P-pure in $\bar{A}$ can be extended to commutative one that is $\psi\pi = \varphi$.

Theorem 4. Direct (inductive) limit of P-pseudo- innovative systems is P-pseudo-innovative system.

The proof of the theorem 3 pass like the corresponding ones in [6].

Let's return to the consideration of external and internal attribute characteristics of the system S with the goal Q. While considering an internal characteristic $P_\alpha, \alpha \in \Lambda$, one obtains P_α – innovative systems, $\alpha \in \Lambda$, describing system S with the goal Q. While considering an external attribute characteristic $P_\beta, \beta \in B$, one get P_β – pseudo-innovative systems, describing system S with the goal Q.

---

[1]Infrastructure is a set of interrelated service structures or objects, components and/or provide the basis for the functioning of the system.

Main result. We obtained the following result: Logic operation of implication explains the principle of duality in classical mathematics, so the rule modus ponens $A \vee \neg A$ beacause the following formula is identically true $\neg p \vee q$.

This raises the following question: what is the principle of duality in multivalued logic?

Let us consider classifications of educational systems as an example.

5 Main Results: Examples of Classification in Learning Process

Education in the wide sense is the process of transmission, as a rule, subsequent generations of accumulated information, followed by quality control. During this transfer process the following conditions must run as a minimum: completeness of transmitted data, transmission how to use transmitted information, the efficiency of transmission, storage, processing and usage of transmitted information, feedback: control of mastering of the information received. If we are speaking about education in the classical school, high school, university education in human society one must add a new condition to this definition that is human factor. The important part in the education as in wide and as in narrow sense of the word is the organization of the process of information transmission.

So we shall consider learning process for system S management. We propose to classify learning process for system S management by it's external and internal attribute characteristics. Thus we have classification by:

- the presence of human factor (common systems of education or peoples' systems of education),
- internal attribute characteristics of a system,
- external attribute characteristics of a system,
- levels of complexity of a system (for example, bachelor, master, Ph.D., etc. for systems with the presence of human factor),
- internal links' management of a system,
- external links' management of a system,
- integrity property P,

Also one can consider classifications in the dynamics while the system is getting new properties.

6 Conclusions

This paper presents the obtained research outcomes of application of the developed approach to a formalization of the concept of a system on the basis of methods of the Theory of Algebraic Systems, Model Theory and Group Theory.

The relevance of this work is defined by:

– a need to improve quality of the control system of students' knowledge;
– a need to optimize the structure and content of the training materials, the need to view the training materials as a system of knowledge;
– a need to develop and introduce learning and knowledge control expert systems in educational process;
– an expansion of network and distance learning.

The novelty of this work includes the following:

– the proposed formalization of the notion of a system based on algebraic systems;
– the characterization of a system by internal and external attribute signs of the system associated with the target of the system, which allows to distinguish between significant and non significant factors determining the possibility of achieving the target of the system;
– in the introducing the concept of pseudo-innovation system and identification of its properties.

The results of this work should be useful in practice of :

– creating and improving expert system training;
– creating and improving expert systems for quality control in education;
– improving the integration of classical learning with learning based on the use of IT technologies [7].

References

1. Mesarovich, M., Takahara, Y.: General system theory: mathematical foundations. In: Mathematics in Science and Engineering, vol. 113. Academic Press, New York, San Francisco, London (1975)
2. Maltcev, A.I.: Algebraic Systems, 392 p. Moscow, Nauka (1970) (in Russian)
3. Kurosh, A.G.: Theory of Groups, 648 c. Moscow, Nauka (1967) (in Russian)
4. Serdyukova, N.A.: On Generalizations of Purities, Algebra & Logic, 30, № 4, pp. 432–456 (1991)
5. Serdyukova, N.A.: Optimization of Tax System of Russia, parts I and II. Budget and Treasury Academy, Rostov State Economic University (2002). (in Russian)
6. Serdyukova, N.A., Serdyukov, V.I.: The new scheme of a formalization of an expert system in teaching. In: Proceedings of the ICEE/ICIT 2014, paper 032, Riga (2014)

7. Monitoring and evaluation of research in learning innovations, Final Report of project HPHA-CT2000-00042 funded under the improving human research potential & the socio-economic knowledge base directorate general science, Research and Development European Commision (Merlin)
8. Serdyukova, N.A., Serdyukov, V.I., Slepov, V.I.: New formalization of the system approach as a tool for study of innovation processes in the IT and Engineering Pedagogy, in print

Forming and Self-evaluation of ICT-Competence of Mathematics Teachers to Be in the Course of Their Professional Training

Irina G. Lipatnikova and Alexander P. Usoltsev

Abstract The article deals with solving the problem of forming and self-evaluation of ICT-competence of mathematics teachers to be, during their professional training. Is made the analysis of preparation's standards of mathematics teachers to be in the foreign countries (Spain, Turkey, North Caroline), which bears evidence of advantages of mastery of ICT as an instrument of searching, interpretation and using information; as a source of knowledge and communication. On the basis of experience of mathematics teachers to be' preparation in foreign countries and recommendations made by UNESCO in the partnership with world leaders in the field of creation of information technologies and leading experts in the field of school's informatics UNESCO's ICT Competency Framework for Teachers (further UNESCO ICT-CFT, formulated demands to ICT competence of mathematics teachers to be and observed ways of its forming. Are proposed methods of realization of self-evaluation ICT- competence by students. Is shown the analysis of results of it's using by the students of pedagogical university –mathematics teachers to be.

Keywords ICT- competence · Professional preparation · Structure of ICT-competence · UNESCO's recommendations · Self-evaluation

1 Introduction

Innovation changes, taking place in modern society and around the world, exerted a big influence on the role and priorities as school as university education, particularly on the mathematics teachers' training. Aptitude for self-development, self-education, mathematics teacher's preparation to the educational work in the

I.G. Lipatnikova (✉) · A.P. Usoltsev
Ural State Pedagogical University, K.Libkhnenta 9, Екаterinburg, Russia
e-mail: lipatnikovaig@mail.ru

A.P. Usoltsev
e-mail: ausolzev@gmail.com

© Springer International Publishing Switzerland 2015
V.L. Uskov et al. (eds.), *Smart Education and Smart e-Learning*,
Smart Innovation, Systems and Technologies 41,
DOI 10.1007/978-3-319-19875-0_34

conditions of information of education; aptitude for using information tools and information and communication technologies for solving professional tasks are becoming the most important components of his professional competence.

However, the analysis of federal state educational standards of higher education, revealed, that the professional preparation of mathematics teachers to be in pedagogical universities is oriented mainly on the forming of his computer-competence as the user, but not as a teacher, which use information communication technologies effectively in professional activities.

Reviewing researches of teachers' readiness in foreign countries (Spain, Turkey, USA) testifies about advantages of mastery of ICT as an instrument of searching, interpretation and using information; as a source of knowledge and communications. Functional importance of ICT is reflected in the following requests to the teachers' readiness, when the teachers can use ICT:

- for work with information, reflection's organization, solving tasks and producing new knowledge;
- for help in self-realization of each student;
- for effective management of his reality's trajectory;
- for students' preparation to the valuable life in modern society;
- for intercultural interaction and resolving conflicts in a peaceful way [1].

Achievement of these demands is impossible without corresponding preparation of teachers, particularly mathematics teachers in the field of using ICT.

2 Fundamental Theory of Demands to ICT-Competence of Mathematics Teacher

It's necessary to underline that determining the level of professional activities in the ICT, the term «information and communication technologies - competence» is used.

Use of different approaches to the definition of the notion «ICT teacher's competence» in modern literature testifies about its various aspects. I.G. Ovchinnikov, underlines that, this competence is a «teacher's information culture» and considers it as a component of general culture, connected with functioning of information in the society [2]. E.K. Henner writes, that ICT is «competence of body of knowledge, skills and abilities, which are formed during the course of education and self-education to the information technologies and the ability to realize professional activities with the use of information technologies» [3].

Dynamism of development of information society doesn't demand the studying of special programmatic instruments, but mastering the essence of ICT by mathematics teachers to be, possibilities of usage these technologies during the course of teaching mathematics and psychological-pedagogical and methodical foundation for their use.

In the frame of professional preparation of future teacher we needed to reveal the demands to the forming ICT- competence of mathematics teacher to be and evaluate it from the point of view of preparation of availability to use it in professional work.

Understanding of the significance of ICT-competence of teacher to be in the modern world has proved the need of creation international recommendations, which set demands to ICT-competence of school teachers or (university lectures), developed by UNESCO in the partnership with world leaders in the field of creation of information technologies and leading experts in the sphere of school's information UNESCO's ICT Competency Framework for teachers (further – UNESCO ICT–CFT, or recommendations). Recommendations include competences, which are needed to the teachers in all aspects of their work:

1. Comprehension the ICT' role in education.
2. Curriculum and evaluation.
3. Pedagogical practical works.
4. Technical and program ICT' tools.
5. Organization and administration of educational process.
6. Professional development [1].

Three approaches (Fig. 1 shows it) to the school information were taken into consideration during creating Recommendations.

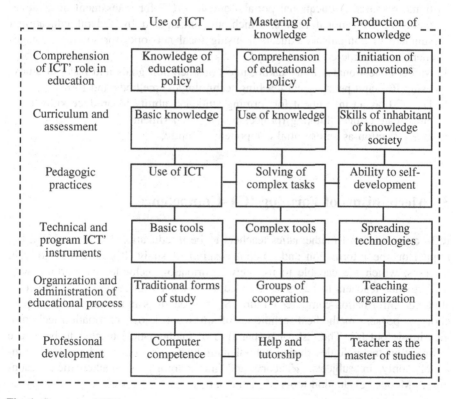

	Use of ICT	Mastering of knowledge	Production of knowledge
Comprehension of ICT' role in education	Knowledge of educational policy	Comprehension of educational policy	Initiation of innovations
Curriculum and assessment	Basic knowledge	Use of knowledge	Skills of inhabitant of knowledge society
Pedagogic practices	Use of ICT	Solving of complex tasks	Ability to self-development
Technical and program ICT' instruments	Basic tools	Complex tools	Spreading technologies
Organization and administration of educational process	Traditional forms of study	Groups of cooperation	Teaching organization
Professional development	Computer competence	Help and tutorship	Teacher as the master of studies

Fig. 1 Structure of ICT -competence of teachers. UNESCO's recommendations

These approaches are connected with the appropriate stages of teachers' professional development, which give opportunity to master practical activities in ICT spread educational environment. The first approach «Use ICT» demands from teachers the ability to help students to use ICT for increasing the efficiency of their cognitive activities. The second approach «Mastering of knowledge» - demands teachers' aptitude for helping students in thorough understanding of content of subjects, using acquired knowledge in solving complex tasks in real world. The third approach «Production of knowledge» demands teacher's aptitude for helping students –citizens and employees to be to produce new knowledge, which is in great demand for balanced development and society's prosperity.

Summing up different approaches to the revealing the notion of ICT – competence of teacher-to-be, it's necessary to define demands to the ICT- competence of mathematics teacher to be.

1. Comprehend role and place ICT in the mathematics course teaching (increasing students' motivation during the course of teaching, increasing of efficiency of study, activation of students' cognitive sphere, providing operative feedback with students.
2. Select, evaluate, create and use ICT' tools for increasing mathematical skills (trainers for trying-out mathematical skills, teaching programs, digital educational resources, educational portals) and use ICT for assessment of achievements of educational results, which are anticipated by federal educational standard of compulsory education (using local networks for solving management tasks in the course of study).
3. Model educational activity taking into consideration goals and tasks of study, using ICT and pedagogical teaching technologies, methods and forms.
4. Use ICT as an instrument for forming students' ability to produce knowledge and create critical thinking (project activity, experiment).
5. Use reflection as an essential component of study.

3 Mechanisms of Forming ICT-Competence

ICT- competence of mathematics teachers to be in educational field, makes a big impact on the information and computerization of study. It's necessary to train teachers, which are capable to use new information technologies working with mathematics teachers to be, give knowledge students with technologies and form students' information-computer culture. Such specialists must know child's psychology, possess methodical techniques of study and know information technologies inside out [4]. The task of preparation of mathematical teacher to be to use information technologies during school mathematics course can be solved completely only in subjects: «Theory and methodology of mathematical teaching», «Modern evaluation tools of mathematics' teaching results».

During studying these courses, it is possible to realize forming of professional competence of mathematics teachers to be in ICT field:

1. On the basis of step-by-step transformation of student's training in quasi-professional activities of mathematics teacher. To achieve this goal is necessary to use computer as a tool of mathematical activity. For example, students are proposed to work out presentation for the lecture, work out and conduct practical lesson on the specific theme, create electronic teaching resources, computer tests for assessment of achievements of educational results [5].

2. Due to the using case study methods, which include modeling, systems analysis, problematic method, game method and suppose mastering ICT' instruments, mathematics teachers to be can see constantly appropriateness and necessity of using ICT' instruments resolving professional situation. Using case study gives opportunity to separate out goals and areas of application ICT for analysis of special professional situation:

   - knowledge retention, acquired during theoretical courses;
   - skills training of practical using of conceptual scheme of theoretical material and explain to the students schemes of analysis of practical situations;
   - skills training of group analyses and making decision during training courses;
   - examination of knowledge, acquired during theoretical course.

   Carrying out laboratory works and using case technologies are created the conditions to acquire knowledge and development of professional qualities of students' personality, which can lead to creative self-realization during student teaching and professional work.

3. Organizing active self-education work, mathematics teacher to be, mastering ICT, can use their potential and find different ways of their application in future professional activities. It can be organized through studying special units of training course «Modern evaluation tools of mathematics' teaching results», using electronic educational resources, doing home assignments and laboratory works, solving professional tasks, working out of easy programs, using systems of computer mathematics.

Besides, information of education includes development and application of the material support of the educational process on the basis of information technologies. The structural components of these technologies consist of technical devices, program and educational maintenance. Using material support during the process of preparation mathematics teachers to be gives opportunity to demonstrate them characteristic properties of computer technologies.

As an informational component, which allows provide a substantial aspect of preparation of mathematics teacher to be in pedagogical university, we propose electronic integrated training course, which includes:

1. Study guide, including instructor's manual, navigation in compulsory and supplementary materials and summary of the units of the course.

2. Manual containing methodical and psychological tools of study, self-assessment and techniques of work with information, complex of professional tasks and problematic situations for each unit, which allow modelling activities of mathematics teacher to be.
3. Supplementary teaching materials, including publications in scientific journals, sites, references to internet resources, internet sites, workshop with using of applied programs.
4. Computer-aided system of assessment and control of knowledge, which is realized as control-training programs. Such programs allow students to evaluate acquired knowledge by themselves.

Offered electronic integrated training course «Modern evaluation tools of mathematics' teaching results» can be presented to mathematics teachers to be. This course can be realized by means of network technologies and other devices such as compact discs. Potential of electronic integrated training courses allow unite different materials on the multimedia basis. These materials can differ in functions, content and form, levels of students' preparation.

4 Self-evaluation of ICT-Competence of Mathematics Teachers

The system approach to the forming of ICT-competence of mathematics teachers to be includes besides substantial and procedural part, well-founded self-evaluation mathematics teachers to be to use ICT in professional work. Self-evaluation readiness to use ICT is the most important tool for checking of mathematics teachers' competences in ICT field. Self-evaluation helps choosing well-founded ways of removal of imperfections of individual educational trajectory and serves as the basis for making effective administrative decisions [6].

For subjective estimation of ICT competence of mathematics teachers to be, was made a questionnaire, were respondents were proposed to evaluate general level of their competences in ICT sphere, according to the scale

(a) quite sufficient
(b) sufficient
(c) can't say
d) insufficient
(e) absolutely insufficient

Formula (1)

$$I_k = \frac{a + 0.5b + 0c - 0.5d - e}{N} \tag{1}$$

where a, b, c, d, e – number of students, which choose the scale's grade, N – the total number of respondents [7, p. 29].

During research were calculated indexes of self-assessment ICT- competence I_k. Using this technique of calculations $+1$ shows maximum, -1 –minimal level of ICT- competence.

51 students took part in research – mathematics teachers to be of institute of mathematics, information and information technologies of Ural state pedagogical university: 19 students of fourth and fifth year, 10 students of the third year, 22 s year students. Quantitative distribution of students according to the subjective level of forming ICT competence is represented in Table 1.

Table 1 Quantitative distribution of students, according to the definition of subjective level of forming ICT- competence

Index	Quite sufficient		Sufficient		Can't say		Insufficient		Absolutely insufficient	
Number of students (5 and 4 years)	5	26.32 %	7	36.84 %	3	15.78 %	2	10.53 %	2	10.53 %
Number of students (3 year)	2	20 %	4	40 %	2	20 %	1	10 %	1	10 %
Number of students (2 year)	2	9.09 %	10	45.45 %	7	31.82 %	2	9.09 %	1	4.55 %

According to the formula (1) were calculated indexes of self-assessment of students' level of forming ICT competence.

$$I_k = \frac{5 + 0.5 \cdot 7 + 0 \cdot 3 - 0.5 \cdot 2 - 2}{19} = 0.29 \quad 5^{th}, 4^{th} \text{ year students} \tag{2}$$

$$I_k = \frac{2 + 0.5 \cdot 4 + 0 \cdot 2 - 0.5 \cdot 1 - 1}{10} = 0.25 \text{ 3th year students} \tag{3}$$

$$I_k = \frac{2 + 0.5 \cdot 10 + 0 \cdot 7 - 0.5 \cdot 2 - 1}{22} = 0.23 \text{ 2}^{nd} \text{ year students} \tag{4}$$

It's evident, that ICT competence of students during the process of studying at the university is increasing without purposeful methodical teachers' training. Self-evaluation's analysis shows that in general self-evaluation of ICT competence of mathematics teachers to be is increasing not so fast, as it was expected according to the objective increasing of the result. It bears evidence that in the beginning of the research, on the second year students demonstrated over-evaluation, at the end of study it became more adequate. Appropriate self-evaluation is main and determinant factor of subsequent self-development of young specialist. Empirical data concerning self-evaluation can be considered as an important evidence of successful work with students.

Besides, mathematics teachers to be were proposed to evaluate functions of information-communication technologies in process of teaching mathematics according to the 5 grades scale:

1. Increasing the student's motivation in process of mathematics teaching.
2. Increasing the efficiency of academic course.
3. Increasing of activation of students' cognitive work.
4. Evaluation and observation study's results.
5. Tool for self-education.
6. Tool for lessons' preparations.

Weigh of each nominated functions was found by the way of calculations of arithmetic average and is represented on the Fig. 2.

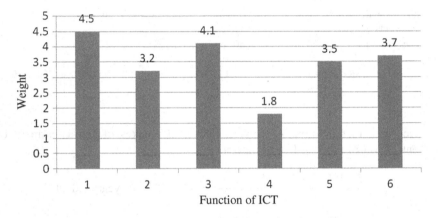

Fig. 2 Average weight of functions of information communication technologies, evaluation of their significance by students

Mathematics teachers to be, consider that the most important information and communication's function is increasing students' motivation during the course of mathematics teaching. It was rather unexpectedly that the functions «evaluation and observation tutorage effects». However, after holding a discussion with students it became evident the problem of lack of coordination of demands to ICT' mastering. Demands are represented in the form of knowledge, abilities, skills of students and (are formulated in author programs in learning ICT in the course of mathematics teaching). Program materials have to satisfy definite demands. This discrepancy makes impossible using all multimedia for control and measuring the level of knowledge in the course of learning. Self-development of new testing programs is complicated work and is not always justified.

5 Conclusions

Forming ICT-competence of mathematics teachers to be is an important task to be solved for training students to the professional activities in modern conditions of education.

Forming this competence can't be achieved only in the course of informatics. The main role in readiness to effective usage ICT in professional pedagogical activity, play such courses as «Theory and methodology mathematic teaching», «Modern evaluation tools of mathematics' teaching results».

Course «Modern evaluation tools of mathematics' teaching results» must be electronic, because it allows organize self-cognitive student's activity, their reflection on the basis of modern methods of organization of study (case-technologies, technologies of forming critical thinking etc.).

Organization of self-evaluation readiness of mathematics teachers to be to use ICT in professional activities is shown: correctness of students' self-evaluation of their ICT-competence; comprehension the significance of getting rid of gaps in professional preparation in ICT field; readiness to make self-improvement in this field.

References

1. UNESCO ICT Competency framework for teacher. Version 2.0. UNESCO. UNESCO ICT. http://ru.iite.unesco.org/publications/3214694/ (2011)
2. Ovchinnikova, I.G.: Information-reflexive approach in the process of development of information culture of students. In: World of Science, Culture, Education, pp. 191–193. Moscow (2009)
3. Henner, E.K.: Forming ICT-Competence of Students and Teachers in System of Lifelong Education, pp. 135–137. BINOM, Laboratory of Knowledge, Moscow (2008)
4. Lipatnikova, I,G.: Modern Evaluation Tools Study's Results. Manual, Ekaterinburg (2010)
5. Lipatnikova, I,G.: Mechanisms of forming information competence of pedagogical university student's in the mathematics teaching course. Education and Science, pp. 104–114. Moscow (2012)
6. Lipatnikova, I.G.: Creation of individual educational trajectory as one of the ways of teaching students to the techniques of making decisions. Fundamental Research, pp. 108–110. Moscow (2010)
7. Diagnostics of conditions of actual problems of mathematics education, pp. 155–156. Collective monograph, Rostov on the Don (2014)

Raising the Level of Future Teachers' Professional Competence in the Conditions of Informational and Educational Environment

Irina V. Rozhina, Anna M. Lozinskaya and Tamara N. Shamalo

Abstract This article discusses the guidelines of the school teachers' professional work in the conditions of informational and educational environment (IEE) and competencies required for its realization, which are grouped into clusters: scientific-theoretical, constructive-designing, organizational-methodical and professional-personal. The ways of increasing of levels of development clusters of professional competence are suggested. They are based on the module-rating technology with techniques of cognitive visualization of teaching material with the use of frames, principles of learning "process is more important than the result" and "learning through teaching", video case lessons which are included in the content of IEE, active learning teaching methods and the technology of mixed teaching.

Keywords Teacher's professional competence · Clusters of competencies · Model of competencies · Development techniques · Technologies of mixed teaching · Competencies formation's assessment

1 Introduction

Teachers' training is associated with the formation and development of professional competencies which are not only multifaceted but multifunctional and dynamically developing items. The achieving of these educational results is directly related to the realization of educational activities which is defined by the use of innovative educational technologies, methods, organizational forms and means of teaching.

I.V. Rozhina (✉) · A.M. Lozinskaya · T.N. Shamalo
Ural State Pedagogical University, Yekaterinburg, Russia
e-mail: irozhina@gmail.com

A.M. Lozinskaya
e-mail: anna-loz@yandex.ru

T.N. Shamalo
e-mail: shamalo@uspu.ru

© Springer International Publishing Switzerland 2015
V.L. Uskov et al. (eds.), *Smart Education and Smart e-Learning*,
Smart Innovation, Systems and Technologies 41,
DOI 10.1007/978-3-319-19875-0_35

393

To improve the efficiency of processes of teaching and evaluation of educational results it is advisable to use the competency model as a basis. This model must meet the following requirements: meet the strategic goals of education; be informative and understandable to all subjects involved in educational process; include an optimal set of components; contain the system of registration and measuring of competencies.

The analysis of the existing practice of professional work of teachers shows that even the high level of subject-methodical training of teachers does not provide the result expected by society. To realize the new goals and content of education school teachers need to be trained to work in the IEE, the new didactic opportunities of which create the conditions for the implementation of innovative educational technologies. According to the law "About Education in the Russian Federation" the organizations engaged in educational activities must form the IEE (wholly or partly by the means of distance learning technologies and e-learning) which will ensure the realization of educational programs and provide the access to the necessary educational resources for all students.

Thus in modern information society the main emphasis in professional education should be given to the mastering of methods of activity in the IEE within the competence approach.

2 Didactic Characteristics of IEE

Following Starichenko B.E. [1], we define the IEE as a set of hardware, software and content which is realized on the basis of modern technological mental models and designed for fully and promptly meet the information needs of all subjects of the educational process related to implementation of the forms and learning activities, as well as organization of information flows associated with teaching and education management.

The modern IEE has a number of didactic characteristics, among them are the following: flexibility; integrity; openness; variability; multifunctionality; interactivity; advanced visualization; operational control of educational achievements; access to the various sources of educational information; the possibility of organizing individual pupils' work, to develop their cognitive independence and creativity by means of ICT, the use of new pedagogical tools for solving educational problems (thus expanding the range of tasks), the transition to a fundamentally new model of the process and objects for analysis, research and experiment.

It is IEE that can attach to the learning process such qualities that will ensure the achievement of educational results demanded by modern society. In this case, the readiness of teachers to work effectively in the IEE is largely determined by their understanding of necessity for significant restructuring of the educational process.

3 Functional Analysis of Professional Work and Competence of the Teacher

The pedagogical activity is inherently multidisciplinary and multifunctional one, which is associated with problems of education, training, and development of competitive personality. Let us briefly consider the guidelines of professional work of school teacher and competencies required for their implementation.

Complex diagnostics pedagogical conditions. The competencies related to this group allow the teacher to identify relevant educational approaches for individual learning and use their observation skills, a variety of media resources (including IEE) and basic knowledge of child development to identify strengths and weaknesses of educational strategies in specific pedagogical conditions.

Development of educational goals and objectives. Skills in this group include the development of education and training purposes in terms of the observed performance and/or behavior on the basis of which will be formulated task and assessed educational results. **Analysis of learning tasks**. Within the formulated aims and defined tasks the teacher should be able to determine the necessary training activities and the adequate teaching methods. **Selection, methodical correction and use of training materials**. The modern teacher must be aware of the wide range of educational media, teaching materials and selection criteria, as well as about methods of modification; be able to attract and adapt to the educational process material from other educational institutions and related areas of knowledge and activity; also be able to use and develop electronic educational resources, have a culture of working with information on the Internet.

Selection, systematic correction and the use of technology and learning strategies. Competencies required for the selection and use of appropriate instructional strategies require awareness of teachers about the various training procedures, methodologies and technologies available for didactic engineering and efficient management of the learning process, as well as possessing the ability to adequately relate the goals and capabilities of students with these strategies for making the most optimal solutions.

Monitoring and evaluation of education. This group of competencies includes the skills of education management, design of test materials based on the existing curriculum, development and application of methods for monitoring and evaluation, including with the use of information and communication technologies and informational-educational environment. **Use of resources**. The competencies related to the ability to use the resources (material, information, technical and technological, human and other) and maintain favorable relationships with them, allow the teacher to develop and use specialized training courses, unique learning technology, learning tools and programming environments; offer and receive educational and related activities with the involvement of relevant resources.

The managing of behavior. The group characteristics of the managing behavior include the mastering of ways to strengthen the desired behavior, the formation of new behaviors and replace them manifestations of undesirable behaviors.

The teacher should own methods of behavior modification, establishing the boundaries of acceptable behavior reactions, the use of developing techniques aimed at creating constructive behaviors. **Professional activities**. The competencies related to professional activities are based on the recognition of the need for constant self-improvement of the individual, professional development and renewal, lifelong learning, participation in professional organizations, and the use of lessons learned and knowledge base in the field of education for empirical research. **Knowledge of modern trends**. The teacher must possess not only the knowledge and skills in their field of education, but also know a lot of the context and the ways in which they can be applied. With its expertise in this area, the teacher more correctly assesses their strengths and weaknesses, as well as its role in the development of the professional field of activity and society. **Subject Education**. Of course, a qualified teacher knows what he teaches others. The content of this group of competencies covers the cognitive, affective and psychomotor areas of expertise.

Interaction with parents and students immediate environment. There is no doubt that parent involvement is very important in promoting the development of their children, parental attitudes and expectations affect the future of students. The competencies in advising parents require skills in interpersonal relationships and knowledge of the dynamics of human interaction.

In accordance with the determination of the structure of pedagogical activity in the work of the teacher in the conditions of the IEE are the following components: gnostic; designing; constructive; organizing; communicative.

4   Clusters of Professional Competence of Future Teachers

Competencies include elements such as knowledge, skills, and abilities and potential, relating to business and personal qualities. At the same time they are habitual patterns of thought, behavior, feelings or speech, the use of which makes a person successful in a particular job or role [2]. Competence can also be combined into clusters - sets of closely related competencies (usually 3–5), the consolidation of related indicators of performance, for the formulation of generalized requirements. In our study identified the main clusters of competencies in accordance with the functional content of teacher's professional activity: scientific and theoretical; structural and engineering; organizational and methodical; professional and personal.

Let us analyze briefly the components of the clusters of professional competence under the IEE, and challenges of their formation.

Scientific-theoretical competence: special education; formed interdisciplinary performances; formed scientific worldview.

One of the factors that determine the quality of education is the content of the special (subject) teacher competencies. They are adapted pedagogical system: scientific knowledge; methods of activity (the ability to act on the model); experience of creative activity; experience emotional and value attitude to nature, society and man. Obviously, the components of professional pedagogical competence of teachers of different disciplines have certain dominant, due to the specificity of the subject and its teaching methodology.

These competencies are manifested in the awareness of teachers about the basic ideas, notions, concepts in the subject areas of knowledge; formation of common (didactic) skills, intellectual abilities in an independent acquisition of new knowledge, tools and methods of cognitive activity and friend. They provide skills activities with the information contained in the academic subjects and educational areas, as well as in the surrounding world.

The teacher demonstrates a need for cognitive activity; ability to navigate and use various sources of information to generate new knowledge; possession of didactic abilities and skills of educational activities; form a holistic view of the world picture, choose their own ideological position; the ability to identify patterns in the basis to study science, norms, rules of social life.

Constructive-designing competence: pedagogical skills forecasting; skills development of modern education systems (innovation); skills development of teaching materials for training.

Engineering competence assume ownership of theoretical methods in the elaboration of holistic process and training sessions on the basis of progressive educational technologies.

Key role in the professional activities of the modern teacher playing skills of designing the educational process in the IEE, as a holistic, reflecting the relationship of all components (taskmgr, research, measurement, monitoring and evaluation of learning outcomes). IEE has its didactic opportunities that did not exist in the arsenal of teachers (flexibility, adaptability, variability of the environment, its transformability from one "version" to another, adjustability for solving various learning tasks). Transition teacher to work in IOS involves the study and analysis capabilities, methods, forms and means of education, characteristic of this environment, as well as learning activities students that achieve new educational results.

Organizational-methodical competence: willingness to teaching activities; willingness to use modern techniques and technology education; ability to work in a team and with the team.

In the context of the implementation of the educational process in IOS changed the nature of the interaction of the participants: students play the role of stakeholders, in contrast to the traditional educational environment, where he served as an object; the teacher's role and the content of his professional activity (in Gnostic, organizational, design, expert, reflexive components).

The content of teacher training to the design of the educational process in the IEE must include the required elements: the concept of information educational environment in the modernization of teacher education; component structure of the educational process in the IEE; technology of designing the educational process in IEE.

Professional and personal competencies: communication skills; management skills; readiness for self-development; possession of professional ethics.

Analysis of studies in the field of teacher training in the IEE showed that the modern teacher must possess the communicative competence, that is, a set of knowledge, skills and personal qualities, allowing to build effective interaction in the digital environment with other entities directly involved in the pedagogical process. Computer-mediated communication is not only limited (non-verbal, emotional), but also a number of didactic advantages, which include flexibility, efficiency, combining information and communication components, personal orientation.

According to a study of Zaslavskaya O.Y. the level of professional competence of a teacher is determined by the degree and level of development of managerial competence, which includes knowledge of management, the ability to implement anticipatory planning, modeling and prediction of the learning process, skills of management personal and student activities, the ability to reflect on and implementation of management functions, to effectively manage the use of information resources, to *implement* the management activities of the informatization of education, to self-improvement of professional competence and individual personality traits of increasing the effectiveness of training.

The teacher must comply with legal, moral and ethical standards: the teacher should be courteous, and attentive to students, tolerance and respect for the customs and traditions of the peoples, to take into account special needs of students and their health, promote the formation of a favorable moral and psychological climate for effective work, their personal behavior to set an example of integrity, impartiality and justice, and create a favorable impression self-disclosure of personal qualities in appearance - style clothing, general care.

Great influence on the quality of the implementation of professional functions teacher provides a manifestation of personal qualities in the system of social relations: a good orientation in the system of social relations and activities, tolerance, empathy, social mobility, integrity and personal identity, its readiness for self-education.

5 Instructional Techniques of Development Clusters of Professional Competencies of Future Teachers in the Conditions of the IEE

5.1 Development of Scientific-Theoretical Competences

In order to improve the learning process, the development of logical thinking and technical methods: (1) should be used problem-based learning (highly effective proved technique M.A. Choshanova based on the implementation approaches "the process is more important than the outcome" and "learning through teaching". The method includes the system of work with video case studies lessons: the task of the decision to view the video case; the task-pause while watching a video of the case; job-reflection after watching the video case) [3]. The use of problem-based learning transforms the received knowledge in the belief that contributes to their implementation, and thus their conscious assimilation; (2) a lot of attention given to the selection and justification of specific cognitive visualization techniques training content (we recommend the use of different models of frame structure and presentation of educational material [4]. Construction of the system data using frames is one of the innovative approaches to the development of teaching materials for students.

Cognitive development goals of educational content are determined using the process of designing educational material frames following models: frame; frame the logical-semantic scheme; frame script.

5.2 Development of Constructive-Designing Competencies

The study by Lapenok M.V. and Rozhina I.V. determined the structure and content of teacher training to the creation and use of the learning process school of electronic educational resources (EER) IEE according to established scientific and methodological approaches based on the use of active forms of learning and activity by: (a) role-playing games that simulate typical pedagogical situations inherent in teaching and cognitive processes using EER IEE; ensuring the formation of the experience of teachers in interactive information interact in real-time and delayed due to the use of instrumental services IEE; (b) teachers create educational resources with the didactic possibilities of instrumental services IEE and didactic, ergonomic and innovative requirements for EER IEE [5].

At the organization of the learning process focused on preparation of teachers to design the educational process in the IEE, it is necessary to take into account that educational process: (a) should be built so that the listener had the opportunity to discover the meaning of personal importance in the formation of professional competencies required him to work in the information educational environment; (b) is carried out in the form of activity-based and differentiated approach to the listeners of different categories and different levels of professional skills, creating the conditions for building individual learning paths in accordance with the professional needs and abilities of teachers; (c) provides the conditions for self-reliance and initiative activity of teachers able to demonstrate having experience with modern teaching tools and updating a variety of reflexive procedures; (d) can be built in a networking of educational institutions offering teacher training and offers a variety of methodological, information and human resources necessary for effective preparation of teachers to design the educational process in the IEE.

5.3 Development of Organizational- Methodological Competencies

In the learning process should be used innovative methods: learning based on information resources, associative methods, methods based on the use of artificial intelligence (the method of "forced assumptions" method precedent educational computer simulation, learning through teleconferencing method reification, and others); different operational modes of interaction, including teaching methods in cooperation (projects, forums, e-seminars), as well as learning technologies, which would create a situation involving students in meaningful professional activities: methodological research project, mini-teaching, information analysis, computer experiment, essays, discussions, etc. The evaluation system of educational achievements can be (Choshanov M.A., and others) through the active participation of students in fulfilling the requirements and tasks: participation in educational discussions on the main themes of the course presented in the IEE; analytical reviews of the Information Resources, which include online resources for learning various subjects school course that students shall be in accordance with the deadlines set out in the calendar of the course; reflection on video case school lessons.

5.4 Development of Professional-Personal Competencies

One method that permits the student to achieve academic goals, record growth dynamics over time, encourage students and for the results to guide the achievement of new and reveal the potential range of works, to ensure continuity of the process of training and personal development by stages of preparation is the method of portfolio. Method of portfolio aims to make the evaluation process manageable, focused. Recognized the high potential of project-based and research method as in improving the cognitive activity and creativity, without which it is impossible to conscious perception of the material.

The inclusion of the method of brainstorming and debate training in the implementation of the cluster allows you to create the ability to: formulate and defend their own point of view, to make conclusions, to build a chain of evidence, identify errors, analyze the information, to concentrate on the essence of the problem, work in a team. The use of active teaching methods helps to develop communication skills and to enable students in the organization of pedagogical process.

Note some of the common questions of construction of methodical system, aimed at improving the professional competence of future teachers in the conditions of IEE. (1) One of the most promising areas of modernization of the training is to use in the classroom modular technology, characterized by a high level of achievement of planned learning outcomes and their reproducibility, structural, content and technological flexibility of modular training programs. (2) The efficiency of modular technology is largely determined by the application of a rating system of education quality control [4]. It is important to note that the problem of development and application of a rating system of control of educational achievements in professional educational institutions in the organization of modular training has now acquired a special significance, including with the rapid development of electronic educational resources and the implementation of competence-based approach to education. (3) When the educational process to enhance professional competencies seems appropriate use of technology blended learning as it combines the advantages of distance learning and compensates for its shortcomings, in which different event-oriented techniques and learning management schemes, such as face-to-face learning, distance learning and on-line learning. In this training is based on the interaction of the listener not only with the computer, but with the teacher in the active form (full-time and distance) when the material studied independently synthesized, analyzed and used for the task. At the heart of our approach to the organization of blended learning is distance learning course and it integrates some of the methods of active learning is implemented on classroom sessions with students, when can be possible to combine group and individual, real and virtual forms as well as focused, intense and controlled independent work of students.

In our study it was found that for the diagnosis of formation of clusters of professional competence of future teachers can be successfully applied the proposed method to measure and evaluate the level of formation of competence in the use of tools IEE services [5], which were identified cognitive and operational (knowledge, skills,) and activity (experience) components of competence developed evaluation scale level of development of each component and the integral indicator of the level of formation of competence."

6 Conclusions

In the age of digital technology society needs not only teachers, but teachers and engineers in one person, who are responsible to integrate knowledge of content of education, educational psychology, resources of information technologies and skills of didactic design. Integration also includes the changing role of the teacher; new educational results can be achieved only in the process of mastering modern learning activities, i.e. in innovative educational process, built in IEE, which inevitably entails a change of the traditional teaching to the research-technique teaching of students. These transformations require the development of following teachers skills: (1) to develop new educational aims, results-oriented in a technologically enhanced information environment that allow students to define their own aims of education, monitor and evaluate progress in the teaching; (2) to design the content of education in the form of interactive content and related practical exercises by selecting and designing tasks, projects and activities with the use of digital resources and information technologies for the formation of learning experiences and the development of research, design and creative abilities of students; (3) to develop a system of monitoring and evaluation in accordance with the objectives and content of training for complex objective diagnosis of learning results, improve the quality of teaching and students' motivation for learning.

Increasing of the level of professional competence is possible due to the use of active methods, techniques and forms of learning activities: problem, heuristics, simulation and gaming, modeling, projective, discussion and others. Active learning promotes the transfer of knowledge and skills of students in the new situation, the study of new problems, the formation of the ability to see alternative solutions, combine the known methods and solutions to create new, original algorithms activities.

Development of pedagogy science towards the use of electronic didactic environments opens up new possibilities for understanding the problems and forms of education in the digital age, the creation of effective teaching methods.

References

1. Starichenko, B.: Conceptual Foundations of Computer Didactics. RIOR-SCIENCE (2013) (in Russian)
2. School Turnarounds Teachers: Competencies for Success. Public Impact for the Center for Comprehensive School Reform and Improvement. (http://www.publicimpact.com/publications/Turnaround_Teacher_Competencies.pdf) (2008)
3. Tchoshanov, M.A.: Engineering of Learning: Conceptualizing E-Didactics. Published by the UNESCO Institute for Information Technologies in Education. (http://iite.unesco.org/pics/publications/en/files/3214730) (2013)
4. Igoshev, B.M., Lozinskaya, A.M., Shamalo, T.N. (eds.): Module-rating technology as a means to improve the effectiveness of teaching physics: monograph. The Humanities. Center VLA-DOS, Moscow (2010) (in Russian)
5. Lapenok, M., Rozhina, I.: Teachers' training and comprehensive assessment of their educability level in the development and use of electronic educational resources. In: The Collection International Scientific-Practical Conference Innovations in Science, Technology and the Integration of Knowledge, pp. 113–122. Berforts Information Press, London (2014)

Pedagogical Practices to Teacher Education for Gerontology Education

Leticia Rocha Machado, Patricia Alejandra Behar and Johannes Doll

Abstract This article discusses possible educational strategies for teaching cyberseniors in Distance Education. The objective of this study was to delineate pedagogical practices that can contribute to teacher training in gerontology education. This need resulted from the need to discuss the increasing longevity of the population. This change in society poses new challenges for education. In this sense, distance learning can become a way to social inclusion, because of its many possibilities. Unfortunately, there are a few related studies, especially considering didactic and pedagogical activity for teachers. Thus, we conducted a study using both qualitative and quantitative approaches. It is based on offering extension courses for the training of professionals and individuals 60 years or older. For data collection we conducted participant observation, interviews, questionnaires and survey of technological productions of participants in a virtual learning environment. From the reports of the participants it is possible to map strategies for teaching and teacher training, including professionals who work or intend to work in distance education with elderly adults.

Keywords Component · Distance learning · Teacher training · Cyberseniors

1 Introduction

Elderly population has been growing in recent years, and this is due mainly by demographic change and an increased attention in aging process. This larger perspective of life amplified new cultural conflicts, and theories of aging process, since

L.R. Machado (✉) · P.A. Behar · J. Doll
School of Education, Federal University of Rio Grande Do Sul, Paulo Gama Avenue
110-12201 Building, 90040-060 Porto Alegre (RS), Brazil
e-mail: leticiarmachado@yahoo.com.br

P.A. Behar
e-mail: pbehar@terra.com.br

J. Doll
e-mail: johannes.doll@ufrgs.br

© Springer International Publishing Switzerland 2015
V.L. Uskov et al. (eds.), *Smart Education and Smart e-Learning*,
Smart Innovation, Systems and Technologies 41,
DOI 10.1007/978-3-319-19875-0_36

elderlies have become a significant part in our society. With this overview appeared differences, questions about rights of seniors and possibilities for quality of life to this population. These questions led to creation of pedagogical strategies that include changes in cultural, social and economic aspects due to increase on life perspective.

The use of technology is not something new. The entire population, in any culture and time, used some sort of technology that facilitated work and daily activities. However, in recent years there has been an exploration and a very rapid evolution of digital technologies that generated a great digital inequality and exclusion in many segments of population, among them older adults.

In this perspective more and more educational resources for use on computer to elderly have been designed to be included to learning process, and can be adapted to different needs of these users. Education, in its turn, is always changing, especially with incorporation of technologies to support learning. In this scenario, new learning environments are created, such as Distance Learning (DL), and new challenges arise for teacher training.

Unfortunately, in respect to use of DL by older adults here are few experiences and research in this field, as far publications of studies carried out in Brazil or internationally.

A study by Kimpeler, Georgieff and Revermann [1] indicates that there is almost no research group with focus on e-learning and elderly. The project, published as a report, points out that there are researches in development stage and with no published results. Paulo and Tijiboy [2] reported in their study that there is still much to be investigated over this subject, especially on teacher training.

Despite limited scientific publications, it is understood that this type of education can meet various demands of audience from targeted content, and a properly use of time and teaching practice to older adults. However, to achieve this purpose there is a need of quality planning and appropriate training.

Thus, this research investigated possible strategies that can be adopted by teachers who work or will work in DL with older adults. In this context, it is important to invest on whole development of human beings as active citizens. To learning process be meaningful it is necessary to consider aging changes in areas like physiological, psychological, cultural, historical, biographical and cognitive, which is provided as a deeper study of gerontology or educational gerontology.

Therefore, it is envisaged an education that fosters creating solutions and transform the society in which we live. Next section, we will address the issue of distance education and teacher training for elderly people.

2   Gerontological Education: The Development of Educational Strategies to Cyberseniors

Learning can happen throughout life (and not just in a period of personal life) and throughout "width" of life (not just in school system) [3–6].

Gerontology or educational gerontology, as a pedagogy for cyberseniors, cares to always stimulate the mind of older adults, encouraging personal interactions, communications, enabling reflection on emotions and work with reminiscence, decreasing thus aging effects [7–9].

Cyberseniors are active elderly on Internet with ease using services offered online like information search, communicate with family and friends, use social networks, perform bill payment for virtual banks etc. This group profile is formed by older people who are most likely to be challenged by different barriers on using of virtual technologies [10, 11].

Education in this scenario can provide offering of activities like reading, interpretation and problem solving that helps to maintain levels of brain acti-vation and in most cases "[...] to recover and to compensate for a lost of contextual or environmental incentives, which occurs particularly after retire-ment" [5]. So, in order to do that it is important to invest in studies of teaching process and student learning to define possible actions to help teachers on didactic-pedagogic activities.

Regarding process of teaching and learning with seniors, some features must be considered and discussed with teachers. Most of older people are more sensitive to disturbances such as external noises in educational environment. Anxiety is one of the main factors that interfere with learning, since this is a distress to learn the new and prove that he/she is still able to accomplish it. In this process, impatience is one of interference in "listening", "ask" and build knowledge by having reference only in education of his/her youth (where learning process was faster and authoritarian). The slow pace derived from neurological process and pronounced by age, it requires using new methods in education as a possibility of a longer time to carry out activities. That is [12], "[...] if content is presented slowly over a relatively long period of time and with breaks, respecting the time of each one" [13].

Socio-educational interventions in this sense contribute a lot to have confidence to face situations, especially new ones, by motivating and enabling a self-assessment and self-esteem [5, 6].

Thus e-learning may be a possibility of continuing education to elderly people. Older students in virtual courses require other strategies for learning and teachers need a different methodology. Thus, everyone should know their roles in order to do the process more effective [15, 16].

In order to meet goals teacher training is crucial: it provides basic skills to deal with speed of information and training in technological tools involved. That is to learn how to use technologies in a consistent, motivating and educational manner. ICT is important for students to be in constant interaction and motivated to continue over a course.

Therefore teacher must be prepared to mediate and provide students to reflect, to ask, to socialize, to interact and to cooperate with colleagues and other educators on virtual learning environment [16–18].

Hence, availability for online courses for cyberseniors is a possibility to envision that they can continue to learn, and that their presence and opinion still make a difference. This presence may generate changes in values and concepts, but it is essential that teachers adopt certain pedagogical practices to work with this kind of audience.

In this scenario, our paper aims to outline strategies for e-learning teaching to cyberseniors from collecting and discussing some indicators during our research. For this purpose, we present methodology adopted in this research.

3 Methodology

Our research was developed from quantitative and qualitative approaches. The research was characterized as theoretical and practical, because in this way important aspects of teacher training in distance education for seniors could be highlighted.

We offered two extension courses. The former was called "FormaDE - Teacher Training: resilience in aging", in which the goal was to train professionals on training or who had already trained and were interested to work with elderly in an e-learning context. This course lasted 20 h and it covered aspects of gerontology and e-learning. We selected nine professionals to attend the course. Of these, six professionals were chosen to serve as teachers in the second extension course (QualiViE) facing seniors, and the three remaining were chosen to act as virtual and face to face monitors in that course. The criteria for selecting of educators were: (a) they accepted to participate as a teacher in a virtual course for older people and, (b) they agreed to teach a subject of interest for elderly (useful in their daily lives).

For this teacher training (FormaDE), were constructed two learning objects by the team of the Center for Digital Technology (NUTED) at Federal University of Rio Grande do Sul: (a) InGeronto: Introduction to Gerontology - which aims to introduce main concepts of gerontology for any educator interested to learn more about this issue (Fig. 1); and (b) GerontoEAD: Gerontology and Distance Learning - which aims to introduce e-learning for teachers who wish to act with elderly population (Fig. 2).

Fig. 1 Learning Objects InGeronto. Source: InGeronto (http://www.nuted.ufrgs.br/objetos_de_aprendizagem/2012/ingeronto_/inicio/ingeronto-inicio.html)

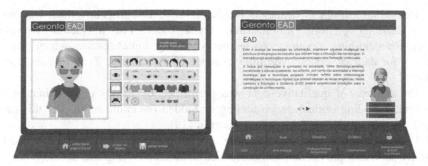

Fig. 2 Screens 1 and 2 of GeronEAD learning objects. Source: GeronEAD (http://www.nuted. ufrgs.br/objetos_de_aprendizagem/2012/gerontoEAD/index.html)

The extension course "FormaDE - Teacher Training: resilience in aging" aimed at training professionals of different areas of knowledge. This course led to discussions on lifelong learning and aging, especially with regard to resilience.

The second course, "QualiViE - Quality of Life: Virtual Workshops for active aging", aimed to discuss gerontological aspects in order to promote quality of life at different levels of aging. The participants of our study were 15 seniors resident in Rio Grande do Sul/Brazil. Inclusion criteria for participation were: (a) age equal or higher than 60 years, (b) be literate (c) be familiar with computer (d) have unrestricted access at home to a computer with internet connection.

All participants were informed in beginning of their activities about intended objectives and methodology, as well about the informed consent form.

The course QualiViE Workshops offered six different themes (proposed by teachers trained over the course FormaDE according to suggestions of elderly participants) focused on cyberseniors: Workshops Soundtracks, History and Memory Workshop, Workshop Colours, Spanish Workshop, Workshop Physiotherapy, Photography Workshop.

The topics of QualiViE workshops may be considered relevant to aging, since it prioritizes well-being and quality of life for seniors. Each teacher developed his/her content and released it in shape of learning objects.

The workshops were offered in a virtual way, with only one face to face meeting. The workshops lasted two weeks, except for Spanish Workshop that need more time for a further deepening on subject. Regular classes were performed at Digital Inclusion Unit for Elderly (UNIDI) in Faculty of Education/UFRGS-Brazil during two hours each.

Assessments by Workshops to elderly at QualiViE enabled data collection of this research, in addition to indicators weighted by teachers during course FormaDE.

For data collection was considered three factors: (a) speaks: semi-structured interviews, (b) written, in which were based on questionnaires at finish of each course, (c) action: from observing participants during courses.

The virtual learning environment ROODA (https://ead.ufrgs.br/rooda/) was used as a learning platform for research development. For interaction between teacher/student/tutor-monitor, we used the following digital features which were a source for data collection: Webfólio; Forum, Chat, Library; Tab "Classes"; and Journal.

The analysis of qualitative data was performed by means of examination of content, including critical understanding or hidden communication. For this purpose, we used the steps suggested by Bardin [19].

Thus, data collected could contribute to teacher training in e-learning aimed to elderly audience, providing a critical reflection about planning and development of educational practices in online courses. Following are outcomes and proposed educational strategies in distance education.

4   Analysis and Discussion of Data

The courses FormaDE and QualiViE, as already mentioned, enabled data collection for design of pedagogical practices for teacher training and his/her feature sat online mode with cyberseniors. Strategies were categorized into five indicators: planning, teacher profile, shape of materials, didactic, and learning objects.

The first indicator is related to planning. Planning corresponds to organize objectives, content and activity and time setting (face to face and virtual) and space.

Planning is essential in virtual courses, because from action plan, teacher or manager may think about the needs of students. Organization creates safety situations with elderly, especially in actions during classes. According to the testimony of a senior student: "The teacher showed plenty of security by exposing content and I felt that the activities were well planned" (SeH1 – Interview Fragment).

It is important to think about the time that will be need to older students participate effectively on online courses. Older students rated that time on virtual classroom was enough, but they showed a need for more time in a face to face class (workshops lasted 2 weeks). In a comment is possible to check this evidence "I wish we had more time together, but I understand that these workshops are more dynamic .... more time together would be good enough to enjoy ourselves" (ShC1E – Interview Fragment).

Organization of time to face to face or virtual classes is also considered in planning of courses. Teachers must define what will be the right format of course in relation to time, if it will be completely virtual or hybrid. Hybrid courses are those that blend face-to-face and online classes. This definition of style of course is held from the moment that we have been outlined needs and profile of target audience. For elderly population according the replies indicated, they preferred format is one which combines face to face and virtual classes.

The second indicator concerns to teacher's profile. In this topic we found relevance of previous training, both technological and gerontological, to who acts, or will act, with older people in virtual space. Elderlies indicated that teachers of Workshops QualiViE were well trained to teach to target audience. Other students have shown that, although teacher knew the contents, there was still a need for most appropriate training for virtual. These data are supported in elderly speech as follows: "It may be that he/she has not shown preparation, but the content, that seemed quite suitable for our age group, it can be said that he/she has prepared for this" (SaH4Q - Questionnaire Fragment).

It was also suggested a pre-survey about knowledge level of students, according to the testimony of one of students: "I thought I needed to make a survey about level of knowledge of the language" (SaE13R –Journal Fragment/ROODA).

All QualiViE Workshop teachers had training course (FormaDE) over geriatric needs of elderly and about technologies involved in process (LO and LMS). Despite this training, contact with audiences requires professional characteristics that cover many affective and communicative aspects. This profile, inclusive, is shown by students themselves, as in following statement: "[...] I think the most important in any workshop for seniors is the patience of teacher with students, because we are slower" (ShC4Q - Questionnaire Fragment).

This observation highlights the importance of investing in gerontological training to teachers and educators in a fast future, audiences will have people with advanced age, both in undergraduate and graduate courses such as formal or informal education.

Training to work with seniors in Brazil has grown since 1990. Since then, different courses of graduate and some undergraduate courses are offered to enable professionals to work with older audiences. In the field of gerontological education, teacher is called social educator, since he/she is considered a support that connects seniors and situations, triggering and stimulating research activities, analysis, creativity, reflection and social organization. Thus, teacher promotes production and social participation, encouraging citizenship [20].

Regarding technological readiness of teachers, all cyberseniors indicated that there was sufficient training to work with virtual modality and technologies. This training in use of technology is critical in virtual courses, because in this way it is possible to assess, to plan and to use resources properly with target audience. This training will enable suggestion of diverse paths to media handling, motivating and considering peculiar characteristics of students during course.

The third indicator refers to formats of available materials. During FormaDE classes, we recommend to teachers to use supplementary materials to assist in development of students' autonomy. Almost all teachers of QualiViE Workshops used such material on which students were questioned and in the answers that material was very useful during Workshops. Other students showed that materials were not enough and they sought other more to deepen content, or that the material was not used. As the testimony of a student: "All support materials helped in the realization of workshop, especially video (movie), because with images is easier to understand the content" (SgH9E – Interview Fragment).

Still considering the content, format stands out as essential for ensuring greater approval of seniors. In this sense, students indicated that content of workshops were appropriate for their audience, but two seniors showed that it was not appropriate, suggesting to use new media.

Even with virtual activities, seniors prefer to receive activities in pdf with a possibility of printing, denoting a need for security on the paper which is a technology well known by them.

In order to do contents enjoyable, elderlies showed that content should be challenging and enabling research and in addition to supplement their learning. These data show that, although older students were educated in a traditional perspective of memorization, they prefer to build activities and knowledge that become meaningful to their lives. Githens [7] reports that older adults are in need of interesting and attractive e-learning materials.

Thereby encouraging research is of great importance to seniors. For them to build knowledge and not just memorize in a mechanical manner. In this perspective and analyzing their learning, participants indicated they prefer challenging content (92 %), followed by issues that would allow greater reflection (8 %).

Fourth indicator relates to didactic. In relation to teachers and their didactics, students indicated a positive assessment by elderly. Some students reported that some teachers failed to return to their messages or have not interacted with them, creating discouraging situations in virtual classes. According to testimony from a student: "Greater interaction with teacher, I had questions and I had no one to ask" (SdE15Q – Questionnaire Fragment).

The profile of a teacher to work with elderly in e-learning should be differentiated of teachers of other virtual courses. Seniors students denoted that, among teachers, many aspects were considered important and that these aspects helped them in their learning. According to seniors, all teachers knew how to work with older audience, but only one that contradicted this assessment information and he/she could not explain why.

In planning activities on virtual scene, there were many suggestions made by interviewees. Elderly people pointed out that they prefer simple activities that use virtual tools and enabling research. The QualiViE Workshops which used unknown tools were evaluated positively by seniors, since they were more practical and were considered useful to everyday lives of students. Long activities were discouraging, not brought great challenges and disregarded the potential of those involved. Confirming testimony from a senior: "I believe that time for activities was good, but content was few challenging and it was very common place" (SaC1Q – Questionnaire Fragment).

Data collected point to different types of activities that most appeal seniors on virtual. Among mentioned activities, those requiring correction by teacher were highlighted, followed by activities which used certain digital resources, writing and research. Contact with and "approval" of teacher are still important to this audience which shows a need for greater interaction of educators with students on virtual.

For older audiences, virtuality suggests a detailed explanation in activities, especially those which use certain technology, as was the case with most virtual QualiViE Workshops. At general, teachers have provided an average of three days to carry out each activity and they have flexibility over time for completion of these activities.

Already the fifth indicator is related to construction of learning objects (LO). Regarding LO was appointed by elderlies that learning objects aided greatly in their learning, and only 20 % said they could not take advantage of LO in everyday life. According to the testimony of one student: "I liked the presentation of the workshop. It was quite objective, direct and easy visually and easy to understand" (SgC1E-Interview Fragment).

At e-learning materials such as LO should be developed striving for a quality, following formatting standards to minimize setbacks, or created by restricting interaction with this type of material. At textual production, for example, it should be noted questions about writing (spelling corrections), readability (font size, contrast), use of hypertext (not recommended to use this type of resource with elderly people), possibility of navigating through links to pictures, videos and other materials to text. Usability issues should also be analyzed, icons and buttons, images and animations, as well as in videos that have now become rich multimedia materials with potential for teaching and learning [14, 21].

Seniors also conducted an evaluation about the interface of LO, where it was possible to realize a necessity of building materials considering gerontological aspects such as biological needs (contrast of colors, text size and suitability of the material to audience).As one participant pointed on an object: "Your page was very good, attractive and without that drawing figure, which I found a kind of ridiculous compared to other workshops, with no sense with the rest. Very well presented, readable, though with black background, everything was clear, large and white fonts when it had to be" (SaH26E– Interview Fragment).

From the indicators categorized based on extracts, testimonials and narratives collected, it was possible to outline possible pedagogical practices for teacher training for older people in e-learning. In this sense, five (5) potential indicators and strategies were outlined in this study: planning, teacher profile, format of materials, didactic, learning objects.

These data suggest a need for more rigorous analysis over didactic-pedagogical aspects used in virtual classes with that audience. Among identified factors are: planning, explanation of content and classroom activities, the doubts that arose in virtual interactions and communications with digital tools and the enthusiasm shown by teacher.

5   Conclusion

Education can be defined as an ongoing process that occurs throughout life. In this process the teacher's performance can help in the research, interact, act, be, and create.

Thinking of elderly population, which each year increases the interest in digital technologies and cyberseniors become, the distance mode can be a great opportunity to qualify them from inclusive and educational activities.

Therefore, the pedagogical strategies that facilitate rapprochement between the teacher and the older student in order to motivate him/her to continue to learn/relearn are extremely important. From the scenario described in this study can reflect on new perspectives for future investigations.

Educational strategies for nursing teacher education, from the categories collected in the survey (planning, teacher profile, the format of materials, teaching, learning objects) should include: gerontological training in the curricula, training for planning, implementation and implementation of educational materials for the elderly population, cases of discussion with teaching and learning situations of the elderly public, interdisciplinary planning and construction integrators, virtual classes and more.

The collected and discussed in this article show a new profile of the elderly who are active and want to participate in continuing education courses. We can also see that are opening new spaces for teaching practice, which requires new practices which may include, effectively, the elderly population in a society that is constantly changing.

References

1. Kimpeler, S., Georgieff, P., Revermann, C.: eLearning for children and elderly people. http://www.tab.fzk.de/en/projekt/zusammenfassung/ab115.htm (2007)
2. Paulo, C.A., Tijiboy, A.V.: Inclusão digital de pessoas da terceira idade através da educação a distância. Novas Tecnologias na Educação 3(1) (2005)
3. Delors, J.: The treasure within: learning to know, learning to do, learning to live together and learning to be. What is the value of that treasure 15 years after its publication? Int. Rev. Educ. 59, pp. 319–330 (2013)
4. Scortegagna, H.M.: Oficinas temáticas como estratégia pedagógica. In: Casara, M.B., Cortelletti, I.A., Both, A. (eds.) Educação e envelhecimento humano. Caxias do Sul, EDUCS (2006)
5. Martín, A.V.: Gerontologia educativa: enquadramento disciplinar para o estudo e intervenção socioeducativo com idosos. In: Osório, A.R., Pinto, F.C. (eds.) As pessoas idosas: contexto social e intervenção educativa. Lisboa, Instituto Piaget (2007)
6. Vallespir, J., Morey, M.A.: A participação dos idosos na sociedade: integração vs. Segregação. In: Osório, A.R., Pinto, F.C. (eds.) As pessoas idosas: contexto social e intervenção educativa. Lisboa, Instituto Piaget (2007)
7. Zimerman, G.I.: Velhice: aspectos biopsicossociais. Porto Alegre, Artmed (2000)
8. Baños, R.M.: Positive mood induction procedures for virtual environments designed for elderly people. Interact. Comput. 24(3), 131–138 (2012)
9. Crowe, M.: Indicators of childhood quality of education in relation to cognitive function in older adulthood. J. Gerontol. A Biol. Sci. Med. Sci. 68(2), 198–204 (2013)
10. McMellon, C.A., Schiffman, L.G.: Cybersenior empowerment: how some older individuals are taking control of their lives. J. Appl. Gerontol. 21, 157–175 (2002)
11. Lee, B.: Cyber behaviors among seniors. Encycl. Cyber Behav. IGI Global web 7, 233–241 (2012)
12. Aldwin, C., Gilmer, D.F.: Health, illness, and optimal aging: biological and psychosocial perspectives. Springer Publishing Company, New York (2013)
13. Azevedo e Souza, V.B.: A motivação do idoso para reaprender a aprender: um desafio para propostas de intervenção educativa. In: Terra, N.L., Dornelles, B. (eds.) Envelhecimento bem-sucedido. Porto Alegre, EDIPUCRS (2002)

14. Monereo, C., Coll, C.: Educação e aprendizagem no século XXI: novas ferramentas, novos cenários, novas finalidades. In: Coll, C., Monereo, C. (eds.) Psicologia da Educação Virtual: Aprender e ensinar com as tecnologias da Informação e da Comunicação. Porto Alegre, Artmed (2010)
15. Moore, M., Kearsley, G.: Distance Education: A Systems View of Online Learning. Wadsworth, Belmont (2012)
16. Mauri, T.T., Onrubia, J.O.: professor em ambientes virtuais: perfil, condições e competências. In: Coll, C., Monereo, C. (eds.) Psicologia da Educação Virtual: Aprender e ensinar com as tecnologias da Informação e da Comunicação. Porto Alegre, Artmed (2010)
17. Palloff, R.M., Pratt, K.: Lições da sala de aula virtual: as realidades do ensino on-line. Porto Alegre, Penso (2015)
18. Bardin, L.: Análise de Conteúdo. Lisboa, Edições 70 (2010)
19. Cachioni, M., Neri, A.L.: Educação e gerontologia: desafios e oportunidades. RBCEH - Revista Brasileira de Ciências do Envelhecimento Humano, pp. 99–115 (2004)
20. Filatro, A.: Design Instrucional na prática. São Paulo, Pearson (2008)
21. Tori, R.: Educação sem distância: as tecnologias interativas na redução de distâncias em ensino e aprendizagem. São Paulo, SENAC (2010)

Part V
Smart Teaching- and Training-related Topics

Investment Competitions on the Current Local Scene from Students' Perspective – Case Study

Libuse Svobodova and Miloslava Cerna

Abstract The contribution deals with the selection of investment competitions for students of the Faculty of Informatics and Management, University of Hradec Králové, who have no experience with investing on the stock exchange and over the counter markets and who would like to get improved in this area. The objective is thus to choose the best simulation game in investing in the Czech Republic. Firstly current trends in e-learning are described and then individual investment games and their assessment are explored and discussed. The actual evaluation of individual investment competitions was conducted via decision support system, specifically in the program Criterium Decision Plus version 3.0. As one of the criteria for evaluation it is also possible to use SMART technology, which is being used by investors when monitoring the development of the financial market and making trades.

Keywords Education · Investments games · Simulators · SMART technologies

1 Introduction

Computer and financial literacy represents an inseparable part of current lives. The increasing importance of electronic education has to reflect advances in technologies. Computer games and simulators go hand in hand with the development of computer technology.

The article deals with utilization of technologies in the educational process; it focuses on on-line investment games. The purpose of computer games development

L. Svobodova (✉) · M. Cerna
Faculty of Informatics and Management, University of Hradec Kralove, Rokitanskeho 62, Hradec Kralove, Czech Republic
e-mail: libuse.svobodova@uhk.cz

M. Cerna
e-mail: miloslava.cerna@uhk.cz

© Springer International Publishing Switzerland 2015
V.L. Uskov et al. (eds.), *Smart Education and Smart e-Learning*,
Smart Innovation, Systems and Technologies 41,
DOI 10.1007/978-3-319-19875-0_37

is not only fun but also demonstration of technical possibilities. Games include some educational elements. Computer games and simulators can positively stimulate different human skills and personal development. Using simulators by investing can help to enhance financial literacy without losing of real money in bad decisions.

During the development of education supported by implementation of e-learning (Computer-based training (CBT) – Web-based training (WBT) – Learning management systems (LMS)) gradual reassessment and criticism of the efficiency and usability of these tools in the process of education started to appear [1].

The Internet has brought crucial changes. The Internet has gradually become a platform where differences between traditional information providers and consumers blur. Technologies such as blogs, wikis, and services of sharing different sources came into existence and established themselves firmly. These technical innovations represent Internet of the second generation and are summed up under the term 'Web 2.0' [2].

The following set of on-line tools represents frequently utilized tools in e-learning: Wiki, Blogs and Weblogs, Podcast, E-books, Digital libraries and data storage, Forums, Synchronous communication tools, Google, M-learning, Social networks and Games and simulations.

Game has been a proved means for educational purposes for centuries just to look back into the 17th century into the first didactics Didactica Magna by Johann Amos Comenius. [3] That was him who formulated key didactic principles like Principle of learner's activity and Principle of fun (game) in the process of education.

Current games which are based on modern technologies, however, have moved into new dimensions in both graphic design and the interactivity; the option to utilize network environment in playing is their unique distinguishing feature. A typical example of game utilization is a 'role-play', when teachers introduce students to the objectives and assign them roles through which they are supposed to achieve the goals. Students must get to know their roles; each role is characterized by some duty, responsibility or motivation. Participants work through various processes corresponding to the role they play so that they could accomplish the task. Games and simulations cater especially for those students who learn best when they can actively work to solve problems, to manipulate things and so on [6].

The investment games and simulators are supposed to be very important for students and instructors. Students can try and test investment theories and this way they are supposed to get the competitive advantage for the next time. It might be beneficial for their economic and personal knowledge. They can apply business and financial know-how or math from their studies and try investment without losing money in the case of bad investment decisions. On the other hand they will not gain any financial profit from their trades.

Literature Review. There are a lot of publications on the local scene dealing with computer games and applications that focus on the education, e.g. [7–10]. When we look at the international scene the offer is enormous. On-line games have been researched by academicians from various perspectives, e.g. as a tool leading to improvement of students' learning performance in the virtual environment of specifically designed on-line game, see more the paper on web based problem solving activities [11]. Beside development of problems solving skills there is another perspective covering a wide research range dealing with design, requirements and potential of on-line games in individual educational subjects or disciplines like languages [12], geography [13] or economics [7, 10].

The following perspective focuses on one of the key trait of games which is fun and joy. Design of educational games is supposed to reflect didactic rules and lead to given objectives in a joyful atmosphere [14]. Van Eck [15] examines the educational potential of computer games where he focuses on the design; essential role of properly designed on-line game.

Online investment games enable students reflect their individual needs and contribute to the development of decision-making skills in investing. On the today market there are plenty of simulators that can be used for investing in securities, currencies, commodities, shares or indexes. Most widely spread games are described in the paper.

2 Materials and Methods

Primary and secondary sources were used. As for secondary sources, they comprised websites of individual competitions and simulators, technical literature, information gathered from professional journals, discussions or participation at professional seminars or conferences. Then it was necessary to select, categorize and update available relevant information from the collected published material so that the basic overview of the subject in the Czech Republic could be provided. Then a testing phase of selected simulators and games followed. Criterium DecisionPlus program was used when processing the final evaluation.

3 Description of Investment Competitions

3.1 Studentbroker

Fio bank has regularly organized Studentbroker competition twice a year for 6 years [11]. Studentbroker is a joint project of Fio Bank and the Czech RM-System, which simulates real trading on world stock exchanges and is included in the curricula of 20 Czech and Slovak universities involved [12]. Students have the opportunity to become acquainted in a fun way with the functioning of capital markets and acquire investment practice. For teachers, the project represents an opportunity simply to monitor the success and investment skills of their students. The exact date of the beginning and end of the competition is never published, but usually this competition begins in spring and autumn, participants can trade almost to the beginning of the next competition. Then a new registration is required. Time of the competition corresponds with summer and winter semesters at universities. The results are evaluated after each semester and the best students are rewarded with valuable prizes and scientific publications on the topic of investing [11].

At the beginning each participant gets a virtual 1 million Czech crowns (CZK) for investments in stocks of the Czech RM-SYSTEM, USA, Germany, Poland and $ 50,000 for transactions with derivatives - futures contracts whose underlying assets are indices, currencies or commodities such as gold, crude oil or wheat and derivates like Index Dow Jones, Index NASDAQ, Index S&P, MINI E-FX. Both amounts are only for trading instruments mentioned above.

For this reason, the competition results are monitored in two separate categories (stocks and derivatives), it is possible to observe firstly the placement of a competitor in the overall ranking in each category, then placement of the competitor within a 'group' (i.e. within students from the school, for which the participant got registered) and finally his/her placement within 'all involved groups'. It is also possible to monitor the overall results in the case of shares of last month and derivatives within the last week. These results, however, are limited to the top 50 players. This does not mean that only these first 50 players know their location. Each competitor can see overall ranking within the framework of derivatives and shares at the top of the competition website.

The actual trading takes place only through the web interface on the competition website www.studentbroker.cz (Fig. 1), where after selecting an investment instrument a page with detailed information about it will appear. The display of the graph cannot be modified and within the technical analysis no indicators which could facilitate decisions on the purchase or sale can be added. The only thing that can be set is the time frame graph, it can span from one day to two years.

The disadvantage in this competition is that the displayed courses are delayed by 20 min, but the order processing is done at prices of real markets. Another disadvantage is the impossibility to enter instructions of protective stop-loss type. The only way to solve this problem is to enter a new order with the opposite position at the price at which the investor wants to close a position. Certain restrictions will occur even if the competitor wants to close the open position within ten minutes after opening. In this case he/she has to wait until the tenth minute elapses since the opening position, because it is not possible to terminate the deal earlier. Other inconvenience for the contestant could be the fact that only three deals can be performed with a single share type. Also for this reason it seems clear that this kind of a competition is designed more for positional traders, not for intraday or scalper ones.

A list of instructions that participant entered into the system can be checked in the browser tab called Instructions. In the Portfolio tab the competitor may follow appreciation of their investments divided into two above mentioned categories, where each category is further divided into individual shares (or derivatives), into which contestants during the game invested (see figure below). Moreover a graphic

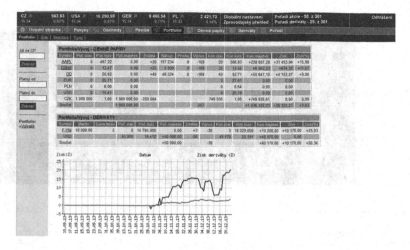

Fig. 1 Student broker portfolio (Source: www.studentbroker.cz)

development of the portfolio can be seen in this tab, this graphic display is updated several times a day, not after the conclusion of each contract.

Fictitious nominal fees for sales made were set to make the competition appear a little more realistic.

Winning this game may not be for everyone so much motivating. The motivation can be seen in the opportunity to try trading without risk or just compete with others.

This competition has no special application for mobile phones or tablets.

3.2 HighSky

Investment Brokers competitions with HighSky are run several times a year. Detailed information, including current positions is available on the www.highsky.cz. website [13]. Any person residing in the Czech or Slovak Republic who has reached the age of majority and has legal capacity can take part in this competition.

At the beginning of the competition, each participant receives one virtual million CZK, which he/she can invest in commodities, stocks, indices, bonds and Forex using MetaTrader free applications available for the PC, tablets and smartphones (iOS and Android).

Prices for individual instruments are not delayed as in the case of Studentbroker or Stock Market Challenge contests, but anyway they may slightly differ from trading on real account. Fees for trade are determined by the amount of spread.

MetaTrader version for PC (Fig. 2) is a trading platform that offers large variety of functions, which might be a problem for a complete beginner. But after a brief familiarization with applications or reading the manual or watching instructional videos, which are usually available for free at respective broker websites, even the

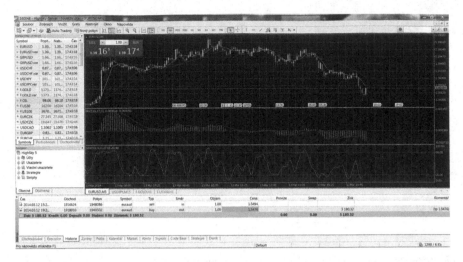

Fig. 2 Metatrader 5 (Source: www.highsky.cz)

beginner finds that trading using this application is very simple. Basically the only activity is just to click on BUY or SELL (see dark blue rectangle in Fig. 2).

This program offers a lot of useful functions. There is a choice of three types of graphs (bar, candlestick, linear) where a different time frame can be selected (from one minute up to 1 month). For better visualization and orientation in the graph it is possible to adjust colors of the graph according to the user's preferences.

It is possible to modify the graph or the window under it; various objects (lines, arrows, comments, pictures, etc.) or indicators that are appropriate for traders using technical analysis can be inserted. Dozens of indicators are included into the installation program. If a trader doesn't select any of them, it is possible to program another or simply download them from the Internet and use them in the application.

Technically more experienced users can use tools to basket strategies and deployment of automated trading systems. The question is whether it's an advantage or a disadvantage, but within the investment competitions from HighSky Brokers, X-Trade Brokers and Bossa the use of automated trading systems is explicitly prohibited in the rules.

Each user can have a report compiled in the programme from all executed trades, where he/she will see a summary of statistics on their trade and graphical representation of the development of his/her assets.

3.3 Bossa

Detailed information about the competition, including investment order is available on the competition website - soutez.bossa.cz. [14] The competition now takes place several times a year and has various conditions.

Usually each participant is given virtual 30 000 CZK, he/she can invest the money into stocks, commodities, indices, bonds and Forex by means of the MetaTrader 4 trading platform. As an example Christmas competition in 2014 with one lap lasting 4 weeks could be presented. The best investor won 100,000 CZK. All four best investors of each week won a Smart-phone worth max. 20,000 CZK. Furthermore, the first ten participants received annual subscription Economic News, Premium or Premium Economist.

As in the case of competition with HighSky Brokers, traded prices are close to those on real markets. Fees for trades are determined by the amount of spread.

Bossa Company provides further education on their sites. Their first language is Polish but there is also a version in Czech language.

3.4 Junior Trader

Junior Trader competition is organized by X-Trade Brokers [15] several times a year. The exact date is not known in advance. Junior Trader is open only to students of Czech and Slovak secondary schools and universities, who have reached the age of 15 years. All information about the contest is available at www.xtb.cz/juniortrader, where the overall standings can be checked.

Initial capital is different every year. In 2014 it was possible to invest € 100,000 into currency pairs, indices and commodities. Usually the top three contestants with the highest appreciation of capital gain various electronic devices, such as tablets, gaming consoles, watches and then each receives a certificate Junior Trader.

Rate of investment instruments are not delayed, they try to copy the real markets, although there may be some limitations. Fees for the execution of the trade are determined by the size of the spread.

Fig. 3 xTab (Source: www.xtb.cz)

Last year trading in this competition was done using applications xMobile, XTab and xStation.

xMobile and XTab applications are trading platforms from X-Trade Brokers for portable devices (xMobile is a version for smart phones, xTab is available for · tablets), they resemble the version of MetaTrader for portable devices with their looks and functions. As mentioned earlier, these applications compared to the version for PCs contain certain limitations so it isn't possible to use them as a fully fledged trading tools. Unlike MetaTrader, we haven't encountered any troubles with translation into the Czech language.

Trading platform xStation looks and functions very much like the version of MetaTrader tablet XTab. Figure 3 presents a screenshot from the xTab application.

The advantage of the xStation platform compared to the MetaTrader version for PC or portable devices could be the fact that it is accessible via a web interface, so there is no need to install any additional software.

4 Results

11 criteria were set for the evaluation purposes, these critera were evaluated via the analytic hierarchy process (AHP). Direct evaluation of criteria was used rather than default set of pair comparisons. The evaluation criteria were as follows:

- Participants
- Periodicity
- Variety of trading
- Smart technologies
- Software
- Graphs

- Instructions
- Languages
- Winnings
- Education
- Real prices
- Maximal number of trades

The following figure illustrates the final evaluation; the higher score the competition reaches, the better (Fig. 4).

As can be seen that the best results were achieved by the investment competition organized by Bossa with a gain of 26.9 % points, followed closely by competition from HighSky (26.7 % points). Junior Trader competition ended up the third with 26.0 % of the points. Studentbroker competition gaining 20.4 % points finished in the last position.

If we focus on all investors (not only students) the best results were achieved by the investment competition organized by HighSky with a gain of 28.8 % points, followed closely by competition from Bossa (27.6 % points). Junior Trader competition ended up the third with 25.3 % of the points. Studentbroker competition gaining 18.3 % points finished in the last position.

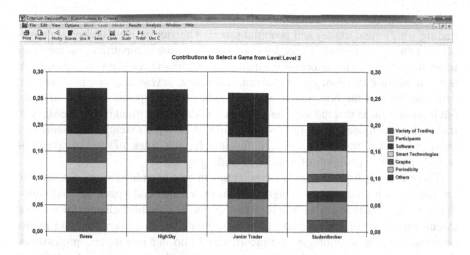

Fig. 4 Evaluation of the investment games (Source: own elaboration)

5 Discussion and Conclusion

Evaluation of various competitions was carried out through the investment decision support system, specifically in the Criterium Decision Plus 3.0 program, in which the very model for the analysis of investment competitions was designed. The goal was to select a suitable investment competition from the perspective of a student of the Faculty of Informatics and Management, University of Hradec Králové.

In total there were four investment competitions (Studentbroker, Junior Trader and competition from companies HighSky Brokers and Bossa) which were rated by eleven evaluation criteria.

Competitions organized by Bossa and HighSky Brokers were evaluated best in this model. These contests are very similar in many aspects. The Bossa competition won especially due to better evaluation in crucial model criteria which was Regularity. The Junior Trader competition was the third; compared with the winner it lost points in two key critera: variety of trading and periodicity. Studentbroker competition came last as it lost points in rating in most of investigated criteria.

Here we are getting to the paradox which might raise a fruitful discussion. If the competitions were judged from the perspective of the teacher the winner would be probably the worst rated Studentbroker competition. Teachers would definitely like to incorporate the competition into classes and profiteer from this unique tool. Studentbroker competition meets the two most important conditions for evaluation from teacher's perspective: firstly it is held twice a year, always during the winter and summer semesters, secondly it offers options to create new divisions by schools and moreover within schools it is possible to divide contestants into other subgroups allowing the teacher an overview of their students.

Psyche of the tradesman represents another considerable problem related to trading with virtual money and potential prizes. The victory itself and prizes might act as a strong motivating factor for participation in the competition; on the other hand, they can force a competitor to businesses that he/she could not afford on a real account because of money management. Therefore, investment competitions and demo accounts at the brokerage company might fit the purposes of trying to trade, but it doesn't mean that the way the contestant trades with virtual money he/she will trade with real money. Factor of psyche could be to some extent circumvented by using automated trading systems, unfortunately, the rules of the competitions usually prohibit utilization of automated trading systems.

There is an indisputable didactic potential in this examined issue. Games and learning/learning via games mingle and form a beneficial space for individual development. Currently when life is in the process of continual change there is one astonishing point of interest; no matter how quickly society develops, no matter how quickly new technologies are incorporated into our private and professional lives where the process of education is no exception there are eternal fundamental principles valid and functioning across centuries like principle of "active involvement" or principal "fun and game".

We believe that investment competition or demo accounts are good for a beginner to try trading and getting familiar with trading platforms, but of course to succeed in trading on real markets much more work and effort are required than. Currently, there are enough investment plays or demo accounts on the market where investor can try risk-free investment options.

This work was supported by the Specific project at the Faculty of Informatics and Management from University of Hradec Kralove.

References

1. Kopecký, K.: E-learning nejen pro pedagogy. 1. vyd. Olomouc: HANEX (2006)
2. O'Reilly, T.: What is Web 2.0: Design Patterns and Business Models for the next generation of software. O'Reilly website, O'Reilly Media Inc. [online], [cit. 2014-12-22]. http://www.oreillynet.com/pub/a/oreilly/tim/news/2005/09/30/what-is-web-20.html
3. Comenius, J.A.: The Great Didactic of John Amos Comenius. Russell and Russell, New York (1967)
4. Valtonen, T., et al.: Net generation at social software: challenging assumptions, clarifying relationships and raising implications for learning. Elsevier B.V. Int. J. Educ. Res. **2010**(49), 210–219 (2011)
5. Greener, S.: How are Web 2.0 technologies affecting the academic roles in higher education? A view from the literature. In: Proceedings of the 11th European Conference on e-Learning, (ECEL 2012), Groningen, pp. 124–132 (2012)
6. Zounek, J.: E-learning—jedna z podob učení v moderní společnosti. 1. vyd. Brno: Masarykova univerzita (2009)
7. Svobodová, L.: Computer games in economics education. J. Technol. Inf. Educ. **4**(2) (2012)
8. Dostál, J.: Výukový software a počítačové hry—nástroje moderního vzdělávání. J. Technol. Inf. Educ. Olomouc - EU, Univerzita Palackého, Ročník 1, Číslo 1, pp. 24 – 28 (2009)

9. Chromý, J.: Využívání počítačových her pro výuku. Media4u Magazine. Praha, Ročník 6, číslo 2, pp. 34–38 (2009)
10. Mls, a Kol, K.: Autonomous Decision Systems Handbook, 240 p. BEN, Praha (2011)
11. Hwang, G.-J., Wu, P.-H., Chen, C.-C.: An online game approach for improving students' learning performance in web-based problem-solving activities. Comput. Educ. **59**, 1246–1256 (2012)
12. Ravenscroft, A.: Promoting thinking and conceptual change with digital dialogue games. J. Comput. Assist. Learn. **23**(6), 453–465 (2007)
13. Tüzün, H., Yılmaz-Soylu, M., Karakus, T., Inal, Y., Kızılkaya, G.: The effects of computer games on primary school students' achievement and motivation in geography learning. Comput. Educ. **52**(1), 68–77 (2009)
14. Kinzie, M.B., Joseph, D.R.D.: Gender differences in game activity preferences of middle school children: implications for educational game design. Educ. Technol. Res. Dev. **56**, 643–663 (2008)
15. van Eck, R.: Six ideas in search of a discipline. In: Shelton, B.E., Wiley, D.A. (eds.) The Design and Use of Simulation Computer Games in Education, pp. 31–60. Sense Publishing, Rotterdam (2007)
16. Studentbroker [online]: [cit. 2014-12-22]. http://www.studentbroker.cz/
17. Fio banka [online]: [cit. 2014-12-22]. http://www.fio.cz/
18. High Sky [online]: [cit. 2014-12-22]. https://www.highsky.cz/registrace#buttonsTop
19. Bossa [online]: [cit. 2014-12-22]. http://bossa.cz/cs/vanocni-soutez/vyhry
20. X-trade Brokers [online]: [cit. 2014-12-22]. http://www.xtb.cz/

The Usefulness of the Virtual Speaking Head, as Well as 3D Visualization Tools in the New Communication, Teaching, Energy Control and Presentation Technologies is Almost Unlimited

Eva Pajorová and Ladislav Hluchý

Abstract The usefulness of the Slovak-speaking virtual head, as well as 3D visualization tools in the new communication, teaching and presentation technologies, as well as in a variety of audiovisual communications software technologies is almost unlimited. One of the options is the learning through technologies for hearing impaired people. Other, which is a most desired technology, is the use of a virtual head in the field of different communication forms. In the field of crisis management, where the virtual head present the warning messages and navigates during the evacuation of people from reproducing the public institutions such as schools, theatres, etc. 3D visualization tools, as well as Slovak-speaking head and Slovak speech visemas have been designed and tested in our the Institute of Informatics, Slovak Academy of Sciences, Bratislava, Slovakia.

Keywords Virtual speaking head · Communication · Audiovisual technologies · 3D visualization · Slovak speech visemas

1 Introduction

In the process of teaching, communication and present technologies have a big role the virtual speaking head, virtual reality, 3D visualization and 3D graphics. For the purposes of the presentation of scientific results in the context of the projects which are solved in our institute. We have designed the virtual, Slovak speaking head and 17 visemas (Fig. 1) of Slovak speech [8] and a number of 3D visualization tools. Utilization of the 3D virtual Slovak speaking head, as well as 3D visualization tools

E. Pajorová (✉) · L. Hluchý
Institute of Informatics, Slovak Academy of Sciences, Bratislava, Slovakia
e-mail: utrrepaj@savba.sk

L. Hluchý
e-mail: ladislav.hluchy@savba.sk

© Springer International Publishing Switzerland 2015
V.L. Uskov et al. (eds.), *Smart Education and Smart e-Learning*,
Smart Innovation, Systems and Technologies 41,
DOI 10.1007/978-3-319-19875-0_38

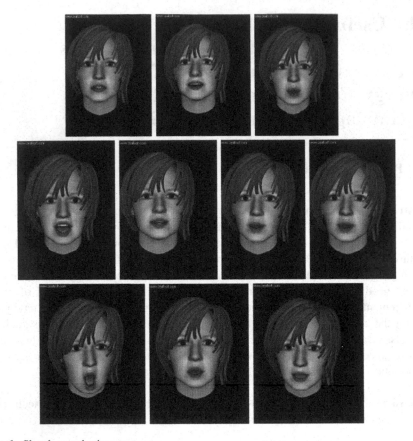

Fig. 1 Slovak speech visemas

in the new teaching, communication and present modern IT technologies is almost unlimited. Slovak language synthesizer was developed in our Institute and he is able convert text to speech and speech to text. Output speech we are using as an input for Slovak speaking head [1, 2].

2 The Use of Audio-Visual Technologies

- *Audio visual technology to generate the warning messages during a crisis situation.*
- *Audio visual technology using in the learning process of the hearing impaired students.*
- *Learning technology for virtual seminars and Conference.*
- *Presentation of the research results from simulation process.*
- *Communication virtual speaking head with robots as agents.*

2.1 Audio - Visual Technology to Generate a Warning Messages During the Crisis Time

Communication in crisis time can be planned for the global public. There are different kinds of group and approach of communication might be defined according to different categories, such as language, age, place, handicapped pupil, hearing impaired pupil, children etc. Mapping environment and crisis zones would help planning information delivery, focus the crisis message better and finally increase the size of the informed public One of critical public group are handicapped and hearing impaired people which are prefer lip reading.

A crisis situation, such as are the large fire, flood, store explosives, a threat to the criminal elements in the area of public institutions, shopping centers, theatres, stadiums, tunnels, etc. are always associated with a great deal of stress situations. To organize the evacuation of endangered places is needful a very good system, which at the moment of crisis will begin broadcasting using the generating and reproduce the warning messages. System must organize whole complicated situation and navigate the pupil. After different test and analyses we are located the warning audio visual system which include in the virtual speaking head. That such a system is helpful to all the people, including the disabled, elderly and children. System must be equipped with monitors and the technology that uses a virtual window on the monitor where the head speaking spoken virtual speech transformed into text [1], the system showing the emergency exits, stairways, informing you whether it is possible for them to use lift and produce needful warnings. In our Institute we have developed such system and it was tested in tunnel and in shopping hall during the unreal large fire (Fig. 2).

Fig. 2 Reproduction warning messages, with transforms the speech to text

preto opustite priestor obchodneho domu. Prosime o pomoc obcanov starsim a hendikepovanym ludom a detom pri rychlom odchode z priestoru obchodneho

2.2 Audio-Visual Technology in the Process of Learning of Hearing Impaired Students

Virtual head makes the computer speakers computer [4, 5]. For the hearing impaired students learning through modern audio visual technology in the current high-technology available, is the main helpful process [3]. In this form of the students forget their handicap. This is mainly because that learning through technology used commonly now as well as other students. Learning to use technology to its fullest accommodates students with disabilities among students without a disability. It is not just about learning the technical aspect, but also from the point of view of psychological, this tutorial is suitable for students with limited hearing. Just such a suitable computer game aimed at teaching. The game brings to the learning process for these students joke, joy, interest, esthetic sensibility. Through the game, the student gets into the virtual world of the other dimensions, can live in a distant past, can live in the cosmos, or in the distant future. In turn, the students can communicate during the game with the computer and work on the tasks in digital form. In the process of teaching joins a collective work over the web. Hearing impaired students can fully participate in the work of this collective is, because they are handy tool, that they were also converted to text when the text and speech. Chatting speech answered the same tool equipped with a synthesizer is willing to convert computer text into speech. A lot of options, that the real world does not allow such students as is the return to the past, find themselves at the bottom of the ocean, the sky, and they can fly in the cosmos, to make a virtual walk distant continents, set foot in the chambers of the human heart and visit the blood vessels of the body and through the many, many other possibilities for them in turn to the virtual world and getting to know things by using the modern technology. Test has been done in school for hearing impaired students (Figs. 3 and 4).

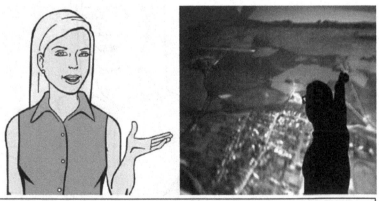

Flood on river Vah caused during the night death of two children. Twenty of houses build near river have been evacuated during the night. For pupil living near the river holds follow warnings:

Fig. 3 Virtual lecture-tutorial suitable for hearing impaired students

Fig. 4 Within the virtual world Students can fly in the cosmos

2.3 Technology for Virtual Seminars and Conferences

Audiovisual systems for the conference very often used 3D technology, virtual reality and virtual talking heads. In this area are increasingly used recordings of lectures where speakers cannot physically attend a real seminar. We have developed system for lecture and often we are using them in our work (Fig. 5).

2.4 Presentation of Scientific Simulations Calculated with Large Data

The results of the various scientific simulations computing with a large data, which are executing on the cluster have been in the past, reproduced by the numbers and graphs. Now the whole process of research can be presented by the audiovisual technology. By the 3D visualization and animation are showing the results of the research. Such presented results are sophisticated, and more understand. The results are understands for a much broader community than just for scientists, dealing with the issue, including for people with disabilities. Visualization tools, which have been developed in our institute, are tested with a large output data from research simulations computed on clusters (Fig. 6).

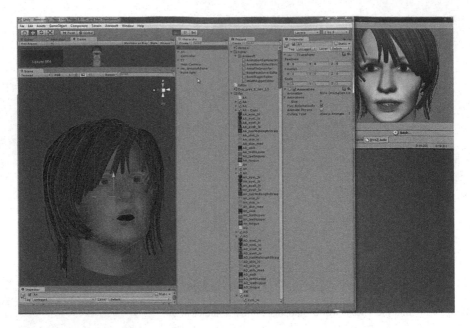

Fig. 5 Virtual conference

Fig. 6 Evolution of prothoplanetary disk

2.5 Use of Virtual Heads in Communication with Robots as Agents

For example it is true, that the social robots do different actions better as pupil. These robots may have ultimately much better properties than the man. They can engage in mutual communication. In the process of communication human - robot, or in the process of communication robot and virtual head, just robot transmit the information for the virtual head. For example, in the monitoring of the environment [6], when in case of detection of a dangerous situation, the robot is able to forward these results to the subsequent process of generation of information or warning messages. The messages are reproducing by the virtual speaking head. Tests have been done in process of testing the new one method developing in our institute, for coordination of a group of mobile agents that can be used for unknown area exploration and monitoring. It was test for communication virtual man - robots as quad - copters. The best form to represent the results form research simulations is 3D visualization (Fig. 7).

Fig. 7 Monitoring infected environment

3 Conclusion

Submitted article presents unlimited use of the virtual, Slovak-speaking head and the main visemas of the Slovak speech. In the future we want to create new one areas of IT technology, or software, which are able to using communications through the virtual speaking head. In the field of viseme creating we want to improve these viseme use of the modern 3D scanner, which are scanned more of a human face, so that we can suggest more points for creating of perfect visemas. We plan working with forty main visemas of Slovak speech. Next, our interest will be concentrate on the field of human-robot and robot - virtual man communication, since it is very grateful field for the use of 3D virtual speaking head.

Acknowledgments This work was supported by project KC-INTELINSYS ITMS 26240220072 and VEGA 2/0054/12 and by the project: Centre of water supply risk of a large city, Nr. 26240220082.

References

1. Sakhia, D., Marián, T., Miloš C., Milan R., Róbert S., Ladislav, H.: HMM speech synthesizer in Slovak. In: 7th International Workshop on Grid Computing for Complex Problems (GCCP), Institute of Informatics SAS, pp. 212–221. Bratislava, ISBN 978-80-970145-5-1 (2011)
2. Sakhia D., Miloš C., Marián T., Milan R., Róbert, S.: Effective triphone mapping for acoustic modeling in speech recognition. In: INTERSPEECH 2011, Speech Science and Technology for Real Life, pp. 1717–1720. ISSN 1990-9772
3. Peter D., Šperka, M.: Face expressions animation in e-learning. lconf06.dei.uc.pt/pdfs/paper14. pdf
4. Kim I.-J., Ko, H.-S.: 3d lip-synch generation with data-faithful machine learning. In: Computer Graphics Forum, Vol. 26(3), EUROGRAPHICS, 2007
5. Albrecht, I., Haber, J., Seidel, H.-P.: Speech Synchronization for Physics based Facial Animation. In: Skala, V. (ed.) Proceedings of 10th International Conference on Computer Graphics, Visualization and Computer Vision (WSCG 2002), pp. 9–16, 2002
6. Marek M., Eva, P.: Cooperative mobile agents for swarm behavior simulation. In: Cooperative design, visualization, and engineering 10th International Conference, CDVE 2013, Luo Y. (ed.) lNCS 8091, Springer, Berlin, pp. 128–136. ISBN 978-3-642-40839-7, ISSN 0302-9743 (2013)
7. Pajorova, E., Hluchy, L.: Visemas and virtual speaking head as a communication form. In: SGEM2014 Conference on Psychology and Psychiatry, Sociology and Healthcare, Education, www.sgemsocial.org, SGEM2014 Conference Proceedings, ISBN 978-619-7105-22-3/ ISSN 2367-5659, Vol. 1, pp. 585–590. 1–9 Sept 2014
8. Pajorova, E., Hluchy, L.: Virtual speaking head for hearing impaired people in crisis time. In: INES 2013 IEEE 17th International Conference on Intelligent Engineering Systems, Proceedings 2013, Article number 6632785, Pp. 69–72. San Jose; Costa Rica; 19 June 2013 through 21 June 2013; Category numberCFP13IES-PRT, Code 101322

A Tool for Developing Instructional Digital Comic Strips with Associated Learning Objectives

Fotis Lazarinis, Vassilios S. Verykios and Chris Panagiotakopoulos

Abstract In this paper a tool, which enables educators to edit comic strips and associate educational goals to each one of them, is presented. Students watch the produced e-comics and the success of the learning objectives can be measured through post-comic activities, such as tests and questionnaires. Various inferences about the learning progress of the participants are also possible through these activities. The edited comics are described using IEEE LOM metadata and they can be packaged as a SCORM object to increase the reusability of the material. In this paper we demonstrate the design of this tool and we present the results of an evaluation experiment.

Keywords Educational comics · Assessment · Edutainment · Educational hypermedia · Reusability

1 Introduction

Children have a natural attraction to comics which therefore may be used as a motivational mechanism in learning [1–3]. Nowadays, children are familiar with various electronic devices and accustomed in reading and watching animated cartoons, either through traditional media (e.g. TV) or new media (e.g. mobile devices). As a result, a growing number of researchers try to utilize digital comics as an educational tool [4].

F. Lazarinis (✉) · V.S. Verykios
School of Science & Technology, Hellenic Open University, Patras, Greece
e-mail: fotis.lazarinis@ac.eap.gr

V.S. Verykios
e-mail: verykios@eap.gr

C. Panagiotakopoulos
Department of Primary Education, University of Patras, Patras, Greece
e-mail: cpanag@upatras.gr

© Springer International Publishing Switzerland 2015
V.L. Uskov et al. (eds.), *Smart Education and Smart e-Learning*,
Smart Innovation, Systems and Technologies 41,
DOI 10.1007/978-3-319-19875-0_39

437

Digital comics can become a useful educational tool, as long as their design and their implementation are based on pedagogical principles and are not used simply as an entertainment tool. There are many challenges that e-comics as educational tools, would have to deal with. E-comics for educational purposes should motivate all students, even students with lower participation in the learning process. Difficult concepts should be presented in a more understandable and concrete way. The learning process should be adapted to each student improving her/his knowledge level and capabilities. Systematic learning planning with the participation of educators should be supported and the learning progress achieved should be evaluated in some way.

Although digital comics have been used to facilitate learning, there is little, if no ability, for teachers to intervene. Thus fact prevents the teachers from adapting the presented educational story to their teaching aims and strategies and their imagination. Therefore in this paper we propose an integrated environment which enables educators to edit comic strips and associate specific learning objectives with each scene. Students watch the produced animations and through post-comic activities, such as tests or questionnaires, we measure the success of the objectives. Various inferences about the learning progress of the participants are also possible through these activities. Further, the data are described using the IEEE LOM [5] metadata scheme and they can be packaged using SCORM [6] to increase data sharing and reusability.

2 Related Work

The idea of using comics in education is based on the concept that comics increase the students' engagement in the educational procedure. Several decades ago it has been observed that comics can motivate poor readers to try harder in order to understand the story [7]. Comics have been used to teach both science and art topics. As reported in [8] children exposed to science comics were able to give scientific explanations for the comics based on their own experiences. Four-frame comic strips were conceptualized from a set of anatomy-related humorous stories [9]. The comics were drawn on paper and then recreated with digital graphics software. These anatomy comic strips were designed to help students learn the complexities of anatomy in a straightforward and humorous way. Comic books have also been used to enhance literature teaching [10].

Digital comics are easier to be developed with the advanced functionality of the comic strip creation tools and the capabilities of modern mobile devices. OurComixGrid is a Web 2.0 e-learning design environment that combines new media creation and online social networking with cyber infrastructure to facilitate multimodal literacy education [11]. According to another study, university students argued that "Adding comic components in digital instructional materials can enhance learning interest" [12]. Mobile devices have been used in a system called PLOG which is a system that time clusters the images and presents them as story vignettes [13]. A preliminary implementation of incorporating digital comics on anticorruption topics through eLearning is examined with positive results in [14].

Comeks is a mobile comic strip creator that enables users to exchange rich, expressive multimedia messages [15]. Pesonen [16] describes the positive outcomes of the use of online interactive comics, by citing that increased motivation and interaction with comics led to a positive learning experience both in vocabulary and terminology. Digital cartoons have been used to represent mathematical concepts where alternative stories branching from a common beginning have been developed [17]. Robin [18] states that "digital storytelling has emerged over the last few years as a powerful teaching and learning tool that engages both teachers and their students". Students have been asked to create their own comics so as to improve motivation, literacy and conceptual understanding [19].

All these studies show that comics have been employed as educational resources for many decades, and, that different approaches on the use and creation of digital comics have already been proposed. Our system which is presented in the next sections, builds on the findings of the previous works. However, we do not simply provide another comic strip creator tool, but we go one step forward supporting the association of specific learning objectives with each sequence of comics. That way the influence of the comic to the learning progress of the students can be measured. More complex inferences about the success of the teaching approach are possible in that way. Additionally, the strips are annotated using learning standards, to increase their findability and reusability.

3 EduComicStrip

The main advantage of using comics in the educational procedure is that comics are fun and that children and older students, really like them. The ability of comics to motivate students and improve their understanding has already been documented. Our proposed system, EduComicStrip, tries to take advantage of this fact to help students learn while they enjoy themselves. However, we need to take into account various parameters to achieve that. The first aim set for our tool is to support teachers so as to be able to adapt and even design the content from scratch, in a quick and easy manner, to support their specific teaching aims. Secondly, in order to measure the pedagogical value of the procedure, there is a need to associate and evaluate specific educational and learning aims. Finally, the system needs to be flexible enough to accommodate various media files and be able to support the reusability of the material.

The EduComicStrip tool is a flexible comic strip creator environment which supports the creation and re-using of comic strips and the association and measurement of specific educational goals. Figure 1 shows the high level components of EduComicStrip. The development of the educational comic strips includes the setting of learning goals and objectives and the association of each comic to specific topics, the editing of the comic strip, the addition of descriptive data about the

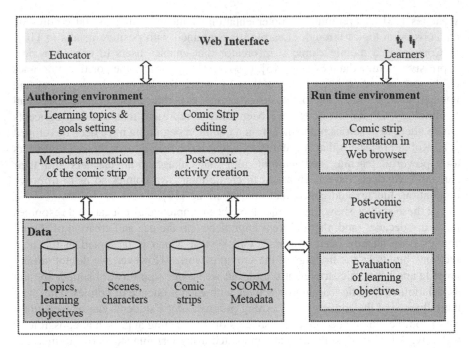

Fig. 1 Components of the EduComicStrip

comic (e.g., educational level, keywords etc.) and finally the creation of a post-comic activity which will be used to assess the learning objectives.

In this first version of the system, the produced comic strip is a set of images embedded in html templates. The html files support navigation through the cartoon frames. At the end of the comic strip the post-comic activity is presented to the student and the learning objectives are evaluated based on the required actions and responses.

The main aim of the proposed design is to provide an alternative learning and assessment tool which can present, through comics, material in a relaxing, inter-active and engaging manner and assess the student understanding. The learning and assessment activities are decided by the educators to support their teaching aims.

3.1 Authoring Environment

The comic editor of the EduComicStrip supports the creation of sequences of images through the selection of scenes, human and animal characters, text captions and other objects (e.g. arrows, stars, boxes, trees). These objects are stored in separate folders in the server's site and are accessible by the client's application.

Fig. 2 EduComicStrip Comic Editor

As seen in Fig. 2, the editing environment is a Web application developed in Adobe Flash where objects can be dragged and dropped or copy-pasted in the central screen to form the scene. Educators can also upload their own images to create more customized material. For example, scenes can be created using any image editor and uploaded to the system as PNG or JPEG images. Presently, the system supports only textual labels for the dialogues between the comic characters but it is technically feasible to associate audio clips to each character to enrich the experience for learners and even support students with physical disabilities, e.g. partially sighted.

Each comic is associated with one or more topics and one or more learning objectives. Topics are hierarchies of concepts and are coded using Topic Maps [20] (see Fig. 3), which is an ISO standard for the representation and interchange of knowledge.

The educational goals can be high level, e.g. "understand concept X" or specific objectives "to achieve specific score in the post-activity". Educators can set their own educational goals by adapting one of the existing predefined rules. For example, they can define the concept X in the pre-defined objective "understand concept X". In general there are three categories of customizable goals supported by the system:

```
...
<topic id="TP_1">
  <baseName>
    <baseNameString>Mathematics</baseNameString>
  </baseName>
</topic>
<topic id="TP_2">
  <instanceOf>
    <topicRef xlink:href="#TP_1"/>
  </instanceOf>
  <baseName>
    <baseNameString>Addition</baseNameString>
  </baseName>
</topic>
...
```

Fig. 3 Part of a Topic Map

1. Related to knowledge (e.g. understand concept X).
2. Related to skills (e.g. be able to perform or to adjust).
3. Related to attitudes (e.g. justify or defend a specific case).

An educator can associate one or more categories of goals to each comic and for each goal one or more specific objectives should be declared. For example, the general goal could be "Understand concept Mathematics/Addition" and the specific objectives are "Achieve score at least 90 % in the post-comic test" and "Answer correctly all questions of difficulty level 1".

In the current version of the system the objectives "Achieve specific score", "Answer correctly specific questions", "Answer correctly questions of specified difficulty" are fully implemented. Educators are able to form a post-comic activity and connect them to the objectives. These activities could in principle be of various types, e.g. tests, questionnaires, problem-solving activities, collaborative activities, etc. Tests contain closed-ended questions. The questions, the correct answers, the difficulty level and a number of other choices can be defined in a graphical way (see Fig. 4). The tests are encoded in IMS QTI XML [21] and are run in the browser by using JavaScript.

The educators are able to provide some keywords and short textual descriptions for each comic. Using a form similar to the previous interfaces, the educator may enter the descriptors which are then encoded in IEEE LOM (see Fig. 5) using the keyword XML element of the standard. These metadata facilitate the reusability of comic strips either by the original creator or by another educator. Our intention is to maximize the reusability of the materials to the benefit of the students and the educators.

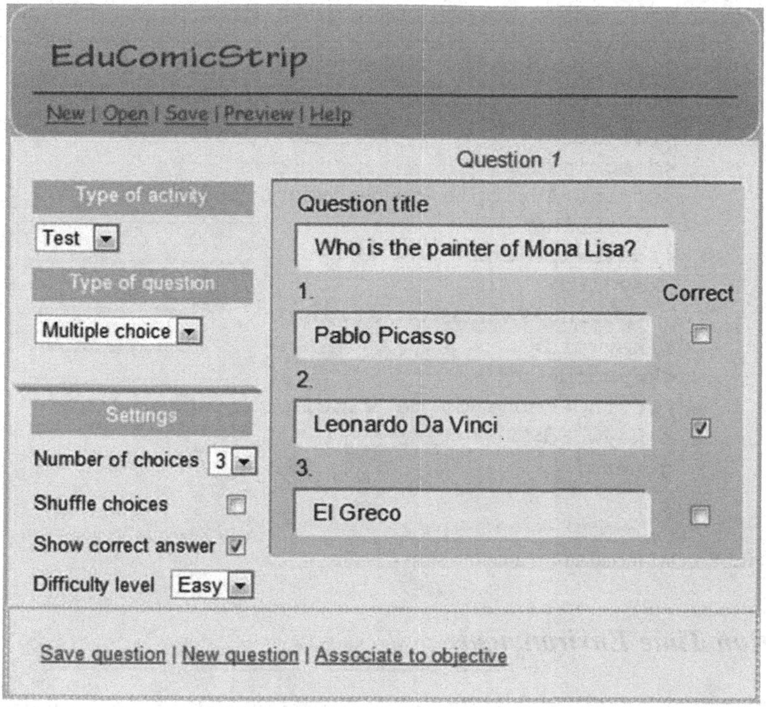

Fig. 4 Post-comic activity

The data can be exported as a SCORM learning object to be reused in other learning platforms. The description in the SCORM package includes the topics, the objectives, the html and PNG files of the digital comics and the IEEE LOM metadata.

It should be noted that the above capabilities are optional. For example, it is not necessary to provide metadata for the digital comic or to create a post-comic activity. The only required action is the creation of a new digital comic or the adaptation of an existing comic. However, in order to provide an intelligent learning environment, educators need to develop some post-comic activity and associate specific objectives. In any case, we tried to keep the authoring environment as simple and intuitive as possible to reduce the required learning and utilization effort. The representation of data in XML format is transparent to the end users in order to hide the technical details.

```
...
<general>
  <title>
    <string>Painters</string>
  </title>
  <description>
  <string>A simple comic about painters.
    </string>
  </description>
  <keyword>
    <string>Painters</string>
  </keyword>
  <keyword>
  <string>Leonardo da Vinci</string>
  </keyword>
</general>
...
```

Fig. 5 IEEE LOM metadata for a comic strip

3.2 Run Time Environment

The resultant sequence of PNG images are embedded in html files and displayed through a regular Web browser (see Fig. 6). Students can navigate through the available pages and watch the strip.

Fig. 6 IEEE LOM metadata for a comic strip

The post-activity runs as JavaScript application which is called at the end of the comic strip. Each student completes the possible assigned task (i.e. the test) and the application checks whether the associated objectives have been satisfied or not. At the end of the post-activity the correct answers are shown to the participant and an email is sent to the educator for each student. This email contains the activity results of the student so as to allow the educator to reach useful conclusions about their students and about the success of the learning goals.

4 Evaluation

In the previous sections we presented the capabilities of the EduComicStrip for creating comic strips, associating specific objectives and measuring their success. A working prototype of the system was evaluated with the aid of 5 educators and 14 young students. The educators were primary school teachers and the students were 10 year-old children who attend the 4th-grade class of the primary school.

The primary aims of this assessment were to realize whether the educators find the tool useful, whether they consider it easy to use, and finally to record how the educators utilize the available options. Therefore, we first explained the ideas of the tool to the educators and we presented the available functions to them. Then each educator had to create a short story. Five short 6 to 10 screen comics were created and a sort sequence of questions was associated with each of the comics.

The creation of the stories lasted from 25 min to 1 hour (avg. time 41 min). Another interesting observation made during this process, is that 25 out of the 28 total post-comic questions, related mainly to the students' attitudes and skills and not on their performance. For example, one question referred to the color of clothing the main character was wearing. The teacher justified her choice saying that she wanted to see how observant the students were. In another question the students were asked to decide whether the answer of the main character on the last action was polite or not. In general, these kinds of questions are positive indications of the utility of the proposed system and the supported knowledge, skill and attitude related objectives.

A short interview with each teacher took place after this step. The initial reactions to the tool were quite positive, as it was considered easy, useful and suitable for different learning cases. All of the educators considered that such a tool could be easily used in their classes. They also enjoyed designing comic strips in such a straightforward mode. Further, the educators made various useful suggestions on how we could exploit and extend the tool. They even suggested that such a tool could be used by the students themselves to create different short stories. This could help them improve their multiple-intelligences.

In the final stage of the evaluation, one of the comic strips was shown to 14 students. This comic strip was a 10 page long sequence of photographs depicting the main means of transport (motorbike, car, boat, train, helicopter and airplane) in various cases. The comic was created using images from the Web. The post-comic activity was a short questionnaire about which means of transportation pollutes the atmosphere more, which is the less safe, etc.

We observed that all the students participated with excitement in our short experiment and at the end of the activity they asked if another comic was available. The teacher who prepared this activity was able to initiate a discussion with the students about the advantages of each medium. The students participated actively in the discussion and the teacher mentioned that the students' involvement was more active than the one she would expect from her experience. She also mentioned that this unexpected students' interest may be due to the visual information her students received prior to the discussion.

Overall, the initial evaluation showed that the system is easy to use and useful for educational purposes and could be employed in a number of learning actions to enhance the understanding of the students and improve their engagement to the activity. However, more assessment experiments need to be performed before it could be used in real world learning activities. Further, a number of improvements should be made to extend its functionality. For example, the set of educational objectives should be broadened. The available set of post-comic activities needs to be enriched with the help of experienced teachers to provide a variety of new and usable options. Finally, the design of customizable templates which could speed up the creation process and the enrichment of the supported media files (e.g. sound), would definitely improve the tool's usability.

5 Conclusions

This paper presents an adaptable learning environment that supports the creation of comic strips. Each digital comic can be associated with learning objectives and post-comic activities to assess the completion of the learning objectives. The interactive environment allows the customization of pre-defined objectives and the quick development of different post-comic activities. All data are encoded in XML using the IEEE LOM metadata scheme, IMS QTI and TopicMasp. This increases the findability and reusability of the data.

The main advantage of the tool is that it supports the measurement of the pedagogical gains of using the tool. Through the execution of post-comic e-activities the educator can realize the influence of the comic to the students. This could support the utilization of a comic strip to larger audiences and the quantification of the results. The initial evaluation showed that the system is useful and flexible and revealed a number of potential system uses. The suggested changes along with the extension of the set of educational objectives and post-comic activities, will be addressed in the future versions of the system.

References

1. McVicker, C.J: Comic strips as a text structure for learning to read. Read. Teach. **61**(1), 85–88 (2007)
2. McTaggart, J.: Using comics and graphic novels to encourage reluctant readers. Read. Today **23**(1), 46–46 (2005)
3. Norton, B.: The motivating power of comic books: insights from Archie comic readers. Read. Teach. **57**(2), 140–148 (2003)
4. Robin, B.: Digital storytelling: a powerful technology tool for the 21st century classroom. Theory Pract. **47**(3), 220–228 (2008)
5. IEEE LOM: Learning Object Metadata, http://ltsc.ieee.org/wg12/ (2006)
6. SCORM: Sharable Content Object Reference Model, http://www.adlnet.gov/capabilities/scorm (2004)
7. Hutchinson, K.: An experiment in the use of comics as instructional material. J. Educ. Sociol. **23**(4), 236–245 (1949)
8. Weitkamp, E., Burnet, F.: The Chemedian brings laughter to the chemistry classroom. Int. J. Sci. Educ. **29**, 1911–1192 (2007)
9. Park, J.S., KiM, D.H., Chung, M.S.: Anatomy comic strips. Anat. Sci. Educ. **4**(5), 275–279 (2011)
10. Versaci, R.: How comic books can change the way our students see literature: one teacher's perspective. Engl. J. **91**(2), 61–67 (2001)
11. Duffy, D., Clark, A.: OurComixGrid: designing a multimodal new media learning environment. In: Proceedings of the 6th International Conference on Networked Learning, Greece, pp. 591–597 (2008)
12. Weng, T.-S.: Interactive e-Books: integrating digital art for educational purposes. In: Proceedings of the 7th WSEAS/IASME International Conference on Educational Technologies (EDUTE'11), Iasi, Romania, pp. 112–117 (2011)
13. Gossweiler, R., Tyler, J.: PLOG: easily create digital picture stories through cell phone cameras. International Workshop on Ubiquitous Computing (IWUC 2004), Porto Portugal, pp. 94–103 (2004)
14. Aryanto, V.D.W.: Using digital comics to enhance elearning on anti-corruption education. Spec. Issue Int. J. Comput. Internet Manage. **15** (2007)
15. Salovaara, A.: Appropriation of a MMS-Based Comic Creator: From System Functionalities to Resources for Action, CHI 2007, pp. 1117–1126. San Jose, California (2007)
16. Pesonen, K.: Using interactive comics for energy education through the web. In: Montgomerie, Viteli, J. (eds.) Proceedings of World Conference on Educational Multimedia, Hypermedia and Telecommunications, 2001, pp. 1482–1488. AACE, Chesapeake, VA (2001)
17. Herbst, P., Chazan, D., Chen, C.-L., Chieu, V.-M., Weiss, M.: Using comics-based representations of teaching, and technology, to bring practice to teacher education courses. Int. J. Math. Educ. ZDM **43**(1), 1–22 (2011)
18. Robin, B.: Digital storytelling: a powerful technology tool for the 21st century classroom. Theory Pract. **47**, 220–228 (2008)
19. Pelton, F.L., Pelton, T., Moore, K.: Learning by communicating concepts through comics. In: R. Carlsen et al. (eds.) Proceedings of Society for Information Technology & Teacher Education International Conference, pp. 1974–1981 (2007)
20. Topic Maps: Topic Maps, http://www.topicmaps.org (2000)
21. IMS QTI: Question and Test Interoperability, http://www.imsglobal.org/question (2006)

Music Coding in Primary School

Luca A. Ludovico and Giuseppina Rita Mangione

Abstract This work stems from the recent reforms made to the Italian education system. On one side, a relevant aspect is the introduction of coding in primary school, where coding is seen as a way to improve learning processes in young students. On the other side, music education is considered a key element to promote the integration of the various aspects of personality: the perceptual-motor, the logic, and the affective-social component. After reviewing the state of the art, this work aims at defining a new discipline - called music coding - presenting the key elements of the two pedagogical approaches. A computer-based tool to foster the development of both coding and music skills in young students will be proposed.

Keywords Coding · Music · Education · Primary school · Pure Data

1 Coding for Young Students

Past research demonstrated that the adoption in primary school of techniques and technologies aiming at active, creative, and participatory training should be encouraged [1, 2]. Despite these evidences, for many years primary school education suffered from the lack of clear guidelines and innovative products able to generate evidence or disprove theories, as well as action scenarios related to this specific domain. Technology in school has been addressed as a form of training, and the educational goal - often enthusiastically emphasized - has become merely teaching how to use a computer.

L. A. Ludovico (✉)
LIM, Dipartimento di Informatica, Università degli Studi di Milano, Via Comelico 39,
20135 Milano, Italy
e-mail: ludovico@di.unimi.it

G. R. Mangione
Indire - Istituto Nazionale di Documentazione, Innovazione e Ricerca Educativa,
Via G. Melisurgo, 4 - 80133 Firenze, Napoli, Italy
e-mail: g.mangione@indire.it

© Springer International Publishing Switzerland 2015

449

V.L. Uskov et al. (eds.), *Smart Education and Smart e-Learning*,
Smart Innovation, Systems and Technologies 41,
DOI 10.1007/978-3-319-19875-0_40

In 2014 the Italian Ministry of Education, University and Research (MIUR) announced the reform plan known as "La Buona Scuola".[1] One of the most relevant features was the introduction of coding as a curricular subject in primary schools. Several Western countries are in the midst of major changes to school curricula, since now the value of introducing topics such as programming, computer science and computational thinking is clear. For example, changes in the curricula have been introduced in 2014/2015 in the USA and UK in order to include basic programming experience in primary schools. These changes are stimulating public debate also in the scientific community, as discussed later.

In Italy, the children-oriented coding initiative - called the *Hour of Code* - is based on Code.org,[2] an international platform that lets young students move their first steps into the world of computer programming through easy tutorials and on line courses, often amusing and playful. Such a Web framework offers a number of courses, ranging from the basic concepts of Computer Science to the implementation and customization of a smartphone app. Lessons can be attended both on line and off line. At the moment of writing, Code.org's homepage states that more than 90 million students have tried the *Hour of Code* and more than 4 billions code lines have been written by young coders. In the framework of the Italian school system, this approach is extremely innovative. The guidelines of the mentioned reform suggest a change of perspective: some concepts and cognitive goals typical of Computer Science are seen as a fundamental asset for the cultural citizenship of the third millennium. The challenge and the bet is that approaching coding at an early age can improve logical and cognitive skills, with a positive impact on the future education and development of students.

The theoretical bases of coding go back to the pedagogical approach of *constructivism* formalized by Piaget, Vygotsky and Bruner [3]. As regards the introduction of coding and computational thinking in primary school, it is worth citing the ideas of Papert: while each discipline claims to push students to think, computer science achieves this result in an operative and concrete way. Papert's theories have been exposed in [4], adopting LOGO as the reference programming language for children's coding activities. An implementation of Papert's ideas requires a deep change in teaching, transforming the mere transmission of knowledge and skills into laboratory activities and structured projects aimed at encouraging collaboration and discussion. Needless to say, this implies not only the redefinition of school curricula but also ad hoc teacher training initiatives.

Opinions about the early teaching of coding spans from an enthusiastic support - problem-solving skills, innovative ways to explore concepts, etc. - to harsh criticism - development of industry-specific skills at the expense of a balanced growth of the individual. An example of the ongoing debate within the scientific community is provided by [5] and [6]. The former work stated the need for new creative and engaging learning experiences for children, with particular reference to job opportunities related to information and computing. However, as argued by the latter work, these

---

[1]Literally: "The Good School", https://labuonascuola.gov.it.
[2]http://code.org.

claims did not convince the entire scientific community. For instance, specific questions were raised about the importance for children to be prepared to work against the opportunity to enjoy a childhood without risks such as alienation from the real world and limitations in the development of imagination.

This debate had an impact on the review of primary school curricula where the so called *Hour of Code* had been introduced as a discipline [7]. A key factor for the updating of school curricula has been identified in [8]: "We teach kids how to use software to write, but not how to write software. This means they have access to the capabilities given to them by others, but not the power to determine the value-creating capabilities of these technologies for themselves".

The mentioned works provide a good overview of pedagogical motivations of coding. Its proponents usually allege one of the following reasons: (i) making students proficient in computer programming in general, and (ii) supporting the computational thinking that will be useful to student also in other educational contexts. When educators and computer researchers are requested to explain the advantages and opportunities that coding offers to young students, they go beyond the consideration that children can become good programmers in the future. Mitchel Resnick, one of the developers of Scratch at MIT in Boston, asserts that children must learn to create through digital technologies, and not only to interact with them [9]. When he describes the motivation that led to the development of Scratch, he claims that the main goal was "to nurture a new generation of creative, systematic thinkers comfortable using programming to express their ideas".

In coding activities, students are exposed to *computational thinking*, a locution which involves the use of computer science concepts such as abstraction, debugging, remixing and iteration to solve problems [10–12]. According to many experts, this form of thinking is fundamental for K-12 students because it requires "thinking at multiple abstractions". Moreover, computational thinking is in line with many aspects of 21st Century competencies such as creativity, critical thinking, and problem solving [13, 14].

It has been demonstrated that computational thinking does not mean only the acquisition of computational skills, even if certain abilities are practiced and trained through programming. For example, [12] argues that programming environments for children can support the development of three dimensions of computational thinking: computational concepts, computational practices, and computational perspectives.

In conclusion, writing code means more than using a computer: it implies finding original ways to achieve a goal, being able to decompose a problem into simpler sub-problems, becoming familiar with concepts such as sequencing, feedback and abstraction. As a consequence, the ability to code is an important skill since it may transform children from passive technology consumers into active technology creators [9].

The word *coding* is currently used by many organizations that promote learning programming skills, e.g. Code.org, Made with Code, Code Club, CoderDojo, Black Girls Code, Codecademy, Code Avengers, CodeHS, and MotherCoders. The increasing interest in coding at primary school is encouraging the creation of new learning environments known as *Initial Learning Environments* (ILEs) [6]. Needless to say, coding environments for children have to present ad hoc features, including gamification and simplified syntax. For example, the adoption of visual rather than traditional programming languages facilitates computational thinking in K–12 contexts because unnecessary syntax is reduced (e.g., the use of semi-colon and curly brackets), commands are closer to spoken languages, and intuitive techniques such as drag-and-snap are employed to link command blocks [15]. In this context it is worth citing visual programming languages such as Scratch, Toontalk, Stagecast Creator, and Alice.

Popular coding environments have made coding accessible to a large number of students and teachers. As a virtuous effect of interdisciplinary contamination, curricular competences and skills related to basic programming are becoming an element in support of other subjects such as mathematics, science and technology. Besides, K-12 programming tools are becoming increasingly important in the context of digital literacy experiences for creating, sharing and remixing digital resources [16–18]. An in-depth review of available products goes beyond the purpose of this work. For further details, [6] provides a critical review of 47 environments used to introduce children to programming concepts, from simple game design to algorithmic challenges.

Most of the environments described in the following support computational thinking through pure self-discovery, without guidance on the cognitive aspects of computational practices and computational perspectives [19]. In this case, programming experience may be non-educative as students are not actively reflecting on their experience. In order to solve this problem in computational practices and computational perspectives there is a need to define new environments where information processing (e.g., metaphor, cognitive conflict and mind-mapping) scaffolding (e.g., explicit marking of causal reasoning) and reflection activities (e.g., peer reviewing and self video recording) could be designed [15].

2 Music and Children Education

Music education and training for children require ad hoc techniques and methods as well as a specific review of school curricula. Recent scientific works show that an integration of multi-modal experiences based on activities such as moving, creating, playing, reflecting [20] support the development of a "symbolically fluent child" [21, 22]. The learning environment should be able to represent activity-oriented musical experiences, where students - properly sustained by scaffold elements - are involved in a process of music construction/deconstruction. For example, according to [23], scaffolding children's early musical experiences and investigations, their engagement in the world of sound, their trans-modal redesign of known literature and song repertoire helps children establish strong, confident, vibrant, and creative identities in learning, communication, and performance.

Currently a new music pedagogy based on an integrated approach is emerging. Recalling the fundamental concepts of pedagogical activism by Dewey [24], the goal is enhancing that educational cross-component able to influence key aspects of the growth such as expressiveness, autonomy and sociality. Music is able to influence the construction of the child's personality because it promotes the integration of perceptual, motor, affective, social and cognitive dimensions [25] by relating basic aspects of human life (e.g. physiological, emotional and mental spheres) with the basic elements of music (e.g. rhythm, melody and harmony).

The abilities of listening, exploration and analysis are fundamental for the development of general meta-cognitive skills of the child, such as attention, concentration, control. In this sense, music is both an opportunity and a crucial educational strategy. For example, through music young students can develop the aspects of analysis and synthesis, problematization, argumentation, evaluation and application of rules. In addition, as regards the ability to read and understand, children have the possibility to train their transcoding skills, moving from the musical domain to the verbal language in order to describe what they heard [26].

The construction of a vertical music curriculum can be the first step in a process that leads to the recognition of the musical dimension as a key element of the training process. The Italian guidelines for school curricula released in 2012 state that music is capable of supporting the achievement of skills essential for citizenship rights. In an official report published by the Italian Ministry of Education, University and Research in 2014,[3] music learning is described as a global experience that must interact with other disciplines and connect to other modes of expression. Learning music is seen as a synthesis of heterogeneous processes of exploration, comprehension and learning, and as a laboratory based on voice practice, conventional and unconventional musical instruments, graphics, gestural and motor activities. The goals to achieve are the integration of different musical languages and the construction of individual and collective identities. Nevertheless, in the Italian education system music teaching is often plagued by numerous factors: traditional lectures, boring subjects, limited instrumental practice, lack of creativity.

In the digital era, new technologies and computer-based approaches can influence music learning and teaching processes. A recent and comprehensive review of this subject can be found in [27], a work that discusses a range of innovative practices in order to highlight the changing nature of schooling and the transformation of music education. Many researchers, experts and music teachers feel a pressing need to provide new ways of thinking about the application of music and technology in schools. It is necessary to explore teaching strategies and approaches able to stimulate different forms of musical experience, meaningful engagement, creativity, teacher-learner interactions, and so on.

The idea of the present work is applying the most recent pedagogical theories to coding and music teaching in primary school through commonly available technologies and free, open-source tools, as detailed in the following sections.

---

[3]Transmission of guidelines relating to ministerial decree D.M. 8/2011.

3 Beyond Instrumental Practice: Music Coding

We have listed a number of scaffolds to provide children with logical and cognitive skills, going beyond the technical capabilities of a good programmer or a skilled music performer. In order to achieve this goal, teaching Computer Science in schools cannot merely imply a passive use of technologies, as well as teaching Music in schools cannot merely imply instrumental practice. We propose to combine the pedagogical advantages of coding and music education in primary school thanks to *music coding*, a new discipline that couples algorithmic thinking, technological tools, and computer interaction with musical experience, creativity, and social processes.

Although less famous than "traditional" programming languages, software tools to achieve music and sound programming are already available and commonly in use. For example, Csound is an *audio domain-specific language* (audio DSL), namely a computer language specialized to the application domain of music. Originally written at MIT by Barry Vercoe and developed over many years, it currently has nearly 1700 unit generators. One of its greatest strengths is that it is completely modular and extensible by the user. Besides, Csound is a free software, available under the GNU Lesser General Public License (LGPL). Csound and its possible applications have been explored in scientific works such as [28] and [29].

Under many points of view, Csound could be a good environment to experiment with music coding. First, this DSL has primitives to implement control flow, e.g. labels and *goto* statements, subroutines, conditional expressions and constructs, loops, etc. Both the compiler and most related software (editors, front ends, etc.) are free, and the environment is cross-platform. Finally, the sound result can be perceived by the programmer even in real time, thus providing a prompt action feedback. Unfortunately, producing sound through this environment is difficult for a child, as well as for a musician without a specific computer training. In fact, Csound takes two formatted text files as input, encoding the *orchestra* and the *score* respectively. The former text file describes the nature of the instruments, whereas the latter sets notes and other parameters along a timeline. Finally Csound compiler processes the instructions in these files and renders an audio file or real-time audio stream as output. Even if graphical interfaces can be implemented, such a programming language is far from visual-aid techniques typical of children-oriented coding environments.

Under the perspective of ease of use and intuitiveness, a formal graphical representation of music processes could help. A possible answer is provided by the extension of Petri nets to music. A Petri net is a mathematical modeling language invented in 1939 by Carl Adam Petri for the description of distributed systems [30]. Extensions of Petri net formalism to the music domain have been described in [31]. Music Petri nets are a formal and graphical tool to represent music processes. Concepts such as sequencing, conditional structures, looping, concurrence, non-determinism can be encoded. Besides, net topology can be changed and the corresponding sound generation can occur in real time, providing a prompt feedback [32]. Nevertheless, this formalism is tricky and can be cumbersome in the description of a music piece when a very fine granularity is required. Moreover, available software tools are not advanced enough to allow easy interaction with music Petri nets.

In summary, the first two approaches present some strengths: (i) They are oriented to music programming and formalization, so they may encourage the computational thinking typical of coding; (ii) They can be used to simulate and teach typical programming concepts; and (iii) Required frameworks are free and cross-platform, so they meet the need of low-cost technologies and availability even in the context of primary school. Unfortunately, these environments have also some relevant drawbacks that make the applicability to children education very difficult. Csound requires writing some text code appropriately formatted, where syntax is difficult to remember and even to be explained to young learners. All this is far from visual programming, drag-and-snap, and other concepts in use in K-12 educational contexts. On the contrary, Petri nets offer a graphical notation and their mechanism is easy to explain (the evolution of a Petri net due to fire rules and token-based marking may even resemble a game), but it is hard to grasp the underlying music meaning.

For the sake of comprehensiveness, another framework supporting visual aids and formalizing connections among functional blocks is the Reactable, namely an electronic musical instrument with a tabletop tangible user interface that has been developed at the Universitat Pompeu Fabra in Barcelona, Spain [33]. In its original implementation, the Reactable is a round translucent table, used in a darkened room, and appears as a backlit display. By placing blocks called *tangibles* on the table, and interfacing with the visual display via the tangibles or fingertips, a virtual modular synthesizer is operated, creating music or sound effects. This tool, used also by professional artists, presents a visual interface that recalls a playful approach. Besides, the Reactable provides a collaborative way of making music, as mentioned in [34]. In addition to licensing fees, this framework cannot be considered a real coding environment, since the function of blocks is predefined and the operating logic is oriented to music production rather than to the formalization of programming concepts.

A compromise solution, presenting all the pros mentioned above and minimizing the cons, is the adoption of Pure Data, a framework commonly in use in the community of musicians, visual artists, performers, researchers, and developers. Pure Data is a free and open-source visual programming language originally developed by Miller Puckette at IRCAM [35] and currently supported by a wide and active community of users and developers. Its applicability to the music-coding domain will be discussed in the next section.

4 Music Coding in Pure Data

Pure Data (Pd) lets users create software graphically, without writing lines of code. Possible applications range from sound processing and generation to 2D/3D graphics and video. As a consequence, Pd is suitable for learning basic multimedia processing and visual programming methods as well as for realizing complex systems for large-scale projects. Pd is a so-called *data flow programming language*, where software modules called *patches* are developed graphically. Algorithmic functions are represented by *objects* placed on a screen called *canvas*. Objects are connected together with cords, and data flows from one object to another through these cords. Each object performs a specific task, from very low level mathematic operations to complex audio or video functions such as reverberation, FFT transform, or video decoding. Moreover, Pd can interface sensors, input devices, and MIDI-compatible music instruments. Thanks to these features, it is possible to design and implement rich, complex, and student-tailored educational environments. As regards the graphical layout of Pd patches, their standard aspect could be hard to understand for children. In this context it is worth citing the GrIPD (Graphical Interface for Pure Data) initiative [36], a cross-platform software aimed to design custom graphical interfaces for patches.

Now let us analyze the points common to a coding environment and to a music-oriented programming language. Coding environments for children should foster the development of three dimensions of computational thinking: computational concepts, practices, and perspectives [15]. *Computational concepts* are the concepts that programmers use, such as variables and loops. *Computational practices* are problem-solving practices that occur during the process of programming. For the sake of clarity, examples include being incremental and iterative, reusing and remixing, abstracting and modularizing. Finally, *Computational perspectives* regard students' understanding of themselves, their relationship to others, and the world around them.

Only the first item is usually covered by general-purpose learning tools and environments for children. On the contrary, a suitable music coding environment can easily fulfill also the other goals. As regards computational practices, in Pd music production as well as sound synthesis can occur in real time, exploring fields such as improvisation and modularization. Besides, the programming environment can be set to encourage reusing and remixing of music materials, like in so-called *step sequencers*. Finally, as regards computational perspectives, let us underline that music production becomes a social activity when students play together. A music coding activity extended to a class not only provides children with self-consciousness, but fosters relationships to other children and the surrounding environment. Moreover, a music coding environment can implement features recalling social-game mechanisms to allow peer-seeking (searching for the helping hand of a friend) and peer-reviewing (asking for other students' comments). Consequently, in our opinion the adoption of Pd - or a similar music programming framework - may provide not only music skills while fostering creativity, but even teach computational thinking better than other "traditional"coding environments.

5 Conclusion and Future Work

Our research aims to define and consolidate a theoretical and applicative framework in order to update school curricula with music coding. This concept implies a comprehensive experience that matches the pedagogical purposes of computational thinking with the development of personality typical of music education and practice. As a result of this theoretical work, a set of Pd patches will be designed and implemented. After their release, a validation stage is needed to ensure that such an educational tool may satisfy the user's needs, here intended both from the teachers' and from the students' standpoint.

In particular, educational validation will be based on the detection of ad hoc indicators and metrics, in order to measure how music coding is able to support the aforementioned axes of computational thinking: computational concepts, computational practices, and computational perspectives. The dimension of the control and experimental group in primary schools is currently under definition, and future agreements will be established with Italian Territorial Centers for School Inclusion (CTI), Territorial Support Centers (CTS) and the Italian Ministry of University and Research (MIUR).

References

1. Kellett, M.: Developing critical thinking skills in 10–12 year-olds through their active engagement in research. Teach. Think. Skills **14**, 32–40 (2004)
2. Schweinhart, L.J.: The high/scope approach: evidence that participatory learning in early childhood contributes to human development (2006)
3. Piaget, J., Beth, E., Brousseau, G.: Learning: creation or re-creation? from constructivism to the theory of didactical situations. A Critique Creativity Complexity: deconstruct. Clichés **25**, 19–33 (2014)
4. Papert, S.: Mindstorms: children, computers, and powerful ideas. Basic Books, Inc. (1980)
5. O'Dell, J.: Why your 8-year-old should be coding. http://venturebeat.com/2013/04/12/why-your-8-year-old-should-be-coding/ (2013)
6. Duncan, C., Bell, T., Tanimoto, S.: Should your 8-year-old learn coding? In: Proceedings of the 9th Workshop in Primary and Secondary Computing Education, ACM, pp. 60–69 (2014)
7. Armoni, M.: Designing a K-12 computing curriculum: the questions. ACM Inroads **4**(2), 34–35 (2013)
8. Rushkoff, D.: Program or be programmed: ten commands for a digital age. Or Books (2010)
9. Resnick, M., Maloney, J., Monroy-Hernández, A., Rusk, N., Eastmond, E., Brennan, K., Millner, A., Rosenbaum, E., Silver, J., Silverman, B., et al.: Scratch: programming for all. Commun. ACM **52**(11), 60–67 (2009)
10. Wing, J.M.: Computational thinking and thinking about computing. Philos. Trans. R. Soc. A: Math. Phys. Eng. Sci. **366**(1881), 3717–3725 (2008)
11. Ioannidou, A., Bennett, V., Repenning, A., Koh, K.H., Basawapatna, A.: Computational thinking patterns. Online Submission (2011)
12. Brennan, K., Resnick, M.: New frameworks for studying and assessing the development of computational thinking. In: Proceedings of the 2012 annual meeting of the American Educational Research Association, Vancouver, Canada, Citeseer (2012)
13. Ananiadou, K., Claro, M.: 21st century skills and competences for new millennium learners in oecd countries (2009)
14. Binkley, M., Erstad, O., Herman, J., Raizen, S., Ripley, M., Miller-Ricci, M., Rumble, M.: Defining twenty-first century skills. In: Assessment and teaching of 21st century skills, pp. 17–66. Springer (2012)
15. Lye, S.Y., Koh, J.H.L.: Review on teaching and learning of computational thinking through programming: what is next for K-12? Comput. Hum. Behav. **41**, 51–61 (2014)
16. Mills, K.A.: A review of the digital turn in the new literacy studies. Rev. Educ. Res. **80**(2), 246–271 (2010)

17. Hague, C., Payton, S.: Digital literacy across the curriculum. Curriculum Leadership **9**(10) (2011)
18. Ng, W.: Can we teach digital natives digital literacy? Comput. Educ. **59**(3), 1065–1078 (2012)
19. Grover, S., Pea, R.: Computational thinking in K-12 a review of the state of the field. Educ. Res. **42**(1), 38–43 (2013)
20. Young, S.: Time-space structuring in spontaneous play on educational percussion instruments among three- and four-year-olds. Br. J. Music Educ. **20**(1), 45–59 (2003)
21. Jorgensen, E.R.: Philosophical issues in curriculum. In: The new handbook of research on music teaching and learning, pp. 48–62 (2002)
22. Barrett, M.S.: Sounding lives in and through music - a narrative inquiry of the everyday musical engagement of a young child. J. Early Child. Res. **7**(2), 115–134 (2009)
23. Tomlinson, M.M.: Literacy and music in early childhood. SAGE Open 3 (2013)
24. Dewey, J.: Art as experience. Penguin (2005)
25. Willems, E., Podcaminsky, E.: Las bases psicológicas de la educación musical. Eudeba (1969)
26. Branca, D.: L'importanza dell'educazione musicale: risvolti pedagogici del fare bene musica insieme. Studi sulla formazione **15**(1), 85–102 (2012)
27. Finney, J., Burnard, P.: Music education with digital technology. Bloomsbury Publishing (2010)
28. Vercoe, B.: Extended Csound. In: Proceedings of the International Computer Music Conference, International Computer Music Association, pp. 141–142 (1996)
29. Boulanger, R.C.: The Csound book: perspectives in software synthesis, sound design, signal processing, and programming. MIT press (2000)
30. Petri, C.A.: Introduction to general net theory. In: Net theory and applications, pp. 1–19. Springer (1980)
31. Haus, G., Rodriguez, A.: Music description and processing by Petri nets. In: Advances in Petri Nets 1988, pp. 175–199. Springer (1988)
32. Baratè, A., Haus, G., Ludovico, L.A.: Real-time music composition through P-timed Petri nets. In: Georgaki, A., Kouroupetroglou, G. (eds.) ICMC|SMC|2014 Proceedings, Athens 14–20 September 2014, pp. 408–415. Greece, Athens (2014)
33. Jorda, S., Kaltenbrunner, M., Geiger, G., Bencina, R.: The Reactable. In: Proceedings of ICMC 2005, pp. 579–582, Barcelona, Spain (2005)
34. Kaltenbranner, M., Jorda, S., Geiger, G., Alonso, M.: The Reactable: A collaborative musical instrument. In: Enabling Technologies: Infrastructure for Collaborative Enterprises, 2006. WETICE'06, IEEE, pp. 406–411 (2006)
35. Puckette, M., et al.: Pure Data: another integrated computer music environment. In: Proceedings of the Second Intercollege Computer Music Concerts, pp. 37–41 (1996)
36. Sarlo, J.A.: GrIPD: A graphical interface editing tool and run-time environment for Pure Data. In: Proceedings of ICMC 2003, International Computer Music Association, pp. 305–307 (2003)

Methods and Technologies for ICT Workers Virtual Mobility

Danguole Rutkauskiene and Daina Gudoniene

Abstract The use of information communication technologies (ICT) in continuing and professional development is one of the possibilities for ICT workers to be oriented to job mobility or virtual mobility. One of the main objectives of the research is to develop prototype of e-Academy for older workers to improve engagement with ICT and benchmark and validate skills and experience against the EQF and to develop a platform for older worker's that will be user friendly and will provide individuals with an opportunity to digitally their skills, experience and qualifications. ICT are important enablers of the new social structure. The aim of the paper is to present a prototype of the platform and virtual mobility, introducing software to the older workers for continuing and professional development. The role of ICT will be analysed for ICT workers in developing communication, cognitive, labour and business skills and competencies, to help them gather information (knowledge), to develop skills and abilities and creatively, to apply them in practice for achieving the teaching (learning) goals via e-Academy platform.

Keywords ICT · Learning platform · Prototype · ICT workers

1 Introduction

Looking from an employers' perspective, the researchers are focusing on policies and practices to foster the older ICT workers' virtual mobility. We are going to start a discussion about what employers and governments could do to support the mobility of older ICT workers in the field of geographic, job and virtual mobility, but in this

D. Rutkauskiene (✉)
Kaunas University of Technology, Kaunas, Lithuania
e-mail: danguole.rutkauskiene@gmail.com

D. Gudoniene
Baltic Education Technology Institute, Kaunas, Lithuania
e-mail: daina.gudoniene@gmail.com

© Springer International Publishing Switzerland 2015   459
V.L. Uskov et al. (eds.), *Smart Education and Smart e-Learning*,
Smart Innovation, Systems and Technologies 41,
DOI 10.1007/978-3-319-19875-0_41

paper we will discuss about the virtual mobility of ICT workers and their continuing and professional development. Currently not only in Lithuania but also in the whole world social and economic changes have evolved, which require more and more skills from the various specialists. In the rapidly changing society a person, who is trying to gain more knowledge and experience in his field of work due to trace of the social situations and adapts to modern life, becomes a successful one. An increasing number of the web-based software packages have been developed to enhance the teaching and design of control systems [5]. The rapid development of a permanent information and communication technology (ICT) changes and diversity also impacts the society changes. The developments and use of ICT is an important element of the European Union's strategy in order to ensure the educational effectiveness and competitiveness. Since 2007 the use of ICT in education has become one of the key issues [8–10]. The digital excellence is recognized as one of the eight skills necessary for all knowledge communities. The aim of the papers is to explore ICT applications, based on open source architecture for e- learning approach and to present the prototype of e-Academy for older workers to improve engagement with ICT and benchmark and validate skills and experience against the EQF.

2 Virtual Mobility

2.1 Overview on ICT and Methods for Virtual Mobility

The virtual mobility refers to ICT workers in continuing education in another institution outside their own country to study for a limited time, without physically leaving their home. The virtual mobility has been defined as an activity that offers an access to courses and study schemes in a foreign country and allows for the communication activities with teachers and fellow students abroad via the new information and communication technologies. The researchers are working on the prototype of the platform and tools using different technologies and allowing learners to communicate and collaborate with teachers, among themselves and with other internet users. On the technological side, trends towards high-quality, converging, mobile and accessible technologies, together with more sophisticated, user-friendly, adaptable and safe applications and services will integrate technology more and more into everyday life. As a consequence, technology will be more smoothly integrated into our daily lives and become a basic commodity [13–15] and e-learning is a general term for a variety of ways of combining different elements of on-line delivery. The use of technology may be integrated within the educational institution, either in class, or within lab work, or the technology use may be restricted to use by the students outside class time, possibly at home. Designing an effective use of technology to support teaching and learning, involves integration three different kinds of knowledge Content Knowledge, Pedagogical Knowledge and Technological Knowledge – as illustrated on Fig. 1 diagram of the TPACK Framework.

Fig. 1 The role of ICT for
future learning strategies [16]

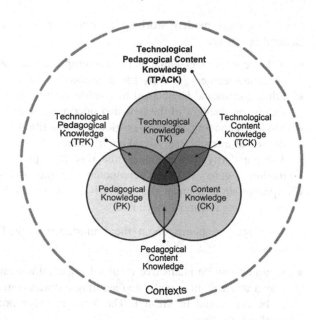

There is a growing evidence that innovative ICT tools can enhance the learning delivery for learners and their engagement with mainstream education. The great strength of such learning tools is their capacity to support the informal learning, which provides a secure environment for acquiring knowledge and rebuilding confidence among learners. The innovation of web technologies will transform the teaching process into a student centred learning process. The technology should allow learners to upgrade themselves from being a passive knowledge consumer (to whom knowledge is pushed) into an active contributor in a social constructionist process (knowledge building) [11, 12]. However for the new technologies of informatics and computer sciences a different approach is needed. For scientific collaboration the possibility to run a software developed by colleagues by internet is an essential. One can test directly results of other researchers by running their software with different data. Therefore the algorithms, software and the results published in the scientific papers can be investigated independently [3, 5].

2.2 Technological Solutions for Virtual Mobility

One can read a lot of articles on innovation, which suggests a set of tools that are considered to be an innovative and thus makes believe that the use of which could be important in education. Nowadays, in the fast moving and changing the world informational communicational technologies play a significant role in the education, industry and business [1, 4]. The adult learners are no exception: they need to adapt to a more modern working environment, improve their ICT skills [2, 4]. General points are:

- ICT becomes an essential as the contemporary education system is focused on preparing learners for daily life & problem solving;
- Education should be adapted to volatile circumstances in a fast moving world: follow the changes of the new and upcoming ICT;
- Modern economy encourages global competition & education should not be limited to a traditional school environment;
- The popularity of ICT change the way how people communicate, find information & gain knowledge (computers, internet, radio, TV in various forms);
- Appropriate & competent usage of ICT improves education system. Nowadays this process is almost inevitable.

More specific points about the importance of ICT in the learning process (studies) are:

- Learning can be interactive & based on communication if ICT are used. This way a wider & more motivated learning environment in an education institution & beyond could be created. The learners solve problems communicating or working together;
- ICT usage allows learning to be applied according to the individual needs, learning content (what do we learn?) & methods (where & how do we learn?);
- Learning can be done anywhere, using a computer, mobile phone etc. In this situation it is important to make individual tasks for learners with special needs (blind, deaf or very talented ones);
- There is no a need to teach all learners at once, as ICT can control the process; this is very convenient for the adult learners who have a job or look after their children.

In addition, there is a subset of educational institutions in the world that have not applied (or just starting) ICT in classes. To tackle this problem, topic about ICT & institution integration should be discussed. Especially: teaching about information communication technologies, using information communication technologies in various lessons or lectures, applying newest technological solutions into the management of educational institution, creating a virtual learning environment. Another important thing about ICT is that it has a lot of useful features, related to continuous learning:

- Flexibility in respect of time & place (learning at home, using a virtual learning environment or distance learning).
- Flexibility in respect of learning material (courses are prepared for example according to organization needs).
- Easy access to information & other people.
- Convenient communication with other people using online resources.
- New approach to organizing learning (individualization of the process, better preparation & control of the learning material).

Despite many advantages of information communication technologies in adult learning, it is necessary to encourage adults to use ICT and show how they can facilitate the learning process. The researchers are working on prototype of e-Academy system that encourages learners' creativity, the ability to work in a team, ability to communicate in the global IT space [6, 7]. ICT can be very effective in developing the communication, cognitive, labour and business skills and competencies, to help students gather information (knowledge), to develop skills and abilities and creatively apply them in practice for achieving the teaching (learning) goals.

3 Prototype of e-Academy Platform

3.1 High Level Description of e-Academy System Design

The main idea how to meet purpose and requirements of the e-Academy is to build universal "My Jobs" plugin for Moodle, which can be installed not only into the e-Platform but also into other already existing Moodle based systems as it is required (Fig. 2).

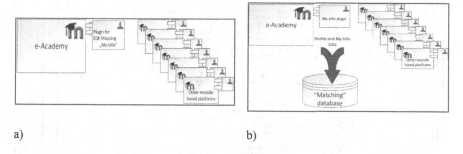

a)   b)

Fig. 2 "My Jobs" plugin can be installed on as many Moodle instances as necessary

The plugin would collect data about user jobs and push it to "Matching" database, based on architecture then we have one "Matching" database and possibly many "My Jobs" plugin installation decisions for "Matching" database to be as a control centre of "My Jobs plugin" was made. All the data from "My Jobs" plugin will be pushed to "Matching" database (Fig. 2(a), (b)). This allows to control all the installed plugins from one place, update questions and other data instantly and without an interruption of Moodle admins. LMS for e-Academy - Moodle 2.7. This Moodle version was chosen because on e-Academy designing period it was latest stable Moodle version.

3.2 Functionality and Design

"My Jobs" plugin is seen as block on the left side of page. The middle part of the page is for courses list. The design is flexible, for example, in the first page can be shown the news, or other information about system, or project if necessary. By pressing "Manage your jobs" the link opens window, where the user can add, edit his jobs, qualification level and how the person meets standardized job description. By pressing "Edit" button on the top right side of qualification the user can change qualification name, level and notes. By pressing "Not sure?" the user can access the test which helps to determine knowledge, skill and qualification level according to EQF description. By pressing "How I fulfil the description of this job profile" that opens standard description of selected job. The user can deselect parts which are not relevant to his/her actual job.

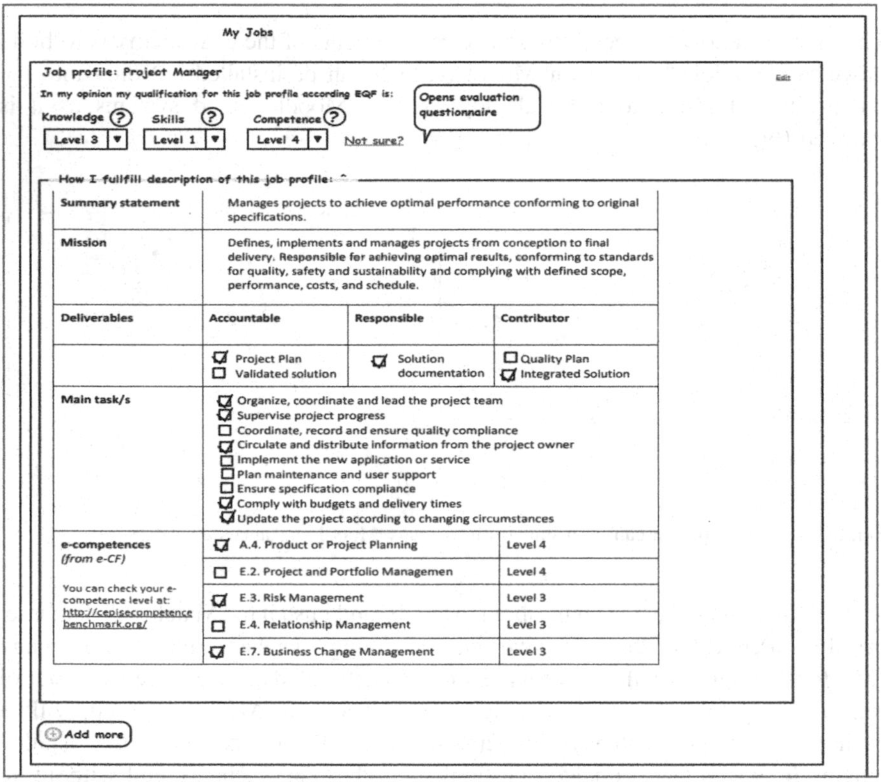

Fig. 3 Functionality and design

EQF Mapping page accessible by pressing "Not sure?" link on My Jobs page. How EQF Mapping page looks can be seen on figure (Figs. 3 and 4).

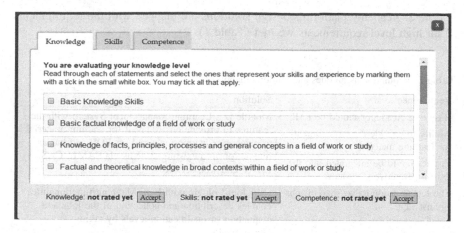

Fig. 4 EQF mapping page

The knowledge, skills and competences tests can be chosen from tabs seen on the top of the page. The user can accept test result by pressing "Accept" button on the bottom of page, after the completion a test. Then the user press "Accept" button and the result of test sets as the qualification level and seen on "My Jobs" page. The e-Academy designed in the way to support many languages. Currently will be installed English, Deutsch, Lithuanian, Polish and Greek languages. "My Jobs" plugin also will support English, Deutsch, Lithuanian, Polish and Greek languages. "My Jobs" plugin related to the database tables in Moodle. The database tables located are located in the Moodle database and related to "My Jobs" plugin as it is shown (Fig. 5).

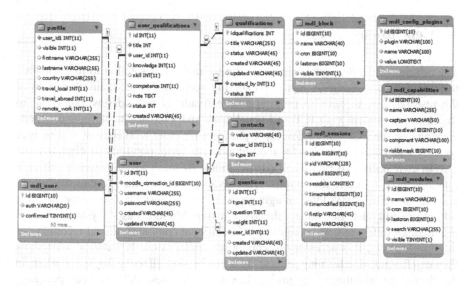

Fig. 5 Database tables located in Moodle database and related to "My Jobs" plugin

The e- Academy platform's design solutions are mapped with the requirements of all high level requirements we met (Table 1).

Table 1 Requirements and design solutions mapping

Requirement	Solution
Base of e-Academy should be LMS	Moodle is a Learning Management System with the features to provide environment for learning material, to categorize learning material for EQF and has an integration into existing systems via SAML, CAS, LDAP and web services
Ability to provide the environment for learning material	
The feature to be integrated into the existing employers' systems	
The feature to digitalize map of jobs against EQF	"My Jobs" plugin with EQF Mapping functionalities enables description and collection of the jobs and evaluation of qualification levels by taking questionnaire
Extra benefit	Solution to use Moodle plugin for EQF Mapping opens extra sources of potential employees providing possibility to install plugin into many Moodle based systems

The implementation of educational concepts/models and the new learning and teaching methods, based on modern ICT, needs an involvement of competitive teachers and support staff. This is particularly very important where the investments to infrastructure and the new equipment should be planned together with the large scale of ICT users on how to use ICT, based on innovative learning and teaching methods in their learning practice.

4 Conclusions

The novelty of the research is shown in the technological solution to use semantic technologies for solving the meta data sharing according to the main objectives of the research, i.e. to develop prototype of e-Academy for older workers and to improve engagement with ICT and benchmark and validate the skills and experience against the EQF and to develop a platform for older workers that will be user friendly and will be provided for individuals with an opportunity to digitalize the map of their skills, experience and qualifications. According to the research, there is intended how to meet the purpose and requirements of the e-Academy where is built universal "My Jobs" plugin, which can be installed not only into the e-Platform but also into other already existing Moodle based systems if it is required. Despite many advantages of informational communicational technologies in adult learning, it is necessary to encourage adults to use ICT and to show how they can facilitate the learning process. ICT can be very effective for ICT workers in developing communication, cognitive, labour and business skills and competencies and to help them gather information (knowledge), to develop skills and abilities and creatively apply them in the practice for achieving the teaching (learning) goals through e-Academy platform.

Acknowledgments This paper has been partially supported by CaMEO/Career Mobility of Europe's Older Workforce: Improving and extending the employability of an ageing workforce in Europe through enhancing mobility, Project Number 539099LLP-1-2013-1-UK-LEONARDO-LNW.

References

1. Redecker, C.: Learning2.0: The Impact of Web2.0 Innovation on Education and Training in Europe. Spain, p. 122 (2009)
2. Burneikaitė, N., Jarienė, R., Jašinauskas, L., Motiejūnienė, E., Neseckienė, I., Vingelienė, S.: Informacinių komunikacinių technologijų taikymo ugdymo procese galimybės. Vilnius, p. 231 (2009)
3. Gray, D.E., Ryan, M. Coulon, A.: The training of teachers and trainers: innovative practices, skills and competencies in the use of eLearning. Eur. J.f Open Dist. E-Learn. http://www.eurodl.org/?p=archives&year=2004&halfyear=2&article=159
4. Rutkauskienė, D., Gudonienė, D.: E. švietimas: tendencijos ir iššūkiai. Konferencijos pranešimų medžiaga: Web 2.0 saitynas. Vilnius, p. 110 (2005)
5. Mockus, J.: Investigation of examples of E-education environment for scientific collaboration and distance graduate studies. Part 2. Informatica **19**(1), 45–62 (2008)
6. Kuzucuoglu, A.E., Gokhan, E.: Development of a web-based control and robotic applications laboratory for control engineering education. Inf. Technol. Control **40**(4) (2011)
7. Vitiutinas, R., Silingas, D., Telksnys, L.: Model-driven plug-in development for UML based modelling systems. Inf. Technol. Control **40**(3) (2011)
8. Pečiuliauskienė, P.: Kompiuterizuoto mokymo metodai pradedančiųjų mokytojų edukacinėje praktikoje, p. 89 (2008)
9. Grodecka, K., Wild, F., Kieslinger, B.: How to Use Social Software in Higher Education, p. 130 (2009)
10. Rutkauskiene, D., Huet, I., Gudoniene, D.: E-Learning in Teachers and Tutors Training Using ICT Based Curriculum. (2010)
11. Turcsányi-Szabó, M.: Aiming at sustainable innovation in teacher education—from theory to practice. Inf. Educ. **11**(1), 115–130 (2012)
12. Bower, M., Hedberg, J.G., Kuswara, A.: A framework for Web 2.0 learning design. Educ. Media Int. **7**(3), 177–198 (2010)
13. Lorenzo, G., Ittelson, J.: An Overview of E-Portfolios, p. 29 (2005)
14. Redecker, C., Leis, M., Leendertse, M., Punie, Y., Gijsbers, G., Kirschner, P., Stoyanov, S. and Hoogveld, B.: The Future of Learning: Preparing for Change, p. 94 (2011)
15. Source: http://tpack.org—Reproduced by permission of the publisher, 2012 by tpack.org

Extensiveness of Manufacturing and Organizational Processes: An Empirical Study on Workers Employed in the European SMEs

Nunzio Casalino, Marco De Marco and Cecilia Rossignoli

Abstract To enhance the competitiveness and the turnover of businesses in the fast-paced world of today means adopting the emerging Automation & Robotics (A&R) equipment that, in turn, require the upskilling and training of employees to enable effective implementation of the business processes. This poses a significant challenge to manufacturing Small and Medium Enterprises (SMEs), which often lack the funds and human resources to invest in staff training, and translates into a distinct need to provide comprehensive A&R upskilling and training programs in the form of smart learning solutions designed to enable SME personnel to pursue their training goals in a flexible, customizable and easy-to-access way. The paper sheds light on the perceived upskilling requirements of SME workers and the broad scope of A&R systems training materials developed by the innovative methodology of ARIALE smart learning system.

Keywords Manufacturing SMEs · Competitiveness · Innovation · e-learning · Managerial skills

1 Introduction

To strengthen European SMEs, it is necessary to customize, integrate, test and validate the emerging innovative technologies and processes. For Europe's competitiveness in manufacturing, it is crucial that advances are taken up in engineering

N. Casalino (✉)
Università Degli Studi Guglielmo Marconi, Rome, Italy
e-mail: n.casalino@unimarconi.it

M. De Marco
Università Telematica Internazionale Uninettuno, Rome, Italy
e-mail: marco.demarco@uninettunouniversity.net

C. Rossignoli
Università Di Verona, Verona, Italy
e-mail: cecilia.rossignoli@univr.it

© Springer International Publishing Switzerland 2015   469
V.L. Uskov et al. (eds.), *Smart Education and Smart e-Learning*,
Smart Innovation, Systems and Technologies 41,
DOI 10.1007/978-3-319-19875-0_42

and manufacturing "at large" as soon as they have the appropriate maturity level. With the advent of online learning methodologies [1], providers of education and training have to create business-learning materials to fulfil the new shrewd demand [2]. Digital technologies combined with suitable learning strategies can help to create open, dynamic and flexible learning environments with implications for countless applications with respect to education and training. Academic institutions, corporations, and government agencies worldwide are increasingly using Internet and digital technologies to deliver instruction and training [3]. What does it take to create a flexible and distributed learning environment for professional learners? Well, a learning environment should be meaningful to all stakeholders, including learners, instructors and support services staff. It is meaningful to learners when it is easily accessible, well designed, learner-centered, affordable, efficient and flexible, and has a facilitated learning environment [4]. When learners display a high level of participation and success in meeting a course's goals and objectives, learning becomes meaningful for instructors. In turn, when learners enjoy all available support services provided in the course without any interruptions, it makes support services staff happy as they strive to provide easy-to-use, and reliable services [5]. Finally, a learning system is meaningful to organizations when it matches the return on investment (ROI), a good level of learner satisfaction, and a low dropout rate. The purpose of this paper is to accompany SMEs through the various factors to adopt open, flexible and distributed smart learning environments. It contains many practical suggestions that it is possible to follow as review criteria that learners should expect [6]. Items in the checklist encompass the critical dimensions of an e-learning environment, including institutional, managerial, pedagogical, and technological aspects. Who can benefit of a smart environment on Automation & Robotics (A&R) fields? First of all a wide range of professionals as:

- Instructors can develop courses for SMEs that can be delivered [7] as: distance education, e-learning, blended learning, online education, web-based training, distributed learning, educational technology, corporate training, etc.
- Instructors, teachers, trainers, training managers, distance education specialists, e-learning specialists, virtual education specialists, e-learning project managers, instructional designers, corporate education specialists, and human resources specialists, can plan, design, and evaluate learning modules for SMEs
- Consultants, human resources managers, department of education staff, Ministry of Education staff, virtual and corporate university [8] teachers can develop strategic plans [9] for designing, evaluating and implementing e-learning initiatives

2 Key Role of SMEs for the EU Economic Development

As described on the European Small Business Survey 2012, small and medium enterprises (SMEs) comprise the 99 % of all enterprises in Europe and take in average more than 60 % of engaged workers. However only 24 % of SMEs provide vocational education and training compared to 80 % of large enterprises (employing over 250 people). SMEs play a key role in generating employment and creating economic wealth, but skill deficiencies in SMEs are adversely affecting their ability to reach their growth potential [10]. By their very nature, SMEs are small, constrained by time and budget and reluctant to engage in learning/training programs. Our empirical study is based on the preliminary results of the ongoing research project "ARIALE - Automatization, Robotization for a New Reindustrialized Europe" (project ID: 2013-1-PL1-LEO05-37568). Its goal is to investigate how SMEs can be engaged in appropriate learning actions [11] to address the automation and robotics fields (www.ariale.eu – Fig. 1). The first SME studies appeared only in the 1960 s. These organizations were largely ignored for a long time, until several reports focused on the significant contribution of SMEs for the European economic development [12]. Since then, research in small business has grown examining their perspectives and inside environments.

Fig. 1 Some screenshots of the "ARIALE - Automatization, Robotization for a New Reindustrialized Europe" smart learning platform - www.ariale.eu

A common theme in this research points to its fragmentation and its failure to provide conclusive evidence about Automation & Robotics development in SMEs and learning needs for their penetration [13]. Another theme emerging points to different deficiencies at the different contextual levels (technological, organizational including managerial, environmental, and individual) and to the lateness of the SMEs in adopting Automation & Robotics. What could be first synthesized from this assessment is that Automation & Robotics are characterized by multi-faceted perspectives and represents fields too large and complex to be encapsulated within one study, one discipline, or one methodology of adoption for SMEs. Still, what makes SMEs decide to adopt effectively automation and robotics technologies is not conclusive and remains the subject of considerable debate among researchers. There is a consensus among researchers and consultants that is possible to develop an effective training system [14] with the appropriate amount of knowledge [15]. How significant is the actual use of traditional courses and e-learning technologies in SMEs? This leads naturally to the question of what makes SMEs adopt or reject e-learning? In addition, how deep is the actual penetration of e-learning in SMEs business environment?

3 Requirements Gathering and Needs Analysis

ARIALE partners have elaborated and distributed among project target groups a detailed questionnaire for requirements gathering and needs analysis regarding the way to plan and improve skills and professional knowledge [16] for SMEs. A total of 103 questionnaires were filled by the target groups: SMEs managers, employees, trainers and consultants involved in automation and manufacturing jobs. The answers were distributed among the partner countries as follows: 36 from Bulgaria, 25 from Italy and 42 from Poland. The answers collected at the end of survey analysis allowed achieving a good level of knowledge about SMEs needs in the automation and robotics training field. The results of the survey illustrate how professional associations and consultants may contribute to the use of e-learning technologies [17] for SMEs. It also shows that – besides significant cost savings – there are further advantages that make the use of e-learning technologies attractive. In fact online learning can provide several good opportunities to SMEs in overcoming part of their technological, environmental, organizational, and managerial inadequacies. Savings of travel or hotel costs are an obvious advantage of e-learning compared to face-to-face sessions. There are, however, also a number of other benefits that have been reported by the target groups:

- **Time independence**. Learning activities can be carried out by managers and employees in the evening or during the weekend
- **Focus**. In their daily work, managers and employees focus on the essentials. This working style is better supported by e-learning rather than face-to-face seminars
- **Learning speeds**. People's learning speeds differ, which makes traditional training courses less effective for those participants who grasp things at a slower pace and who are often reluctant to keep asking for explanations
- **Advanced learning culture in SMEs**. The application of e-learning as part of the learning opportunities in a SME company leads to develop a new learning culture

4 Lessons Learned and Learning Principles

The survey included 103 people in Bulgaria, Italy and Poland. Below is a summary of the results of the analysis.

4.1 Profile of the Respondents and Preferences

The majority of the respondents to the survey in all partner countries were within the age group 26–45 years, which was not surprising as all of the respondents are employed in SMEs or are trainers/teachers. In terms of gender the naturally for this economic sector the majority of the respondents in all partner countries are male. There were no respondents below Upper secondary level of education. The most predominant group were respondents with a master degree. However, bachelor's and PhD degrees were also not rare. The job positions of the interviewed varied a lot among the different countries. However, in the aggregated results there was balanced distribution between all levels within the organization: Manager (17); Executive (13); Middle manager (13); Consultant (7); Office-worker (15); Skilled worker (24); Unskilled worker (0); External supervisor (3); Other (15). The largest group of the respondents (63) declared that they have been working for more than 5 years – which means very experienced people have answered to the questions regarding the course. As for the computer skills, almost all respondents stated that they are Good or Very good. The close percentages show a positive trend that more and more people are taking online courses and see the benefits [18] from such kind of training. With regard to the possible methods for course subscription, a convincing majority of 62 % have voted for online registration form, while the second best choice selected by 32 % was via e-mail. Therefore, papers and phones should be limited in the registration process. The most preferred examination time for 32 % of the respondents were after the class/course and before, in the middle and after the class/course. The 25 % pointed that they prefer to be examined during the class/course. All respondents except one have chosen some form of examination, so the mobilizing effect of the examination process should not be underestimated in the ARIALE course. The most desired form of exam is via practical assignment (49 respondents), followed by online based exam at home and/or work without supervision (37). In order to answer to these requirements the ARIALE course may give a choice to the learners at the end of the training whether to prepare some practical project or take an online exam.

4.2 Organizational Aspects of the Virtual Smart Learning Environment

The respondents considered all aspects of the virtual smart learning environment important, leading are the Overall functionality (37 %) and Ease of use (32 %). Navigation has seen as important by 14 %, while equal number of people perceive Design/layout and Information glossary as significant. Therefore, in the development of the ARIALE course special attention should be paid to the overall functionality of the smart platform. The communication channels to be used with the other learners are the following: Online forum - 50 respondents; E-mail - 40 respondents; Chat - 23 respondents; Video conference - 16 respondents; Audio conference - 6 respondents. About 75 % of the interviewed have taken a classroom course in the last 3 years. However, only 31 % have taken an online course. 62 % of the workers (mainly in Bulgaria and Poland) believe that an online course could be effective for professional training [19] as a classroom course. Almost equal percentages of respondents think that an online course will allow to cover a larger variety of topics that are not covered in a regular classroom course. About 83 % of the respondents require a short introduction to the course presenting the main goals. Analogically, 86 % want to be acquainted on the course's framework about its duration, requirements, etc.

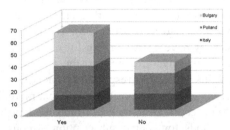

Number of people interviewed who think or not that an online training can be used as effectively as classroom training as far as professionals are concerned.

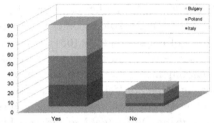

Number of people interviewed that need or not a short introduction to the course in the beginning acquainting them with its goals.

About the 81 % of the respondents prefer to have simplified interface and an easy workload of the e-tasks. In addition, 70 % of the respondents consider that it should be a compulsory precondition that the test questions are correct before to continue studying. A particularly high number - 74 % of the respondents stated that in their previous training experiences the content of the materials explained the knowledge and concepts well. In addition, a high number of respondents think that e-collaboration and teamwork activities can improve the learning results [20]. E-tasks and online assignments are the most preferred approach for the respondents (54 %). Also not surprisingly 90 % of the respondents think that a short presentation about the training contents by the teacher or instructor can positively affect participant understanding ability. Analysis of the questionnaires resulted in 70 % of the respondents stating that the previous courses in which they took part had met at least some of their objectives, and only 58 % stating that they would be able to apply their learning in their personal and professional life within 12 months. There were 92 % of participants who agreed that it is useful to share with others and 66 % felt that they could become managers that are more effective. It was interesting to note that whilst there was a poor response to questions about the on-line collaboration area, 85 % of participants found the help of the tutors/facilitators very useful. The 78 % of the respondents want to be in contact with other training participants through a virtual classroom [21]. For many of the respondents the availability of the teacher or instructor to provide help when needed (by email, phone, etc.) is considered very important (89 %). A substantial number of the respondents (80 %) confirmed that they would like a course available off-line. The topics included and evaluated are the following:

- ICT based means for automation and innovation
- Sensors in industrial automation
- Actuators in industrial automation
- Application of PLC in industrial automation
- Industrial networks and interfaces in industrial automation systems
- Industrial robots in automation systems

5 Traditional Classroom Course: Respondents' Perceived Requirements

According to the survey respondents, many different requirements and factors contribute to the success of the traditional classroom course:

Participant

- Preliminary learning background
- Motivated participants
- Small group of participants
- Possibility to interact with other course participants
- Opportunity of networking and making new contacts

Teacher/Trainer

- Motivation of the lecturer
- Skilled lecturer able to present the training material in an interesting way
- The attitude and charisma of the lecturer to give participants the impression that what they learn is important and will be practically applicable in their workplace
- To have good oratory skills, to attract participants' attention

Organizational Environment

- Comfortable working environment
- The number of trainees should be considered respect to a traditional classroom
- Focus and concentration
- Good relations and interaction between teachers and trainees
- Possibility of exchange of opinions and experiences

Learning Materials and Facilities

- Availability of business cases and concrete trials
- Ensuring conditions for practical assignments

- Short video and audio clips
- Well-structured presentation materials and training accuracy
- Clear learning tasks and interface easy to use
- Actual and practically useful topic for the trainees
- Self-check test for checking the learned material
- Opportunity to provide feedback and communicate with the mentor/trainer
- Availability of links to external contents and websites
- Clearly formulated topics and tasks
- Use of images to make easier to memorize the contents
- Flexibility and possibility to learn from home

5.1 Perceived Barriers to Online Training and Upskilling Courses

Most of the participants in the survey have not taken online courses and therefore could not share first hands experience. However, the opinions of the respondents who answered to this question tended to converge around several common barriers:

- Lack of personal contact with the trainer and of possibility for discussions
- In a classroom there is the chance to participate and can be a motivating factor
- The learner has to have very strong internal motivation to take the course as well as systematic work is necessary

Some of the respondents were very fond of online courses as they encourage self-learning and ensure opportunity to learn at any time [22]. Also online courses allow covering more diverse topics, which are not covered in the traditional classroom training and shortens the time for preparation. There were various recommendations regarding the form and presentation of the online course. Most prevailing opinion was that the lessons should be no longer than 45 min. One of the respondents suggested that all the lessons are presented as video clips and are accompanied by a pdf file explaining the material presented in the clip. The distribution of the learning material should be considered along with the time allotment for the course – over flooding with information should be avoided. Participants should have the opportunity to put into practice what they learn [23]. There should be practical tasks envisaged after each module as well as conditions where the participants can perform these tasks. Learners should be given the opportunity to choose for themselves which modules to study. A common suggestion was that the course should have a short theoretical part and significant practical part. Multiple choice tests and short open questions were the most preferred forms of examination suggested by the respondents in the open-end questions of the survey. Simply tests without open-end questions were seen as an easier way to cheat, while open-end questions would reduce this risk. An interesting observation was that the questions before the course give the trainer an idea about the knowledge level [24] of the participants. Another observation was that short questions are a good idea during the course but it should be carefully used to not discourage participants. Some of the respondents expressed doubts in the self-assessment as a tool and that results from tests could be not very reliable. One of the respondents recommended that there should be tests after each chapter/module. At the end of the course, one general test is recommended to check the overall understanding [25] of the learning contents.

6 Requirements, Considerations and Conclusions

From the analysis of the collected ARIALE questionnaires on end users requirements, many considerations emerged:

- The target group involvement in competences' development and contents' improvement can have a very positive effect on SME's competitiveness [26]
- Formal methods of teaching and learning are not necessarily the most appropriate way of engaging, motivating and transferring knowledge to SMEs workforce. So formal training is not the best way of learning for SMEs. Instead, non-formal and informal learning can develop skills and competencies required at workplace
- Training activities have to be focused on the specific needs of the SMEs for example giving the possibility to assemble learning contents in the smart platform
- The SMEs' heads frequently own a negative attitude to change and learning. In many cases, time devoted to learning activities is considered as lost time. To get effective motivation the learner [27] should be put in the center of learning activities
- SMEs are driven primarily by profit and they are focused especially on bottom line. The role of promotion is very essential. No matter how good the training and support material, it has to be carefully promoted from the chief
- Learning for many SMEs' heads has seen unfortunately as a cost, and they do not always consider it as an investment for the future. The curricula should have a measurable impact within the organization [28] and should be affordable
- SMEs use a short-term approach; they only set up a training plan on Automation or Robotics only when they face meet problems. Approaches to learning and training in small firms needs to take account of the shorter planning periods they adopt
- An informal environment should be built to encourage networking. It can also strengthen and connect SMEs in order to build or enhance their business at the regional, national, or international level

Other target groups' comments highlighted some successful and not successful aspects to evaluate for the development of a smart LMS such as:

- Enjoyable and useful live face-to-face sessions can encourage virtual collaboration
- Some contents have to be specifically focused on management issues
- Structured and easy tools can be very helpful for inexperienced online participants
- Expert participants contribution as platform facilitators and tutors
- Collaborative aspects of the smart learning platform are confirmed as essential
- 24 h/24 and access to external resources (such as e-libraries) look valuable
- Previous experiences showed that simulations were unusable and not intuitive
- Not well-structured presentation of materials or graphs
- Lack of collaboration and of direct contact with other people
- Clearer structure at the beginning and tasks with clear deadlines

As conclusion, it is possible to involve successfully workers by using an engagement strategy that communicates real needs and addresses their needs. The combination of face-to-face and virtual action learning (blended learning) can operate well, and enable SMEs to incentive competitive growth and business turnover [29]. The need for a clear structure of the curricula was underestimated and more attention should be given to informing participants on tasks. Frequently organizations are too focused on the issue of addressing business and neglect the opportunity that the workplace learning and training can provide. Although workplace learning and training may be seen as something that is "good in of itself", in a business context, it is primarily a way for enhancing business performance in order to become competitive. Ultimately, SMEs can meet the challenges of competitiveness and growth if they invest on learning and training, gathering their organizational and business objectives.

References

1. Uskov, V., Casalino, N.: New Means of organizational governance to reduce the effects of European economic crisis and improve the competitiveness of SMEs. Law Econ. Yrly. Rev. J. **1**(part 1), 149–179 (2012) (Queen Mary University, London, UK)
2. Attwell, G.: Exploring models and partnerships for E-learning in SMEs. Developing New Pedagogies and e-Learning in SMEs, report, Stirling, Brussels, Belgium (2003)
3. Observatory of European SMEs: Competence Development in SMEs. Enterprise DG of the European Commission. Publications of the European Communities Luxembourg (2003)
4. Birchall, D., Hender, J., Alexander, G.: Virtual Action Learning for SMEs. A review of experiences gained through the ESeN Project, Henley Management College (2004)
5. Alexander, G.: SMEs Learning Needs. Henley Management College, Oxfordshire (2006)
6. Bolden, R.: Leadership Development in Small and Medium Sized Enterprises. Final Report, Centre for Leadership Studies, University of Exeter (2001)
7. Casalino, N., Buonocore, F., Rossignoli, C., Ricciardi, F.: Transparency, openness and knowledge sharing for rebuilding and strengthening government institutions. In: Klement, E. P., Borutzky, W., Fahringer, T., Hamza, M.H., Uskov, V. (eds.) Proceedings of Web-Based Education—WBE 2013, IASTED-ACTA Press Zurich, Innsbruck, Austria (2013)
8. Blackburn, R., Hankinson, A.: Training in the smaller business: investment or expense? Ind. Commer. Train. **21**, 2 (1989)

9. Casalino, N.: Piccole e Medie Imprese e Risorse Umane nell'Era della Globalizzazione, n. 90, pp. 1–273. Wolters Kluwer, Cedam (2012)
10. Casalino, N.: Simulations and collective environments: new boundaries of inclusiveness for organizations? Int. J. Adv. Psychol 3(4), pp. 103–110 (2014) (Science and Engineering Publishing, USA)
11. Hilton, M., Smith, D.: Professional education and training for sustainable development in SMEs, European Foundation (2001)
12. Casalino, N., Cavallari, M., De Marco, M., Gatti, M.: Defining a model for effective e-Government services and an inter-organizational cooperation in public sector. In: Proceedings of 16th International Conference on Enterprise Information Systems—ICEIS 2014, INSTICC, vol. 2, pp. 400–408. Lisbon, Portugal (2014)
13. Eraut, M.: Non-formal learning, implicit learning and tacit knowledge in professional work. The Necessity of Informal Learning. Policy Press, Bristol (2000)
14. Casalino, N., Ivanov, S., Nenov, T.: Innovation's governance and investments for enhancing competitiveness of manufacturing SMEs. Law Econ. Yrly. Rev. J. 3(part 1), 72–97 (2014) (Queen Mary University, London, UK)
15. Hodgson, V., Watland, P.: Researching networked management learning. Manage. Learn. 35(2), 99–116 (2004)
16. Stewart, J.A., Alexander, G.: Virtual Action learning: Experiences from a study of an SME e-Learning Programme. Henley Management College, Oxfordshire (2012)
17. Casalino, N.: Innovazione e Organizzazione nella Formazione Aziendale, n. 10, pp. 1–212. Cacucci, Bari (2006)
18. Kirby, D.: Management education and small business development: an exploratory study of small firms in the UK. J. Small Bus. Manage. 28(4), 78 (1990)
19. Learning and Skills Development Agency LSDA: Working towards skills: Perspectives on workforce development in SMEs. Research report (2002)
20. Observatory of European SMEs: SMEs in Focus. Main Results from the 2002. (KPMG/ENSR). Enterprise Directorate-General of the European Commission. Office for Official Publications of the European Communities, Luxembourg (2002)
21. Armenia, S., Canini, D., Casalino, N.: A System Dynamics approach to the paper dematerialization process in the Italian Public Administration. Interdisciplinary Aspects of Information Systems Studies, pp. 399–408. Physica-Verlag (2008)
22. Small Business Service: SME statistics UK 2004, SBS (2004)
23. Casalino, N., Ciarlo, M., De Marco, M., Gatti, M.: ICT Adoption and organizational change. An innovative training system on industrial automation systems for enhancing competitiveness of SMEs. In: Maciaszek L., Cuzzocrea A., Cordeiro J. (eds.) Proceedings of 14th International Conference on Enterprise Information Systems—ICEIS 2012, Wroclaw, Poland, INSTICC, pp. 236–241. Setubal, Portugal (2012)
24. Unisys: Final Report: E-learning in Continuing Vocational Training, Particularly at the Workplace, with Emphasis on Small and Medium Enterprises. Directorate General for Education and Culture EAC-REP-003 (2005)
25. Casalino, N., D'Atri, A., Braccini, A.M.: A management training system on ISO Standards for organisational change in SMEs. Int. J. Prod. Qual. Manage. 9(1), 25–45 (2012) (Inderscience Publishers, USA)
26. Stanworth, J., Purdy, D., Kirby, D.A.: The Management of Success in "Growth Corridor" Small Firms, Şmall Business Research Trust Monograph (1992)
27. Lave, J., Wenger, E.: Situated Learning: Legitimate Peripheral Participation. Cambridge University Press, New York (1991)
28. Stewart, J.A., Alexander, G.: Engaging SMEs in an E-learning trial. In: Networked Learning Conference, Lancaster University (2006)
29. Rossignoli, C.: The contribution of transaction cost theory and other network-oriented techniques to digital markets. Inf. Syst. E-Bus. Manage. 7(1), 57–79 (2009)

Smart Control System of Human Resources Potential of the Region

Olga A. Ivashchuk, Igor S. Konstantinov and Irina V. Udovenko

Abstract In this paper we present the results of our research and modeling of the Smart Control System for the human resources potential of the region (SCS of HRP) which has to become an integral part of the general intellectual network of management of the Smart City functions. We construct the conceptual model of human resources potential as an object of automated control. As a result the object of control is considered as a complicated dynamic informational, social and economic system. We develop the generalized structural model of SCS of HRP, which is endowed with both traditional functions of collecting and processing information and functions of intellectual analysis of data based on situational and mathematical modeling of the process of interaction of components of human resources potential between each other and environment. This approach provides the possibility of adaptive balanced management as well as development of rational productive managing decisions.

Keywords Smart control system · Human resources · Modeling

1 Introduction

The level of social and economic as well as scientific and technical development of regions of a modern state is considerably defined by not only available raw and material and technical base, but also by the condition of their human resources potential, which the competitiveness of enterprises and organizations as well as the

O.A. Ivashchuk (✉) · I.S. Konstantinov · I.V. Udovenko
Belgorod National Research University, Belgorod, Russia
e-mail: ivaschuk@bsu.edu.ru

I.S. Konstantinov
e-mail: konstantinov@bsu.edu.ru

I.V. Udovenko
e-mail: udovenko@bsu.edu.ru

© Springer International Publishing Switzerland 2015
V.L. Uskov et al. (eds.), *Smart Education and Smart e-Learning*,
Smart Innovation, Systems and Technologies 41,
DOI 10.1007/978-3-319-19875-0_43

481

effect of investments into region economy, the result of introduction of innovations, the development of services sector depend on.

In this paper we will consider the term "human resources potential of the region" (HRP) as the complex of: formed and constantly developing professional knowledge and skills of the available human resources of a region (including the registered in an employment office); accumulated knowledge and skills of future human resources, who are students of educational professional institutions now; as well as a future knowledge, skills and abilities of minor part of the population of a region.

In the conditions of the innovation-focused development of economy formation of HRP is characterized by the following key features:

– all the above-stated structural components of HRP as well as the factors, which influence them directly, for example, a demographic situation, a condition of modern educational services and regional labor markets, are changing continuously;
– the processes of their interaction between each other and environment are characterized by high dynamics and complexity;
– the level of state regulation is extremely low in the field.

Many researchers consider a problem of the balanced development of the countries and regions on the basis of effective management of their personnel potential with the relevant decision of tasks of overcoming of a disproportion between the markets of workplaces and labor. However the following approaches are generally used:

– the formation of an active state policy [2, 6, 7];
– the formation of the optimal structure of the business community [1, 3–5, 9, 10];
– improvement of the educational processes in institutions of professional education [8].

At the same time, the general methodological approach to building effective management systems which will be adaptable to changes in the HRP (a control object) and in the external environment, as well as an instrumental mechanism to ensure the functioning of such systems to date have not been established.

It should be noted that the effectiveness of control processes such complex dynamic objects as HRP, connected with the necessity of collecting and processing large volume of heterogeneous information, creation and implementation of predictive models. Thus, for the creation of such control systems, it is necessary to use IT, technologies of automation, advanced methods of situational and mathematical modeling. This direction, as shown by various studies [4], is at an early stage of development.

The authors set the task to model a Smart Control System for the human resources potential of the region (SCS of HRP) which has to meet the following main requirements:

– to provide the process of effective and productive control of human resources potential in various spheres of economy,

- to be universal for any level of hierarchy of administrative and territorial division,
- to be adaptive to current changes in social and economic sphere on the considered territory as well as in environment.

2 Modeling of the HRP as Object of Smart Control

The analysis of interaction between HRP and social and economic and environment of the regions was carried out; it allowed us to determine its state parameters as a modeled operated system, to distinguish controlled and operated ones from them as well as to define the borders of this system with the environment. As a result we constructed a conceptual model of HRP as an object of automated control, which is not considered as a traditional "black box", but as a multicomponent system with concrete structure and information flows reflecting interaction of its systems between each other and between its systems and environment. Schematically this model is shown in Fig. 1.

Fig. 1 HRP model as an object of Smart control

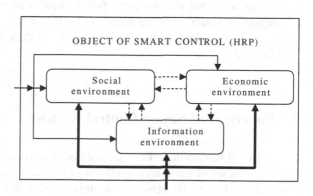

The system includes the following three main classes of components: system "Social environment"; system "Economic environment"; system "Information environment". Thus, HRP is presented as a complex dynamic information-social-economic system. On the scheme the dashed lines show the flows reflecting the interaction of the main systems of the object of control between themselves; the continuous thin lines – influence of environment; continuous bold lines – control actions directed at the main systems.

The example of the corresponding classification of the parameters of the state of the object of control (its main components) is shown in Fig. 2. More concrete structure of the considered controlled system, classification of the parameters of the state and the corresponding mechanisms of control of HRP depend on the specifics

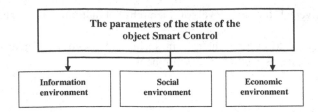

Fig. 2 The parameters defining the state of the human resources potential of the region, as an object of smart control

of the region and features of those branches of economy, which are priority for this territory.

For example, consider the parameters of the information environment:

- parameters connected with structuration and completeness of the contents of labor markets and educational services of the region;
- availability and quality of the information environment integrated with other regions;
- number of students using distance learning;
- parameters of the technology of public opinion forming;
- number of available PC to access to the Internet;
- number and availability of radio-, TV centers in Institutions of Higher Education.

3 Functions of Smart Control System

As a result of functioning of the modeled SCS of HRP the steady state of the latter in the conditions of environment influence *(realization of the objective of the system functioning)* has to be achieved; in this case at any moment quantitative and qualitative parameters of the labor force of the region have to be the most closely resembling the demanded target state *(realization of the main objective of the human resources policy)* that corresponds to the sustainable innovation-focused development of the region, its competitiveness *(realization of the main objective of social and economic development of the region)*. Further specification of the purpose of the system (objectives tree derivation) and the corresponding systematization of ways of its achievement (systems tree derivation) allow us to reveal the main systems, the processes of their interrelated functioning.

The main functions of the SCS of HRP and the corresponding systems realizing these functions are presented schematically in Fig. 3.

Fig. 3 The main functions of the smart control system of HRP and its the subsystems

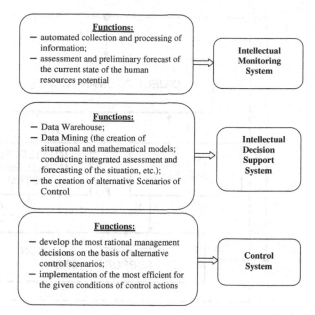

Functions:
— automated collection and processing of information;
— assessment and preliminary forecast of the current state of the human resources potential

→ Intellectual Monitoring System

Functions:
— Data Warehouse;
— Data Mining (the creation of situational and mathematical models; conducting integrated assessment and forecasting of the situation, etc.);
— the creation of alternative Scenarios of Control

→ Intellectual Decision Support System

Functions:
— develop the most rational management decisions on the basis of alternative control scenarios;
— implementation of the most efficient for the given conditions of control actions

→ Control System

4 Structural Model of the Smart Control System

From the point of view of the set-theoretic approach the model of the system of the considered class can be presented by the following way:

$$S_{SCS} = \langle \Sigma_{SCS}, X_{SCS}, Y_{SCS}, Z_{SCS}, \Omega_{SCS}, F_{SCS}, \Theta_{SCS} \rangle, \tag{1}$$

where Σ_{SCS} – set of components-systems of SCS;
X_{SCS} – set of states of elements (entrances) Σ_{SCS};
Y_{SCS} – set of states of elements (exits) Σ_{SCS};
Z_{SCS} – set of states of systems Σ_{SCS};
Ω_{SCS} – set of influences of environment on Σ_{SCS};
F_{SCS} – set of the displays carried out on $\Sigma_{SCS}, \Omega_{SCS}$ and Y_{SCS};
Θ_{OC} – set of the relations over elements $\Sigma_{SCS}, \Omega_{SCS}$ and Y_{SCS};

According to the scheme of the figure $\Sigma_{SCS} = \{S_{OC}, S_{CS}, S_{IM}, S_{IS}\}$, where S_{OC}–object of SCS;

S_{CS}–Control System; S_{IM} – Intellectual Monitoring System; S_{IS}– Intellectual Decision Support System (IDSS). Respectively, the set of states of components of SCS of HRP can be presented in the form $Y_{SCS} = \{Y_{OC}, Y_{CS}, Y_{IM}, Y_{IS}\}$; set of influences of environment $\Omega_{SCS} = \{\Omega_{OC}, \Omega_{CS}, \Omega_{IM}, \Omega_{IS}\}$.

The structural model of SCS of HRP is schematically presented in Fig. 4. During its construction the approaches stated in [11–13] were used.

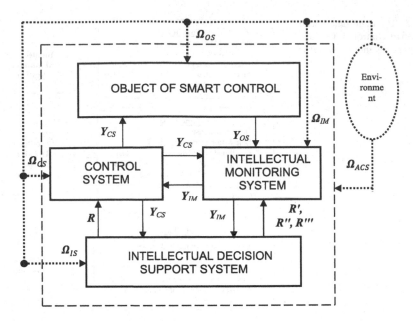

Fig. 4 The structural model of SCS of HRP

It should be noted that the components of set Y_{IS} are controlling signals both for the object of control of SCS, and for the control system itself as well as for the system of intellectual monitoring. They regulate the work of these components of SCS according to the current changes in the object of control and environment.

Thus, the inputted system of IDSS provides formation of internal contours of control in the SCS of HRP, in each contour there is an internal subject of control – IDSS, and internal objects of control – either a control system or a system of intellectual monitoring. The inputted contours provide the possibility of self-adjustment of the system that corresponds to the principle of adaptability in HRP control. The set Y_{IS} includes the following Y_{IS} components $= \{R, R', R'', R'''\}$, where R – a set of alternative scenarios of control to regulate the state of the object of control, R' – a set of models to carry out assessments of different levels (the state of the components of the object of control and intellectual system of monitoring, the productivity of the managing influences, etc.), R'' – a set of forecast models, R''' – a set of the influences regulating the structure of the system of intellectual monitoring. The controlling signals coming from ISSDM in the course of internal control are formed as a result of the following displays:

$f_{IS}: \Omega_{IS} \times Y_{CS} \times Y_{IM} \rightarrow R$ – formation of alternative scenarios of control to regulate the work of the control system of SCS of HRP;

$f'_{IS}: \Omega_{IS} \times Y_{CS} \times Y_{IM} \rightarrow R'$ – formation of models for to estimate and regulate the structure of the intellectual monitoring system and the parameters of monitoring, to estimate the current state of HRP;

$f''_{IS}: \Omega_{IS} \times Y_{CS} \times Y_{IM} \to R''$ –formation of models for preliminary forecasting of any change of the current state of HRP;

$f'''_{IS}: \Omega_{IS} \times Y_{CS} \times Y_{IM} \to R'''$ – formation of the control influences to regulate the structure of the system of intellectual monitoring.

Specified in the generalized SCS model sets are filled with concrete contents depending on the tasks being solved in the field of the balanced management of HRP as well as the features of social and economic development of the territory.

5 Modeling of Intellectual Monitoring System and Intellectual Decision Support System

To choose and carry out concrete rational actions for the qualitative and quantitative balanced regulation of HRP the control system has to have, firstly, reliable information about the current state of the object of control, and, secondly, rather full set of alternative scenarios of control created on the basis of demographic, production, social and financial forecasts. These problems are solved at the level of the introduced above specialized intellectual systems of SCS of HRP: the Intellectual Monitoring System and the Intellectual Decision Support System.

Specification of the structure of the intellectual monitoring system is shown in Fig. 5. Its main components: a system of information collecting (about the parameters of all elements of the object of control, the control system and IDSS as well as about the parameters of the controlled external influences); the system of assessment of the current state of the object of control; the system of the preliminary forecast of the current state change of the object of control. The last two systems are the ones, which provide intellectualization of the monitoring system and for their effective functioning it is required to use specially developed situational and mathematical models (the controlling signals R' and R''). Besides, the structure of the system of intellectual monitoring has to adapt according to the changes in the object of control and environment (signal R'''). Necessary models are formed in IDSS. $Y_{IM} = \{X, X', X''\}$, where X – results of information collecting in the system of monitoring, X', X'' – sets of results of the model assessment and the preliminary forecast of the state of HRP.

The structure of IDSS is defined by its functions (see Fig. 3). In Fig. 6 the IDSS model that includes the following main systems: *Knowledge Base*; *Modeling*; *Formation of Alternative Scenarios of Control; Regulation of the Structure of Monitoring System* is schematically presented. The Knowledge Base contains: information, collected and previously processed in the monitoring system, which is used inside IDSS for carrying out spatiotemporal analysis, modeling, imitating experiments in the form of elements of the set D; the models developed and used both inside IDSS and for formation of sets R, R', R'', R''' (elements of the set M); rules P necessary for formation of models.

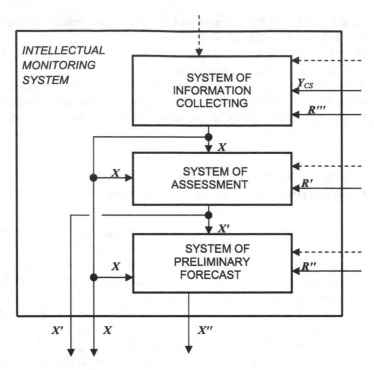

Fig. 5 Model of the intellectual monitoring system as a part of SCS

Modeling system using the information of the knowledge base forms the models necessary for the solution of a full range of problems of SCS of HRP. During identification of new cause-effect links in the course of modeling new rules are also formulated (P').

In the system of formation of alternative control scenarios on the basis of the carried out estimations and forecasts possible control influences are defined and alternative scenarios of control of the human resources potential of the region are formed; they are transferred to the control system of SCS of HRP (besides they are stored in the knowledge base).

In the figure there are M_1, M_2 – models constructed by the modeling system for formation of alternative scenarios of controls of HRP and regulation of the system of monitoring respectively.

6 Summary, Conclusions and Implications

Construction and organization of a Smart Control System with use of the presented models will enable to provide effective intellectual decision support and on this basis realization of productive managing decisions on the balanced formation and development of the human resources potential of the region.

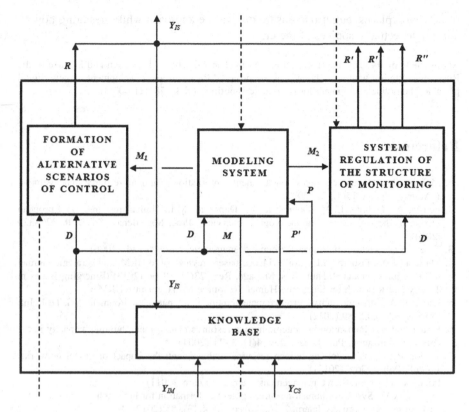

Fig. 6 Structural model of Intellectual decision support system

It should be particularly noted that such systems have to become an integral part of the general intellectual network of control of "Smart City". The ideology of "Smart City" was created and is developing actively for the purpose of providing real favorable conditions for sustainable social and economic development of the territories of modern states, self-sufficiency of their sociocultural space, attraction and preservation of human resources potential.

It is necessary to emphasize that for real implication of investments, innovations and talented people from various spheres of economy and science, it is essential that "Smart City" not simply meet the modern requirements on natural and energy resources use, environmental standards, improvements and safety of its territory. It has to provide the most favorable climate for a person's development (his intelligence, creativity, cultural level, etc.), possibility of comfortable effective work, minimum time and human expenditures on receiving services of any kind, which, in turn, will propel the economy of a city and a country as a whole to the totally new level of development. Thus, it is necessary not only to introduce innovations while constructing "Smart City", arranging its essential services, education and creation of new workplaces, but also to ensure innovative approach while arranging how to apply intellectual resources of the city.

Acknowledgments This paper was supported by The Ministry of Education and Science of the Russian Federation within the bounds of the project "Research and development of new digital portable photo/video equipment for panoramic shooting" (N 14 581 21 000 3).

References

1. McBain, R.: Human resources management: international human resource management. J. Manag. Update **13**(4), 21 (2002)
2. Ecirlia, A., Dobreb, E.-M., Dobrescuc, E., Danetiud, M.I.: Human resources in European market in the past decade—a sociological overview. Proc. Soc. Behav. Sci. **150**, 320–329 (2014)
3. Tyson, S.: Essentials of Human Resource Management. Elsevier Ltd. (2006)
4. Marlera, J.H., Fisherb, S.L.: An evidence-based review of e-HRM and strategic human resource management. J. Hum. Res. Manage. Rev. **23**(1), 18–36 (2013) (Emerging Issues in Theory and Research on Electronic Human Resource Management (eHRM))
5. Esmaeili, A.: Strategic human-resource management in a dynamic environment. Sci. Tech. Inf. Process. **39**(2), 85–89 (2012)
6. Farazmand, A.: Innovation in strategic human resource management: building capacity in the age of globalization. Pub. Organ. Rev. **4**(1), 3–24 (2004)
7. Taylor, R.: China's evolving human resource management: the impact of global networks. J. Glob. Policy Gov. (2013)
8. Price, A.: Human resource management. Cengage Learn. (2011)
9. Sidorin, A.V.: System of human resources potential formation for high tech industries on the basic of cluster approach. Intern. J. Naukovedenie **4**(13), 55 (2012)
10. Asrorova, M.A.: Formation and development of human resources potential in the region: problems and ways of perfection. Herald TSULBP **3**(55), 82–88 (2013)
11. Konstantinov, I.S., Ivashchuk, O.A.: Tool means of organization and support of electronic services to the population. Comput. Sci. J. Moldova **19**(2), 56 (2011)
12. Arhipov, O.P., Ivashchuk, O.A., Konstantinov, I.S., et al.: Way of creation automated control system of innovative "Smart City". Inf. Syst. Technol. **6**(68), 85–95 (2011)
13. Ivashchuk, O.A., Udovenko, I.V.: Support decision-making in the control of personnel potential of the construction cluster. Sci. Statements BelSU 2014 **15**(186, 31/1), 108–114 (2014)

Smart Learning Environments Using Social Network, Gamification and Recommender System Approaches in e-Health Contexts

Pierpaolo Di Bitonto, Enrica Pesare, Veronica Rossano
and Teresa Roselli

Abstract The fundamental role that ICT plays in the process of modernization of education and training, whether in formal or informal, is now universally recognized. The lifelong learning become increasingly urgent in many areas such as, for example, adult education, innovation and learning in the workplaces, health & wellbeing, cultural heritage, and so on. This requires the creation of learning paths cantered on the specific needs of the individual, for the development of skills and abilities as well as for the acquisition of content. Re-inventing the ecosystem training and re-strengthen the teaching and learning in the digital age, through practices more open and innovative in order to create learning experiences richer, engaging and motivating, is a priority. The article presents some solutions of smart learning environment in e-health domain that combines pedagogical approaches of social learning and game-based learning with technological approaches of the social network and recommender systems in order to provide engaging learning experiences.

Keywords e-health · Gamification · Game based learning · Social learning

1 Introduction

The growing interest in ICT, as pointed out in the Horizon 2020 Workprogram [30], combined with the increasing availability of technologies in education processes, poses new challenges to the current world of education: creating a digital ecosystem

P. Di Bitonto · E. Pesare (✉) · V. Rossano · T. Roselli
Department of Computer Science, University of Bari, Via Orabona, 4 – 70125 Bari, Italy
e-mail: enrica.pesare@uniba.it

V. Rossano
e-mail: veronica.rossano@uniba.it

T. Roselli
e-mail: teresa.roselli@uniba.it

© Springer International Publishing Switzerland 2015   491
V.L. Uskov et al. (eds.), *Smart Education and Smart e-Learning*,
Smart Innovation, Systems and Technologies 41,
DOI 10.1007/978-3-319-19875-0_44

that goes beyond traditional learning environments and that is able to support informal learning processes.

E-learning research has to define new methods and techniques for creating environments addressed to the skills acquisition, the user empowerment and the social inclusion. In particular, in the field of e-health and chronic diseases the use of new technologies can be effective in therapeutic education, as well as in professional training. Therapeutic education and professional training require also new forms of teaching/learning that stimulate the interest and motivate the user to a faster learning. For this reason, the use of learning environments, which are able to personalize the learning paths and make the learning process easier for the user, is essential in these contexts.

As stated by Hwang [14] the definition of intelligent learning environments has been the subject of research since the 80 s, from the Intelligent Tutoring Systems (ITS) to the Adaptive Hypermedia and Web-based Learning System [12, 15, 19, 24]. Each of them is characterized by specific methods and techniques that can support the user through the learning process and customize both the content and the educational processes.

In recent years, the spread of "smart technologies" has fostered the birth of the new term "Smart Learning Environment" (SLE): the debate is still open, but the term "smart" is now meant not only as the ability to provide adaptive and personalised pathways but also as combined use of smart technologies and smart pedagogies. In other words the SLEs are not only able to provide the most appropriate service but can combine different pedagogical approaches (ubiquitous learning, situated learning, social learning, game-based learning, etc.) and the most current technological methods than those used in the past (recommender system, multi-agent architectures etc.) to make learning more effective and efficient [25].

The numerous definitions, however, leave still some open issues: the literature has shown that the personalised paths are necessary elements but no longer sufficient in traditional learning environments and, even more, in informal learning contexts. In our opinion, the challenge is to define learning models that are able to encourage the engagement and motivation of the user, that represent the key factors for successful learning. In addition, the current models should be able, not only to support the transfer of knowledge in declarative form, as required by traditional training, but also to meet the emerging needs of informal training by supporting the acquisition of specific skills.

For this, the Smart Learning Environments, as intelligent learning environments have to meet the demands from the current educational landscape and make the learning path an experience that meets and satisfies user expectations. For this reason, motivation and user engagement play key roles in making the learning process effective and efficient. Experimental evidences have already widely demonstrated how this is true in the context of traditional education and, even more, in the context of informal education, where to attract the user to promote significant and deeper learning becomes a priority.

In this context, our research aims to design and implement Smart Learning Environments able to encourage the user's engagement and motivation. The proposed solutions combine pedagogical approaches of social learning and game-based learning with technological approaches of the social network and recommender systems in order to provide personalized learning experiences based on the specific needs of individual actors. The systems developed were tested to measure the effectiveness of learning skills in contexts of chronic diseases where the acquisition of specific behaviours is essential.

The paper is organized as follows: the next section describes the whole technological and pedagogical approaches that represent the fundament of our research work; Sect. 3 describes two different SLEs defined in two different settings. Finally, some conclusions and future works are described.

2 Technological and Pedagogical Approaches

The approaches herein described are often proposed separately. The challenge addressed in this research is the definition of a SLE able to combine different pedagogical approaches and technological methods that have contributed over the years to define intelligent learning environments. In particular, the approaches used are the following: the social learning to foster the active participation of student in the learning process; the situated learning and recommender systems to support the learner in the selection of the content and the personalization of the learning paths; the gamification and game-based learning to promote users' engagement and motivation.

The social learning puts the user in the middle of the learning process and allows her/him to become the active part of her/his own education and training process. Social learning, in fact, takes place through observation, imitation and copying of actions and behaviours. The works of Wenger [29] and Besana [3, 4] relies on the social learning theory proposed by Bandura [1]. In particular, Wenger focuses on learning in communities of practice, where conversation and discussion allow formal or informal learning, without a conscious decision to learn and without realizing that learning is taking place. On the other side, Besana focuses on the use of social networks and web 2.0 tools to foster learning in different contexts. The use of social network, in fact, in informal learning is strategic in communities of practice where the professional growth is based on sharing experiences, identifying best practices and providing mutual help in dealing with the daily problems rather than on fixed training program [26]. The user takes then an increasingly active role, becoming leading actor and content producer making his/her knowledge and experience available to the community.

Moreover, to support user in contents and activities selection according to his/her learning goals is basic for successful learning. In this context, the situated learning [17] is one of the most interesting pedagogical approaches. In the situated learning, in fact, learning is conceived as a social process in which knowledge is co-constructed, and the learning process occurs in a specific context in which the knowledge should be applied. Therefore, the learning of a specific concept has to be contextualized into physical, social and cultural environment in which it takes place.

This approach is very interesting in medical training, both for the empowerment of the patient and for medical workers training. In both cases, the contextualization of learning reduces training time and produces more effective results [16].

From the technological point of view, we applied the situated learning by means of recommender systems that can offer the most adequate resource to the user, taking into account different dimensions of the context in which learning takes place [27]. In fact, recommender systems are widely used in different domains, but their use in learning poses new challenges. Their use is often limited to the recommendation of resources or contents, in other words they usually provide learning resources, resource sequences or people to work with in a learning activity while in recent years are emerging recommender systems able to predict learning performance, to suggest learning activities and/or learning tasks [11]. In this case, the recommender system has to take into account learner history, environment, timing, resources availability and so on, but the context dimensions should be selected properly depending on the final goals.

In order to make the learning efficient, aside from to supply the best suitable resources and/or activities, it is important to engage the student in the learning process. The game-based learning and gamification approaches, relying on the desires of success, competition, status recognition and so on, may promote user engagement: in fact, allowing the player to deal with the distance between desires and impossibility of satisfying them immediately, they represent a response to unmet needs [28].

On the one hand, game-based learning can support different pedagogical principles: game can support multi-sensory, active, experiential and problem-based learning; it promotes the recovery of prior knowledge, because players have to use previously learned information to progress in the game; it provides instant feedback that allows players to test different hypotheses and to learn from their own actions; it provides self-assessment tools through the use of score and different levels achievement [22, 23]. Furthermore it can allow skills acquisition in different areas: personal and social development, language and communication development, collaboration abilities, logical-mathematical and critical thinking, problem-solving abilities, etc. [20].

On the other, the Gamification approach, proposed by Deterding [10, 18, 31], involves the use of game mechanics (points, levels, lives, charts, etc.) and rules to enhance the users experience and actively engage them in areas not directly related to the game and fun (e.g. marketing, politics, health and e-learning). The gamification approach can improve motivation and keep user engaged in order to make the learning process more effective. In the e-health field, for example, there are situations in which both patients and caregivers are forced to acquire knowledge and skills to improve their life-style. Moreover, the gamification approach could help practitioners in acquiring new skills.

In the following sections two examples of SLEs developed for the e-health context have been described. Both of them use the described approaches in order to foster therapeutic education and professional training by sustaining the user engagement and motivation.

3 Smart Learning Environments for Medical Purposes

3.1 Ubicare

The UBICARE project (Ubiquitous knowledge- oriented Healthcare) is a project co-funded by Region Apulia in the U.E. - FESR P.O. 2007–2013, which aims to promote the de-hospitalization of patients suffering from chronic diseases. In particular, the project is addressed to patients suffering from chronic heart failure and peritoneal dialysis. The goal is the creation of a technological infrastructure to enable the sharing of clinical data, to support medical and paramedical staff in diagnosis and monitoring of patients, to train and inform them about diagnostic procedures, therapeutic interventions and follow-up of patients, and, finally, to empower the patient [2, 5].

The UBICARE system is based on the Social network paradigm in order to allow the various actors, involved in the patient management, to share information and clinical data in real time. The project is addressed to hospital and general practitioners, nurses, patients and caregivers who need to keep track of the clinical history of the patient in order to properly manage the patient in the hospital and at home. Furthermore, the network of operators and patients becomes a training place at different levels: for patients and caregivers that have to learn how to manage the disease at home; for nurses and general practitioners that have to acquire specific skills and knowledge on disease management; for specialist physicians who can share and discuss best practices and treatment protocols in the disease management with colleagues.

A recommender system has been used to foster the training of different users and provide them with the right resources. In particular, hybrid methods [7] have been adopted to build different functions:

- Recommended for you is a push service that suggests proactively the best resources based on the user profile, the usage context and the device used;
- Customized Search is a service for specific requests, based on semantic correlation between tags used to describe the resources.

Furthermore, the resources list is refined by selecting those ones that have been appreciated by similar users.

In addition, to promote deeper learning and motivate users, the social learning and game based learning approaches have been used. A component for sharing clinical experiences has been developed to implement the social learning approach: it provides tools for managing and sharing resources for professional training and update (international guidelines, interesting sites, etc.) and for discussing controversial cases and best practices. Resources are organized into three sections:

- Links that allow the user to share links to various sources for professional update;
- Experiences that allow the user to share best practices, guidelines for diagnosis and treatment protocols, and successful clinical cases;

- Question & Answer that allow the user to share problems to be discussed in the community; the answers of other users will be evaluated by the community and will become new resources to enrich the knowledge base of the system.

The game based learning approach has been used both for the training of nurses and general practitioners, who have to acquire skills about the patients and critical events management, and for the training of patients and caregivers who have to learn the daily care and management of the disease. Simulation of clinical cases and edu-games use the gamification approach to enhance the interactive user experience and engage them in the learning process. In both cases, different clinical scenarios and different learning goals for each professional role are offered.

In the first case, a component proposes simulation of clinical cases created from real clinical data stored in patient's records. The player is asked to solve the patient's problem in a given slot of time. For each intervention a feedback and a score is provided. In addition, to ensure the scalability of the system, a trainer can enter new relationships between clinical data and enable the creation of new problems [6]. The simulation component, interacts with a decision support service that using data mining algorithms and ontologies in order to support user in clinical decision and patient monitoring and provides interesting cases to be used in the training process [9].

In the second case we developed an edugame component in order to help the patients to acquire skills required for the therapy success, such as: adopt an healthy lifestyle compatible with the disease, manage daily tasks and recognize abnormalities of their health status. The component includes a series of scenarios and interventions that can be combined in the most suitable way by the training staff to provide support for the training needs of each patient.

3.2 DiabeteNetwork

The DiabeteNetwork is a project that aims to support young patients and their families from the diagnosis to the awareness that good management of diabete type I guarantees a high quality of life.

The diagnosis of type I diabetes in children and young patients finds often unprepared not only the parents but also teachers and other figures responsible for young patients during the day (teachers, coaches, etc.). The main goal of the social learning network is to provide parents and caregivers with the right information, to promote knowledge sharing among caregivers and young patients themselves, and to promote right behaviours and healthy lifestyle adoption.

In particular, the main aims of the defined environment are: to support knowledge sharing through different types of training and information resources such as learning objects, links, recipes and games; to encourage sharing of personal experiences about life with diabetes and to promote virtual and/or face to face meeting among the community members.

DiabeteNetwork has been designed to be a landmark for information on the disease, but it aims to support young patients and people around them in their day life with diabetes. Young patients could discuss and share feelings, interests, ambitions, and so on; parents can discuss with medical or paramedical operators and other parents. The professional staff in the social network is important for the training of patients and caregivers in order to transfer basic information about healthy lifestyle and disease management.

Since different actors are involved in the DiabetesNetwork, a recommender system has been used with a dual purpose: first, to improve the users grouping based on their interests and relationship in common, and, secondly, to foster the acquisition of knowledge and the achievement of educational goals. In the first case, a system based on the user's published contents and the relationships with other users allows to discover new users on the basis of common interests [8, 13]. In the second case, a hybrid recommender system will suggest not only educational resources and personal experiences but also activities suitable to the achievement of educational goals.

Social learning in this case will take place at different levels, more or less formal, through the sharing of information, training resources and personal experiences. For the first purpose a section has been designed to allow sharing of information and resources about the disease and the appropriate lifestyle to adopt: different kinds of resources will be available (learning objects, documents, links, recipes etc.) and all the community can comment on them and discuss about the issues contained in the resources. For the second purpose each user can share personal experiences, more or less related to the disease or to the daily life with the disease. Other users can comment it or offer some advice starting a debate able to promote an informal learning process.

Moreover, in order to engage and motivate the user, the social network has been gamified. The social gaming approach can be seen as a social expansion of the individual game and can encourage spontaneous interaction among users and involve them to be more active in the game and, then, in the social network [21].

The system uses different mechanics:

- levels to measure the status reputation and to reward the most active users and participants in the community and those with higher influence;
- badges to recognize particular merits such as winning a competition;
- challenges that may take place both in the virtual and in the real world to sustain the multidimensional continuity of SLE.

In particular, new badges and challenges can be proposed by the operators according to specific requirements to promote or encourage the development of right behaviours or create opportunities to enhance interaction and collaboration among users.

4 Conclusions and Future Work

The increasing use of ICT in traditional and informal learning is rapidly changing to meet the new challenges of skills acquisition, user empowerment, therapeutic education and professional training. Especially in the field of e-health, it is important to define engaging environments in order to ensure the effectiveness of the learning process. The paper presents two different solutions of smart learning environments that combine pedagogical approaches of social learning and game-based learning with technological approaches of social network and recommender systems in order to provide engaging learning experiences for knowledge and skills acquisition.

The UBICARE project has been evaluated with a pilot study in two experimental contexts: the care of patients suffering from chronic heart failure and peritoneal dialysis. The pilot study was firstly carried out at the Departments of Nephrology and Cardiology "Policlinico" hospital in Bari, involving 16 physicians and 6 nurses, then it was extended to 20 patients and caregivers at home. The data collected registered a good appreciation in terms of knowledge gain and user satisfaction among the practitioners, while the experimental with patients and caregivers was to short to appreciate changes in users knowledge acquisition. This will require a long term study.

The DiabeteNetwork, is still under development. Only a preliminary pilot study is ongoing with a sample that involves users at the Departments of metabolic diseases at the "Giovanni XXIII" hospital in Bari. The work has already gained the appreciation of the medical staff and association of young diabetes involved in the project. In the next future the results coming from the users involved in the experiment will be available.

Acknowledgments This work was supported in part by the Project UBICARE (UBIquitous knowledge-oriented HealthCARE) - EU-FESR P.O. Puglia Region 2007–2013 Grant in Support of Regional Partnerships for Innovation - Investing in your future (UE-FESR P.O. Regione Puglia 2007–2013 – Asse I – Linea 1.2 - Azione 1.2.4 - Bando Aiuti a Sostegno dei Partenariati Regionali per l'Innovazione - Investiamo nel vostro futuro).

References

1. Bandura, A.: Social learning theory. Prentice-Hall, Oxford (1977)
2. Berni, F., et al.: A knowledge management service for e-health. In: ICERI2013 Proceedings, pp. 488–493 (2013)
3. Besana, S.: Schoology: il Learning Management System diventa "social". TD-Tecnologie Didatt. **20**, 51–53 (2012)
4. Besana, S.: Social network e apprendimento informale: un contributo. TD-Tecnologie Didatt. **20**, 17–23 (2012)
5. Di Bitonto, P., et al.: Distance education and social learning in e-health. Int. J. Inf. Educ. Technol. **4**(1), 71–75 (2014)

6. Di Bitonto, P., et al.: Training and learning in e-health using the gamification approach: the trainer interaction. In: Lecture Notes in Computer Science, pp. 228–237. Springer Verlag (2014)
7. Burke, R.: Hybrid web recommender systems. In: The Adaptive Web, pp. 377–408 (2007)
8. Chen, J., et al.: Make new friends, but keep the old. In: Proceedings of the 27th international conference on Human factors in computing systems—CHI 09, p. 201. ACM Press, New York, New York, USA (2009)
9. Corriero, N., et al.: Simulations of clinical cases for learning in e-Health. Int. J. Inf. Educ. Technol. 4(4), 378–382 (2014)
10. Deterding, S., et al.: From game design elements to gamefulness. In: Proceedings of the 15th International Academic MindTrek Conference on Envisioning Future Media Environments—MindTrek '11, pp. 9–11 (2011)
11. Drachsler, H., et al.: Panorama of recommender systems to support learning. In: Handbook on Recommender Systems, 2nd edn. (2015)
12. Graf, S., et al.: A flexible mechanism for providing adaptivity based on learning styles in learning management systems. In: Proceedings of the IEEE International Conference on Advanced Learning Technologies (ICALT 2010), pp. 30–34 (2010)
13. Hannon, J., et al.: Recommending twitter users to follow using content and collaborative filtering approaches. In: Proceedings of the fourth ACM conference on recommender systems —RecSys '10, p. 199. ACM Press, New York, New York, USA (2010)
14. Hwang, G.: Definition, framework and research issues of smart learning environments-a context-aware ubiquitous learning perspective. Smart Learn. Environ. pp. 1–14 (2014)
15. Kinshuk, L.T.: user exploration based adaptation in adaptive learning systems. Int. J. Inf. Syst. Educ. 1(1), 22–31 (2003)
16. Kneebone, R.: Simulation in surgical training: educational issues and practical implications. Med. Educ. 37(3), 267–277 (2003)
17. Lave, J., Wenger, E.: Situated learning: legitimate peripheral participation. Cambridge university press (1991)
18. Lee, J.J., Hammer, J.: Gamification in education: what, how, why bother? Acad. Exch. Q. 15 (2), 146 (2011)
19. Martens, A., Uhrmacher, A.M.: Adaptive tutor processes and mental plans. Lect. Notes Comput. Sci. 2363, 71–80 (2002)
20. McFarlane, A., et al.: Report on the Educational Use of Games. TEEM (Teachers evaluating educational multimedia), Cambridge (2002)
21. Montola, M.: Exploring the edge of the magic circle: defining pervasive games. In: Proceedings of DAC, p. 103 (2005)
22. Oblinger, D.G.: The next generation of educational engagement. J. Interact. Media Educ. 2004 (8), 1–18 (2004)
23. Papastergiou, M.: Digital game-based learning in high school computer science education: impact on educational effectiveness and student motivation. Comput. Educ. 52(1), 1–12 (2009)
24. Van Seters, J.R., et al.: The influence of student characteristics on the use of adaptive e-learning material. Comput. Educ. 58, 942–952 (2012)
25. Spector, J.M.: Conceptualizing the emerging field of smart learning environments. Smart Learn. Environ. 1(1), 1–10 (2014)
26. Trentin, G.: Apprendimento in rete e condivisione delle conoscenze. Ruolo, dinamiche e tecnologie delle comunità professionali on-line. FrancoAngeli (2005)
27. Verbert, K., et al.: Context-aware recommender systems for learning: a survey and future challenges. IEEE Trans. Learn. Technol. 5(4), 318–335 (2012)
28. Vygotsky, L.S.: Play and its role in the mental development of the child. J. Russ. East Eur. Psychol. 5(3), 6–18 (1967)
29. Wenger, E.: Communities of practice and social learning systems. Organization 7(2), 225–246 (2000)

30. Workprogramme Horizon 2020, ICT 2014—Information and communications technologies. http://ec.europa.eu/research/participants/portal/desktop/en/opportunities/h2020/topics/90-ict-21-2014.html (2014)
31. Zichermann, G., Cunningham, C.: Gamification by Design: Implementing Game Mechanics in Web and Mobile Apps. O'Reilly Media Inc., Sebastopol (2011)

Real-Time Feedback During Colonoscopy to Improve Quality: How Often to Improve Inspection?

Piet C. De Groen, Michael Szewczynski, Felicity Enders,
Wallapak Tavanapong, JungHwan Oh and Johnny Wong

Abstract Colorectal cancer (CRC) is the second leading cause of cancer deaths in the US despite wide use of colonoscopy to prevent CRC-related death. The current explanation for the failure of colonoscopy to prevent most CRC-related death is that lesions are not detected or completely removed. Real-time feedback during colonoscopy has the potential to alert endoscopists that sub-optimal visualization of the colon mucosa is occurring. To determine what type and frequency of feedback most likely will improve colonoscopy, we studied a set of randomly obtained video files for four features associated with quality of visualization: clear or blurry frames, camera speed, amount of remaining debris and effort of the endoscopist to inspect in a circumferential fashion all of the mucosa. Our results show that the two types of feedback most frequently needed to improve visualization are reminders to obtain clear frames or to inspect all of the mucosa in circumferential fashion.

Keywords Video stream analysis · Quality features · Real-time feedback · Colonoscopy · Colorectal cancer · Education

1 Introduction

Colorectal cancer (CRC) is a preventable cancer that is diagnosed in around 150,000 people each year in the US. Despite the fact that it can be prevented, about 50,000 patients die from CRC annually [1]. There is wide-spread consensus that the current preventive strategies in place in the US should drastically reduce the incidence and mortality of CRC, yet for a multitude of reasons this has yet to happen [2, 3].

P.C. De Groen (✉) · M. Szewczynski · F. Enders
Mayo Clinic, Rochester, MN, USA

W. Tavanapong · J. Wong
Iowa State University, Ames, IA, USA

J. Oh
University of North Texas, Denton, TX, USA

© Springer International Publishing Switzerland 2015
V.L. Uskov et al. (eds.), *Smart Education and Smart e-Learning*,
Smart Innovation, Systems and Technologies 41,
DOI 10.1007/978-3-319-19875-0_45

Of all the methods to prevent death from CRC, colonoscopy holds most promise; it is a technique that allows detailed inspection of the entire colon and at the same time removal of all premalignant lesions. The latter is commonly performed during the withdrawal phase of the procedure. Colonoscopy is also readily available in most geographical areas of the US with wide-spread coverage of the procedure by payers.

The main problem with colonoscopy is the relatively limited CRC protective effect it currently provides. Several studies have shown limited or even total absence of a protective effect (i.e., in the right colon) against mortality of CRC, especially outside carefully controlled trials [2, 3]. More recent studies have shown a definite protective effect, in particular for CRC of the left colon [4, 5]. CRC of the right colon appears to be more difficult to prevent and numerous explanations for the relative failure of colonoscopy have been proposed. In general, these explanations can be divided into two sets. One set focuses on patient and biology related factors; these include a poor preparation, an inability of the patient to cooperate during the procedure, an abnormal anatomy, flat polyp morphology or an unfavorable polyp or tumor biology. Indeed, these factors all may be present more in right-sided CRC: frequently bile and small bowel content covers the right colon, the deep folds of the right colon make inspection difficult, and flat, more rapidly progressing tumor biology (CIMP pathway) is much more likely in neoplasia of the right colon. The other set of explanations focuses on procedure and endoscopist related factors: suboptimal equipment, no removal of remaining debris, not reaching the cecum, fast withdrawal, no effort at inspection of areas behind folds and angulations, and inadequate polyp removal technique.

A key study supporting this concept was published in 2014 and shows that for every 1 % increase in adenoma detection rate (ADR), there was a 3 % decrease in interval CRCs; the lowest interval CRC rate was observed among endoscopists with an ADR > 33.5 % [6]. Proponents of the first set of factors may point to the patient responsibility for a clean colon, the type of preparation and patient compliance, outline the benefits of propofol sedation and believe that interval CRC is a result of rapid growth. Proponents of the second set of factors are of the opinion that gastroenterology-trained endoscopists provide better quality than other endoscopists, believe that removal of debris, complete inspection and total removal of all neoplasia can be achieved and should lead to nearly complete protection against CRC if screening and surveillance guidelines are followed; interval cancers are considered a result of missed lesions (polyp or small cancer) or incomplete resection of identified lesions at prior colonoscopy. In reality there is not a strict separation into two sets of opinions but a gradual range of opinions.

Several years ago the general opinion within gastroenterology, in particular related to interval cancers, was more along the first set of explanations with a focus on tumor biology. Lately, research has shown serious gaps in procedure quality, suggesting that improvement in this area would dramatically improve patient outcomes. The most important observations that favor procedure and endoscopist are reports that show

(1) vast differences in interval CRCs among endoscopists, [7]
(2) a very low interval CRC rate in endoscopists with an adenoma detection rate (ADR) of at least 20 % in Poland or > 33.5 % in the US, [6] and
(3) a very high CRC mortality reduction of 89 % when implementing a colonoscopy protocol that enforces high quality and results in 34 % ADR [8].

The question, then, is how to improve endoscopist technique during the colonoscopy procedure within today's challenging medical environment in which physicians often feel pressure to see more patients in less time. Our hypothesis is that providing feedback during the critical phase of the procedure − the withdrawal phase - will result in improved endoscopist technique and higher ADRs.

2 Related Work

2.1 Endoscopic Multimedia Information System (EMIS)

Since 2003 our group has worked on creating an automated system to capture, analyze and summarize video files representing an entire endoscopic procedure [9]. We have called our system EMIS for Endoscopic Multimedia Information System. We have focused our efforts on colonoscopy. Our work has shown that our manual EMIS annotation technique is reproducible among annotators with fair to good inter-operator agreement; inter-operator agreement is best for very low and very high quality procedures, but varies when quality is average. Our automated EMIS technology results correlate with our manual annotation results, and both manual and automated annotations correlate with ADR − the most widely accepted main determinant of colonoscopy quality − for a set of video files representing the work of a single endoscopist or an endoscopy group.

2.2 Commonly Used Indicators of Quality

All commonly used indicators of colonoscopy quality, such as cecal intubation rate, average withdrawal time, ADR, polyp detection rate, and interval cancer rate, are averages and provide no information about a single procedure [10]. Instead, these quality parameters are summary data that reflect a group of procedures performed by an individual endoscopist or a group of endoscopists over a specific time period. An inherent feature of summary data is that a few really poor procedures combined with a larger set of higher quality procedures will result in acceptable overall quality scores. Intuitively it does not make sense to set as goal a specific withdrawal time (several specific times have been proposed) or a specific number of cases in which polyps should be detected. Instead, it would make sense to measure features that directly define quality of each procedure. However, an accepted method to measure quality of colonoscopy *during* the procedure does not exist at the present time. Therefore, our automated annotation, if it could be performed in real-time with real-time reporting of measured quality, has the potential to provide real-time feedback about quality, and thereby influence the outcome of the procedure.

2.3 Direct Indicators of Quality

Three things need to happen at the same time in order for a colonoscopy to be of high quality. First the colon needs to be well prepared (Clean). Second, most if not all of the mucosa needs to be inspected (Look Everywhere). And third, all neo-plastic lesions, where possible, need to be completely removed (Abnormality Removal) [11]. We have combined these three features into the CLEAR acronym. EMIS uses computer-based algorithms to analyze the image stream generated during – not after – colonoscopy for specific metrics based on the CLEAR principle. EMIS does not interfere with actual colonoscopy as the same image stream is displayed on a monitor allowing the endoscopist to view the colonic mucosa and perform diagnostic and therapeutic procedures as indicated. To allow streaming video file analysis for CLEAR features we created SAPPHIRE: middleware that handles multiple simultaneous real-time algorithms and automatically distributes these to either one or more threads, CPUs or GPUs, all the time making sure that all single frame-related algorithms are completed before the next frame becomes available [12, 13]. With a video frame rate of 30 frames/s this means SAPPHIRE must complete all single frame-related analyses within a time span much shorter than 33 ms in order to process the results and generate feedback information.

2.4 Features of EMIS

EMIS can detect whether the colon is clean, whether the endoscopist removes remaining debris, whether the endoscopist tries to inspect the entire colon and whether polyps are removed. Using a graphics card attached to a MS Windows OS7 workstation we automatically capture the video stream from the endoscope image processor; algorithms process the video stream for many features related to quality. For each algorithm we went through a similar multi-step process. First, we decided what new or existing features needed to be derived to measure the desired quality metrics. Next we created a training set of images that incorporated the presence or absence of the features; for some training sets we used a binary approach (biopsy cable in frame: present/absent), for others a continual range (stool pixels per frame: 0–100 %). The third step consisted of creating algorithms that measured the features of interest. Initially we developed those in high level language such as MatLab. Then we would use Machine Learning techniques to train our software on the training set; next we determined sensitivity and specificity using the test set.

Our goal is to achieve around 95 % sensitivity and specificity. Once we achieved these marks, we rewrite our code in either C/C++ to increase speed of execution, or when this does not result in fast enough execution, assembly language.

2.5 Real-Time Feedback

In addition to automated real-time analysis, EMIS allows real-time feedback. For this we developed within SAPPHIRE a reporting module that summarizes from all algorithms a summary state that is continuously updated in real-time; this summary state can be sent for display on a monitor. Recently we described all the features required to allow EMIS to function in real-time within a healthcare network; we encountered numerous challenges yet eventually we solved all [14]. Thus we are ready to test our first real-time feedback modules.

3 Methods

3.1 Features Ready for Real-Time Feedback

The protective effect of colonoscopy is directly related to removal of all precancerous lesions; as lesions can only be removed if identified, the first thing that needs to occur during colonoscopy is inspection of all mucosa, and where mucosa is covered with remaining debris, this debris needs to be removed. This is not as simple as it seems as the colon is a convoluted, moving tubular organ constantly receiving bile-stained digestive juices and food particles from the small bowel. Our algorithms for features that measure colon preparation and mucosal inspection have gone through a number of iterations aimed at improving accuracy and speed; at present they are fast enough to provide real-time analysis and the results have been coupled to a feedback module. Thus we are ready to start providing feedback related to colonic preparation and mucosal inspection.

3.2 In-Person Real-Time Feedback

Real-time feedback is commonly provided to physicians in training; indeed, nowadays endoscopy is taught first by acquisition of basic endoscope handling techniques on endoscope simulators. These simulators provide real-time feedback about the force that is used to manipulate the endoscope within a simulated patient and at the end of the simulation may summarize amount of mucosa and the number of simulated polyps seen as well as extent of intubation and duration of the withdrawal phase in a report. Next, after having observed experienced endoscopists during colonoscopy on real patients, trainees themselves start performing colonoscopy on patients under continuous in person supervision of experienced endoscopists. Especially in the first days and weeks of hands-on patient endoscopy, the trainee receives continues feedback about how to insert the endoscope, how to clean the lens of the endoscope and the mucosa of the patient from remaining debris, and how to maneuver the tip of the endoscope in order to achieve optimal inspection of as much of the colon mucosa as possible.

3.3 Automated Real-Time Feedback

Our real-time feedback is targeted at experienced endoscopists as well as relatively advanced trainees. Therefore, the intention is only to provide feedback when endoscopic actions are seen that ideally are not present in endoscopies by experienced gastroenterologists performing high quality colonoscopy. We are unaware of an existing system that currently provides real-time analysis with real-time feedback during endoscopic or surgical procedures; thus there are no examples of when and how to provide real-time feedback. Based on domain expertise we defined four feature/time feedback triggers during the withdrawal phase. First, we decided that whenever the image was not clear or blurry due to debris covering the lens or the tip of the scope being stuck within mucosa, the endoscopist should show efforts at obtaining a clear image within 15 s. Second, we expected that with debris, easily removable, an effort at removal should occur within 30 s. Third, we determined whether removal speed was rather fast over 30 s or more. And last but not least, we determined if there was obvious circumferential inspection activity within a 30 s withdrawal segment. Whenever polyps were removed, the latter two could not be evaluated as absent. For each 30 s segment we scored expected behavior as "0" and behavior that would trigger feedback as "1". Table 1 lists the feature annotations and scoring. A single annotator with over 3 years of video file annotation experience annotating thousands of colonoscopy video files manually annotated all video files. This annotator was trained in colonoscopy video file annotation and then bench-marked to a set of 10 video files, annotated by a group of experienced endoscopists.

Table 1 Features and annotation triggers

Name	Feature	Expected (no feedback)	Feedback
Blur	Clear image	Clear within 15 s	Blurry for at least 15 s
Stool	Remaining debris	Cleared within 30 s	Not cleared within 30 s
Speed	Withdrawal speed	Moderate during 30 s	Fast for at least 30 s
Spiral	Circumferential withdrawal	Present within 30 s	Not present within 30 s

4 Results

Video files were selected from a large set of video files obtained automatically as part of our quality studies in January and February of 2014. Video files were captured at a rate of approximately 30 frames/s. A total of 100 video files were annotated for this study. There were 4 video files (4 %) where not a single feedback trigger annotation was made; an example is shown in Table 2. This example is of interest as the total withdrawal time, 3 min and 12 s, is far below the recommended minimum withdrawal time of 6 min, yet not a single time was any of the manual trigger thresholds reached. The remaining 96 video files each included from 1 to 44 feedback triggers. Table 3 summarizes the results for the 100 video files.

As can be seen from Table 3, there was a wide fluctuation within each feature trigger with a large range for clear images and circumferential withdrawal. The expected average number of feedback triggers given the current feature definition

Table 2 Example of a colonoscopy video file without any feedback triggers

Video file name	Actual time	Frame	Time (s)	Blur	Stool	Speed	Spiral
ColonoscopyVideoFileName	6:00	10818	15	0	0	0	0
	6:15	11239	30	0			
	6:30	11689	45	0	0	0	0
	6:45	12138	60	0			
	7:00	12588	75	0	0	0	0
	7:15	13037	90	0			
	7:30	13487	105	0	0	0	0
	7:45	13937	120	0			
	8:00	14386	135	0	0	0	0
	8:15	14836	150	0			
	8:30	15285	165	0	0	0	0
	8:45	15735	180	0			
	9:00	16184	195	0	0	0	0
	9:12	16557	207	0			

Table 3 Summary of results for four inspection features. Withdrawal Time and Trigger Interval in seconds

	Withdrawal frames	Withdrawal time	Blur	Stool	Speed	Spiral	All alerts	Trigger interval
Average	20518	685	4	1.01	0.02	3.09	8.13	114
SD	10689	357	4	1.34	0.14	3.56	7.23	93
Minimal value	5661	189	0	0.00	0.00	0.00	0.00	35
Maximal value	69362	2314	24	5.00	1.00	27.00	44.00	692

was 8 ± 7 (Mean ± SD) triggers per colonoscopy. The average interval between triggers was nearly 2 min with a very wide range from once every 35 s to once in nearly 12 min. Feedback triggers were not randomly distributed throughout video files. Table 4 shows part of the feedback trigger annotations for the colonoscopy with the largest number of triggers, 44. As can be seen many triggers were grouped together between minute 30 and 33 after the start of the procedure.

5 Educational Analysis

The two examples shown in Tables 2 and 4 were further analyzed by careful review of the events by an expert endoscopist with greater than 20 years of colonoscopy experience to determine the potential educational value of the feedback. Annotations were found to be accurate. In the procedure shown in Table 2 the colon was well cleaned and most remaining debris was removed during insertion and withdrawal. The image was always clear and speed during withdrawal was constantly in the direction of the anus. Spiral activity was present each 30 s segment. However, not all of the mucosa was seen, there was no inspection behind large folds and flexures and the endoscopist did not go back to inspect mucosal areas missed. Automated analysis of the video file (data not shown) revealed a very low total spiral score, the number of completed 360 degree inspections, of 4; the absence of back and forth movement with inspection of folds explains the low spiral score and the very short withdrawal time. Thus the manual annotation was accurate but the manual spiral annotation as performed not sufficient to generate feedback. The automated analysis which can summate the complete circumferential inspection activity over the entire withdrawal phase and provide an update of the time spent inspecting the mucosa however would provide real-time feedback of the score and withdrawal time, and inform the endoscopist that more effort towards inspection is warranted.

In the procedure shown in Table 4, the colon was fairly clean except for sporadic seeds, likely from a fruit or vegetable ingested in the days prior to colonoscopy. Around 20 min a polyp was resected which resulted in some bleeding and while inspecting the bleeding site, the endoscopist lost track of the polyp specimen. Around minute 27.5 the polyp specimen was found and suctioned into the instrument together with a number of seeds. Seeds are known to occlude the suction system. More seeds were seen and by minute 30 the endoscopist had lost suction; the endoscope tip was submerged in remaining debris, water used to wash the lumen and seeds. After three minutes the suction ability for water, debris and air was restored. The withdrawal phase was long, 30 min, but at least 10 min were spent inspecting the bleeding polypectomy site, looking for the polyp specimen and trying to regain suction. The automated analysis showed a spiral score of 19, which seems correct, but is inflated due to the efforts at inspecting the bleeding site and trying to find the polyp specimen. Another algorithm, determining forward and

Table 4 Example of video file annotation with triggers (red); not all triggers are shown

Video file name	Actual time	Frame	Time (s)	Blur	Stool	Speed	Spiral
Colonoscopy video file name	9:30	17085	15	0	0	0	0
	9:45	17533	30	0			
	10:00	17982	45	0	0	0	0
	10:15	18432	60	0			
	10:30	18882	75	1	0	0	1
	10:45	19331	90	0			
	11:00	19781	105	0	0	0	0
	11:15	20230	120	0			
	11:30	20680	135	0	0	0	0
	11:45	21129	150	0			
	12:00	21579	165	0	0	0	0
	12:15	22028	180	1			
	12:30	22478	195	0	0	0	0
	12:45	22928	210	1			
	13:00	23377	225	1	0	0	0
	13:15	23827	240	0			
	13:30	24276	255	0	0	0	0
	13:45	24726	270	1			
	14:00	25175	285	0	0	0	0
	14:15	25625	300	0			
	29;00	52148	1185	0	0	0	0
	29;15	52598	1200	0			
	29;30	53047	1215	0	0	0	0
	29;45	53497	1230	0			
	30;00	53946	1245	1	0	0	1
	30;15	54396	1260	1			
	30;30	54846	1275	1	1	0	1
	30;45	55295	1290	1			
	31;00	55745	1305	1	1	0	1
	31;15	56194	1320	1			
	31;30	56644	1335	1	1	0	1
	31;45	57093	1350	1			
	32;00	57543	1365	1	0	0	0
	32;15	57992	1380	1			
	32;30	58442	1395	1	1	0	0
	32;45	58892	1410	1			
	33;00	59341	1425	0	0	0	1
	33;15	59791	1440	0			

(continued)

Table 4 (continued)

Video file name	Actual time	Frame	Time (s)	Blur	Stool	Speed	Spiral
	33;30	60240	1455	0	0	0	1
	33;45	60690	1470	0			
	34;00	61139	1485	1	0	0	1
	34;15	61589	1500	1			
	34;30	62038	1515	0	0	0	0
	34;45	62488	1530	0			
	35;00	62937	1545	1	0	0	0
	35;15	63387	1560	0			
	35;30	63837	1575	0	0	0	0
	35;45	64286	1590	0			
	36;00	64736	1605	0	0	0	0
	36;15	65185	1620	0			
	36;30	65635	1635	0	0	0	0
	36;45	66084	1650	0			
	37;00	66534	1665	0	0	0	1
	37;15	66983	1680	0			
	37;30	67433	1695	0	0	0	0
	37;45	67883	1710	0			
	38;00	68332	1725	0	0	0	0
	38;15	68782	1740	0			
	38;30	69231	1755	0	0	0	1
	38;45	69681	1770	0			
	39;00	70130	1785	0	0	0	0
	39;15	70580	1800	0			
	39;22	70791	1807	0	0	0	0

backward motion, can detect that the endoscope is not moving in meaningful forward or backward direction, and the lack of this movement can be included in determining circumferential withdrawal without movement in anal direction.

6 Discussion

Improving quality of colonoscopy with real-time feedback during the procedure has never been done. Therefore we are investigating each step in the development of real-time feedback. Here we determined how often feedback related to blurred images, failure to remove remaining debris, high speed of withdrawal and absence of circumferential withdrawal would occur if we used 15, 30, 30 and 30 s respectively as feature triggers activating feedback. Our results support the following conclusions. First, real-time feedback during colonoscopy is needed as quality fluctuates among procedures. Second, as shown in Table 3 real-time feedback is more likely to be beneficial for blurry images and absence of circumferential withdrawal. Third, the thresholds used in this study, which were based on domain expertise, for blurry frames and circumferential withdrawal with trigger time periods of 15 and 30 s respectively seem ready for testing in the clinical environment. Fourth, to our surprise, real-time feedback related to absence of removal of remaining debris and withdrawal speed is unlikely to greatly influence quality of colonoscopy. Fifth, feedback related to withdrawal time and cumulative circumferential withdrawal should be included with blurry image and circumferential withdrawal feedback. Sixth, the average interval between triggers of nearly 2 min seems acceptable; moreover, any improvement in technique due to real-time feedback will – automatically – decrease the number of trigger events and thereby increase the interval between triggers.

Our in depth analysis of two outlier cases, one without any triggers, and one with the maximal number of triggers revealed that meaningful real-time feedback during colonoscopy will have to include a large number of features. Relative position of the endoscopy, determined by delta forward-backward movement, will allow us to determine whether circumferential inspection is occurring at more or less the same location or during a gradual withdrawal of the instrument. 3D mapping of the colon mucosa from 2D images will provide a second means of determining endoscope movement direction anus [15]. Without any doubt additional features will need to be included or developed once we have tested real-time feedback in clinical practice.

Use of real-time feedback by endoscopy trainees for only spiral score and cumulative spiral score was associated with significantly higher inspection technique [16]. Thus EMIS has the potential to educate "on the job" and likely will be useful for objectively measuring and improving endoscopic technique of staff endoscopists. Yet although these findings are very encouraging, several questions remain. First, will EMIS be acceptable to staff endoscopists, especially those who have been accustomed for many years to practice endoscopy without any peer review? Second, will the effect of EMIS wear off over time? Third, more thorough inspection in general means longer inspection per procedure and fewer procedures per day; thus higher colonoscopy quality may lower endoscopist income. Lastly, EMIS itself comes with costs which further may decrease endoscopist income.

In conclusion, we are one step closer to testing real-time feedback in the clinical practice of experienced, staff endoscopists. Our results shown here support testing a number of features and suggest that a moving feature trigger period of 15–30 s will provide meaningful feedback without too many feedback triggers per procedure. Additional studies are required to determine whether "on the job" education via real-time feedback is acceptable and will lead to persistent improvement in endoscopist technique and decreases CRC mortality.

References

1. Siegel, R., et al.: Cancer statistics, 2013. CA Cancer J. Clin. **63**, 11–30 (2013)
2. Brenner, H., et al.: Protection from right- and left-sided colorectal neoplasms after colonoscopy: population-based study. J. Natl. Cancer Inst. **102**, 89–95 (2010)
3. Baxter, N.N., et al.: Association of colonoscopy and death from colorectal cancer. Ann. Intern. Med. **150**, 1–8 (2009)
4. Zauber, A.G., et al.: Colonoscopic polypectomy and long-term prevention of colorectal-cancer deaths. N. Engl. J. Med. **366**, 687–696 (2012)
5. Nishihara, R., et al.: Long-term colorectal-cancer incidence and mortality after lower endoscopy. N. Engl. J. Med. **369**, 1095–1105 (2013)
6. Corley, D.A., et al.: Adenoma detection rate and risk of colorectal cancer and death. N. Engl. J. Med. **370**, 1298–1306 (2014)
7. Gupta, R., et al.: Colon cancer not prevented by colonoscopy. Am. J. Gastroenterol. **103**, S551–S552 (2008)
8. Xirasagar, S. et al.: Colorectal cancer prevention by an optimized colonoscopy protocol in routine practice. Int. J. Cancer (Journal international du cancer) (2014)
9. Oh, J., et al.: Measuring objective quality of colonoscopy. IEEE Trans. Bio-med. Eng. **56**, 2190–2196 (2009)
10. Rex, D.K., et al.: Quality indicators for colonoscopy. Am. J. Gastroenterol. **101**, 873–885 (2006)
11. de Groen, P.C.: Advanced systems to assess colonoscopy. Gastrointest. Endosc. Clin. N. Am. **20**, 699–716 (2010)
12. Stanek, S.R., et al.: SAPPHIRE: a toolkit for building efficient stream programs for medical video analysis. Comput. Methods Programs Biomed. **112**, 407–421 (2013)
13. Stanek, S.R., et al.: Automatic real-time detection of endoscopic procedures using temporal features. Comput. Methods Programs Biomed. **108**, 524–535 (2012)
14. de Groen, P.C., et al.: Challenges associated with introduction of real-time quality monitoring and feedback during colonoscopy within a secure, tightly regulated healthcare network. Frontiers Artif. Intell. Appl. **262**, 592–601 (2014)
15. Hong, D., et al.: 3D Reconstruction of virtual colon structures from colonoscopy images. Comput. Med. Imaging Graph. **38**, 22–33 (2014)
16. Srinivasan, N., et al.: Real-time feedback improves the quality of colonoscopy by trainees: a controlled clinical trial. Am. J. Gastroenterol. **107**, S596 (2012)

Author Index

© Springer International Publishing Switzerland 2015
V.L. Uskov et al. (eds.), *Smart Education and Smart e-Learning*,
Smart Innovation, Systems and Technologies 41,
DOI 10.1007/978-3-319-19875-0

Printed in the United States
By Bookmasters